Acoustics: Basic Physics, Theory and Methods

Acoustics: Basic Physics, Theory and Methods

Paul Filippi
Dominique Habault
Jean-Pierre Lefebvre
Aimé Bergassoli

Laboratoire de Mécanique et d'Acoustique
13402 Marseille cedex 20
France

ACADEMIC PRESS
San Diego London Boston
New York Sydney Tokyo Toronto

This book is printed on acid-free paper

Academic Press
24–28 Oval Road, London NW1 7DX, UK
http://www.hbuk.co.uk/ap/

Academic Press
a division of Harcourt Brace & Company
525 B Street, Suite 1900, San Diego, California 92101-4495, USA
http://www.apnet.com

ISBN 0-12-256190-2

A catalogue record for this book is available from the British Library

Typeset by Mathematical Composition Setters Ltd, Salisbury, Wiltshire
Printed in Great Britain by MPG Books Ltd, Bodmin, Cornwall

99 00 01 02 03 04 MP 9 8 7 6 5 4 3 2 1

Contents

Foreword xii

Preface xiii

Chapter 1. Physical Basis of Acoustics 1
Jean-Pierre Lefebvre

Introduction 1
1.1. Review of mechanics of continua 1
  1.1.1. Conservation equations 2
  1.1.2. State equation 4
  1.1.3. Constitutive equations 8
  1.1.4. Equations at discontinuities 10
1.2. Elementary acoustics 11
  1.2.1. Linearization for a lossless homogeneous steady simple fluid . . . 12
  1.2.2. Equations for entropy and vorticity: Fundamental character of acoustic motion 13
  1.2.3. Equations for pressure and other acoustic quantities: Wave equations 14
  1.2.4. Velocity potential 16
  1.2.5. Acoustic energy, acoustic intensity 18
  1.2.6. General solutions of the wave equation in free space 20
  1.2.7. Harmonic waves 25
  1.2.8. Acoustic sources 27
  1.2.9. Boundary conditions 27
  1.2.10. Units, orders of magnitude 33
  1.2.11. Perfect gas 34
1.3. Elementary acoustics of solids: Elementary elastic waves 34
  1.3.1. Linearization for an isotropic, homogeneous, purely elastic solid 34
  1.3.2. Compression/expansion waves and shear/distorsion waves 35
  1.3.3. Plane waves: Longitudinal and transverse waves 37
  1.3.4. Orders of magnitude 37
  1.3.5. General behaviour 38
1.4. Conclusion 38
Bibliography 38

Chapter 2. Acoustics of Enclosures 41
Paul J.T. Filippi

Introduction 41
2.1. General statement of the problem 42
2.1.1. The wave equation 42
2.1.2. The Helmholtz equation 43
2.1.3. Boundary condition for harmonic regimes 44
2.1.4. Eigenmodes and eigenfrequencies, condition for existence and uniqueness of the solution 47
2.2. Sound field inside a parallelepipedic enclosure: free oscillations and eigenmodes 48
2.2.1. Eigenfrequencies and eigenmodes for the Neumann problem 49
2.2.2. Resonance frequencies and resonance modes for the Robin problem 52
2.2.3. Forced regime for the Neumann problem: eigenmodes series expansion of the solution 54
2.2.4. Forced regime for the Robin problem: expansion into a series of the eigenmodes of the Laplace operator 55
2.3. Transient phenomena – reverberation time 58
2.3.1. A simple one-dimensional example: statement of the problem . . . 58
2.3.2. Eigenmodes expansion of the solution 60
2.3.3. The 'boundary sources' method 63
2.3.4. Description of sound establishment by a series of successive reflections 66
2.3.5. Sound decay – reverberation time 67
2.3.6. Reverberation time in a room 68
2.3.7. The formula of Sabine 69
2.4. Acoustic field inside a circular enclosure: introduction to the method of separation of variables 71
2.4.1. Determination of the eigenmodes of the problem 71
2.4.2. Representation of the solution of equation (2.77) as a series of the eigenmodes 74
2.4.3. The method of separation of variables 75
2.5. Enclosures bounded by plane surfaces; introduction to the method of images 77
2.5.1. Reflection of a spherical wave by an infinite plane: the 'plane wave' and the 'geometrical acoustics' approximations 77
2.5.2. Forced regime in a polyhedral enclosure: series representation of the response by the image method 80
2.6. General case: introduction to the Green's representation of acoustic fields . . 81
2.6.1. Green's representation of the acoustic pressure 81
2.6.2. Simple layer and double layer potentials 85
2.6.3. Boundary integral equations associated with the Green's representation of the pressure field 86
Bibliography 87

Chapter 3. Diffraction of Acoustic Waves and Boundary Integral Equations . . 89
Paul J.T. Filippi

Introduction . . . 89
3.1. Radiation of simple sources in free space . . . 90
3.1.1. Elementary solution of the Helmholtz equation . . . 90
3.1.2. Point sources . . . 93
3.1.3. Cylindrical and spherical harmonics . . . 95
3.1.4. Surface sources . . . 97
3.1.5. Properties of the simple and double layer potentials . . . 98
3.2. Green's representation of the solution of linear acoustics boundary value problems . . . 102
3.2.1. Statement of the problem . . . 102
3.2.2. Green's representation of the acoustic pressure . . . 104
3.2.3. Transmission problems and non-local boundary condition . . . 105
3.3. Representation of a diffracted field by a layer potential . . . 107
3.3.1. Representation of the field diffracted by a closed surface in $\mathbb{R}^3$ or a closed curve in $\mathbb{R}^2$ . . . 107
3.3.2. Diffraction by an infinitely thin screen . . . 108
3.4. Boundary integral equations . . . 110
3.4.1. Integral equations deduced from the Green's representation of the pressure field . . . 110
3.4.2. Existence and uniqueness of the solutions . . . 112
3.4.3. Boundary integral equations deduced from the representation of the diffracted field by a hybrid layer potential . . . 114
3.5. Two-dimensional Neumann problem for a circular boundary . . . 115
3.5.1. Interior problem and Green's representation . . . 115
3.5.2. Exterior problem and Green's representation . . . 117
3.5.3. Exterior problem and hybrid layer potential . . . 118
Bibliography . . . 120

Chapter 4. Outdoor Sound Propagation . . . 121
Dominique Habault

Introduction . . . 121
4.1. Ground effect in a homogeneous atmosphere . . . 122
4.1.1. Introduction . . . 122
4.1.2. Propagation above a homogeneous plane ground . . . 122
4.1.3. Propagation above an inhomogeneous plane ground . . . 132
4.2. Diffraction by an obstacle in homogeneous atmosphere . . . 136
4.2.1. Introduction . . . 136
4.2.2. Diffraction of a plane wave by a circular cylinder . . . 136
4.2.3. Geometrical theory of diffraction: diffraction by a convex cylinder . . . 137
4.2.4. Diffraction by screens . . . 143

4.3. Sound propagation in an inhomogeneous medium 149
4.3.1. Introduction 149
4.3.2. Plane wave propagation 150
4.3.3. Propagation of cylindrical and spherical waves 153
Bibliography 156

Chapter 5. Analytic Expansions and Approximation Methods 159
Dominique Habault

Introduction 159
5.1. Asymptotic expansions obtained from integral expressions 161
5.1.1. Elementary kernels 161
5.1.2. Integration by parts, Watson's lemma 163
5.1.3. The method of stationary phase 164
5.1.4. The method of steepest descent 166
5.2. Kirchhoff approximation 169
5.3. Neumann series 170
5.4. W.K.B. method. Born and Rytov approximations 171
5.4.1. W.K.B. method 171
5.4.2. Born approximation and Rytov approximations 173
5.5. Image method, ray method, geometrical theory of diffraction 174
5.5.1. Image method 174
5.5.2. Ray method 176
5.5.3. Geometrical theory of diffraction 177
5.6. Parabolic approximation 179
5.6.1. Replacing the Helmholtz equation by a parabolic equation 180
5.6.2. Solution techniques 181
5.6.3. A word about starters 182
5.6.4. Error estimates 182
5.7. Wiener–Hopf method 182
5.7.1. Description 182
5.7.2. Obtaining a Wiener–Hopf equation 182
5.7.3. The decomposition theorem 186
5.7.4. The main difficulties of the method 186
Bibliography 187

Chapter 6. Boundary Integral Equation Methods – Numerical Techniques . . 189
Dominique Habault

Introduction 189
6.1. Techniques of solution of integral equations 191
6.1.1. Collocation method 191
6.1.2. Galerkin method 194
6.1.3. Method of Galerkin-collocation 195
6.2. Eigenvalue problems 195
6.2.1. Interior problems 195
6.2.2. Exterior problems 198

6.3. Singularities . . . 200
6.3.1. Singularity of the kernel . . . 200
6.3.2. Domains with corners, polygons . . . 200
Bibliography . . . 202

Chapter 7. Introduction to Guided Waves . . . 203
Aimé Bergassoli

Introduction . . . 203
7.1. Definitions and general remarks . . . 203
7.1.1. Guided waves . . . 203
7.1.2. Boundaries . . . 204
7.2. The problem of the waveguide . . . 213
7.2.1. General remarks . . . 213
7.2.2. The condition for mode propagation . . . 213
7.2.3. Solution of some simple problems . . . 214
7.3. Radiation of sources in ducts with 'sharp' interfaces . . . 221
7.3.1. General remarks . . . 221
7.3.2. Point source and Green's function in the duct . . . 221
7.3.3. Radiation impedance of the source in the duct . . . 223
7.3.4. Extended sources – pistons . . . 224
7.3.5. Interference pattern in a duct . . . 227
7.3.6. Impulse response in a duct . . . 228
7.4. Shallow water guide . . . 229
7.4.1. Properties of the shallow water guide . . . 229
7.4.2. The Pekeris model . . . 230
7.4.3. Solutions $\psi_j(z)$ . . . 231
7.4.4. Eigenvalues . . . 231
7.4.5. Eigenmodes . . . 233
7.4.6. Solutions $R(r)$ . . . 233
7.4.7. Essential remark . . . 233
7.4.8. Radiation of a harmonic point source . . . 233
7.4.9. Propagation of a pulse . . . 234
7.4.10. Extension to a stratified medium . . . 236
7.5. Duct with absorbing walls . . . 236
7.5.1. Duct with plane, absorbing walls . . . 237
7.5.2. Orthogonality of the modes . . . 238
7.5.3. Losses on the walls of the duct . . . 239
7.6. Ducts with varying cross section . . . 240
7.6.1. Linear duct with discontinuous cross section . . . 240
7.6.2. Rectangular and circular cross sections . . . 242
7.6.3. Connection between an absorbing duct and a lossless duct . . . 242
7.6.4. Changes in direction . . . 244
7.7. Conclusion . . . 246
Bibliography . . . 246

Chapter 8. Transmission and Radiation of Sound by Thin Plates 247
Paul J.T. Filippi

Introduction 247
8.1. A simple one-dimensional example 248
8.1.1. Governing equations 248
8.1.2. Fourier-transformed equations and response of the system to a harmonic excitation 249
8.1.3. Transient response of the system to a force acting on the panel 255
8.2. Equation governing the normal displacement of a thin elastic plate . . . 256
8.3. Infinite fluid-loaded thin plate 260
8.3.1. Free plane waves in the plate 261
8.3.2. Reflection and transmission of a harmonic plane acoustic wave 265
8.3.3. Infinite plate excited by a point harmonic force 268
8.4. Finite-dimension baffled plate: expansions of the solution into a series of eigenmodes and resonance modes 271
8.4.1. Expansion of the solution in terms of the fluid-loaded plate eigenmodes 273
8.4.2. Expansion of the solution as a series of the resonance modes . . 274
8.4.3. The *light fluid* approximation 276
8.5. Finite-dimension baffled plate: boundary integrals representation of the solution and boundary integral equations 279
8.5.1. Green's kernel of the *in vacuo* infinite plate 280
8.5.2. Green's representation of the fluid-loaded plate displacement . . 281
8.5.3. Boundary Integral Equations and polynomial approximation . . 283
8.5.4. Response of a rectangular baffled plate to the wall pressure exerted by a turbulent boundary layer 286
8.6. Conclusion 289
Bibliography 289

Chapter 9. Problems 291

Mathematical Appendix: Notations and Definitions 301
Dominique Habault and Paul J.T. Filippi

Introduction 301
A.1. Notations used in this book 302
A.2. Classical definitions 303
A.3. Function spaces 304
A.3.1. Space $\mathcal{D}$ and space $\mathcal{D}'$ 304
A.3.2. Space $\mathcal{S}$ 305
A.3.3. Hilbert spaces 305
A.3.4. Sobolev spaces 307
A.4. Distributions or generalized functions 309
A.4.1. Distributions 310

A.4.2. Derivation of a distribution 311
A.4.3. Tensor product of distributions 313
A.5. Green's kernels and integral equations 314
Bibliography 315

Index 316

Foreword

At the invitation of the authors, it is my privilege to contribute a Foreword to this book. It is an English-language version of the French-language *Acoustique Générale*, published in 1994 (which I reviewed in JSV 186(5)), translated into English and revised by the authors to make its contents more accessible to English speakers.

Its contents are what is important. No other English-language book on acoustics contains as much concise information, both physically and mathematically rigorous, on the basic physical and theoretical concepts of acoustics and the methods of obtaining solutions applicable to practical problems of the mathematical problems of acoustics.

Although acoustics is presented in the book as a collection of mathematical problems and how to solve them, the physical bases of these problems are initially spelled out in generality in the first chapter, in the context of continuum mechanics (which is not done in any other books on acoustics). In the subsequent seven chapters, both mathematically rigorous and approximate methods of solving the acoustical equations thus described are presented, including both analytical and numerical methods. The presentation throughout is very concise, and so the book is like a diamond mine for those willing to dig deep enough to find diamonds which encapsulate the problems they wish to solve. Since the authors, however, have all been brought up in the French tradition of 'rational mechanics', rather than the 'empirical mechanics' tradition favoured among most English speakers, I have to conclude by saying that 'English-speakers' must be prepared to dig deep into the book if they want to find their diamonds. If they do so, I am sure that they will be duly rewarded.

P.E. Doak

Preface

These lecture notes were first written in French, while the authors were teaching Acoustics at the University of Aix-Marseille, France. They correspond to a 6-month course given to postgraduate students before they begin a Ph.D. thesis.

The reason for writing this book was, at the time we began, the lack of textbooks. Most of the books were either too specialized, and thus almost incomprehensible for students, or too exhaustive. We needed a basic course which could be presented within a rather short period of time to students who had a fair background in Mathematics and Mechanics but often no knowledge of Acoustics.

The purpose of this book is to present the main basis of modelling in Acoustics. The expression 'modelling' used here includes the procedures used to describe a physical phenomenon by a system of equations and then to solve this system by analytical and/or numerical methods. The unknown function of the equations is generally the sound pressure itself or some related function. First (and most) of all we wish to stress that when modelling a phenomenon it is necessary to have a good knowledge of the approximations introduced at each step, the kind of neglected phenomena and the conditions of validity of the solution methods. In wave propagation, for example, the unit length that is well adapted to the description of the phenomenon is the wavelength. Then there are 'low frequency' and 'high frequency' methods, and all dimensions must be compared with the wavelength. It is always essential to choose 'reasonable' approximations and the methods adapted to the problem. It is obviously a nonsense to develop or run a big routine in order to compute results which can be obtained simply through analytical methods. The book tries to give a clear overview on these aspects.

Throughout the book we insist also on the properties of the solutions – mainly existence and uniqueness. One and only one solution corresponds to a stable physical phenomenon. It is therefore both natural and essential that the corresponding system of equations has one and only one solution. It must not be believed that because almost all problems must be solved with a computer, there is no need to examine the properties of the solutions. We attempt to show that mathematical theorems cannot be ignored when modelling physical phenomena and solving equations.

Another purpose of this book is to provide the reader with basic and main concepts in order that he could read more specialized publications.

Apart from corrections or improvements, the content has only been slightly modified from the French version. Chapter 1 has been reduced and Chapter 8, originally written by C. Lesueur and J.L. Guyader, is replaced by a chapter on the same subject written by P. Filippi.

The book is split into nine chapters. The aim of Chapter 1 is to establish wave equations in fluids and solids by linearizing the equations of conservation of mass, momentum and energy. The equations are obtained for homogeneous and heterogeneous media, boundary conditions, moving sources, etc.

Chapter 2 is devoted to Acoustics in bounded domains. Methods to solve the Helmholtz equation are presented: method of separation of variables, image method and Green's representations. They are applied to several examples. The respective roles of eigenmodes and resonance modes are also shown and detailed in an example. The results are applied to the description of the establishment and extinction of sound.

Chapter 3 is dedicated to boundary value problems solved by methods based on Green's kernels and integral representations. The radiation of simple sources in free space is expressed by using Green's kernels. Then Green's representations of the sound field, layer potentials and boundary integral equations are examined in detail. As an example of application, the Neumann problem for a circular boundary is considered.

Chapter 4 is devoted to outdoor Acoustics. Phenomena of outdoor sound propagation are divided into three parts: propagation above ground, diffraction by an obstacle and propagation in an heterogeneous medium. For each of them a general survey of solution techniques is presented.

Chapter 5 is a general presentation of (partly) analytical methods used in Acoustics. They are all based on analytical representations or approximations. Most of them are related to asymptotic expansions. They are all characterized by particular conditions of validity which must not be forgotten when using them. These conditions are described for each method.

Chapter 6 contains the numerical aspects of the methods presented in chapter 3. It is dedicated to boundary integral equations: solution techniques, eigenvalue problems and singularities.

Chapter 7 is dedicated to Acoustics in waveguides. Various aspects of sound propagation in ducts are examined. Most of the results are obtained through the method of separation of variables which leads to modal representations.

Chapter 8 is devoted to a simple problem of interaction between a fluid and a vibrating structure – namely, sound transmission and radiation by thin plates. This is a coupled problem for the sound pressure in the fluid and the displacement on the plate. The different aspects of the physical phenomenon are examined on various examples.

Exercises are gathered in Chapter 9. They provide complementary notions and illustrative examples of the course.

Because in most chapters expressions related to Mathematical Analysis are used, a mathematical appendix is added to present in some detail the general interest of this theory when applied to Acoustics. This appendix also contains many of the notations used in the book.

Each chapter as well as the appendix is ended by a list of books and publications in which more detailed information can be found, and the reader is highly recommended to refer to them. These lists are not exhaustive. They include most of the references used by the authors when preparing the present text.

Publishing this manuscript has been a long story. The first discussions between the authors date back to 1986. It was first an internal publication of the Laboratoire de Mécanique et d'Acoustique. Then, with the strong help of Bernard Poirée, it was published by the Société Française d'Acoustique and the Editions de Physique in 1994. The present edition is mainly due to Professor Doak, who has been kind enough to appreciate the French version. He has dedicated himself to convincing Academic Press to publish it, and he does us the honour of writing the Foreword of this book.

Finally, we are happy to take the opportunity to thank all the persons who contributed to the successive versions of the manuscript. We would particularly like to thank all the secretaries who helped us for the early drafts; they probably had the worst part of the work. We are also very grateful to the Academic Press staff for their very efficient help and courtesy and, finally, thanks again to Phil Doak for his friendly enthusiasm.

Dominique Habault, Paul Filippi, Jean-Pierre Lefebvre and Aimé Bergassoli
Marseille, May 1998

CHAPTER 1

Physical Basis of Acoustics

J.P. Lefebvre

Introduction

Acoustics is concerned with the generation and space-time evolution of small mechanical perturbations in a fluid (sound waves) or in a solid (elastic waves). Then equations of acoustics are simply obtained by linearization of the equations of the mechanics of continua.

The main phenomenon encountered in acoustics is wave propagation. This phenomenon is the only one that occurs in an infinite homogeneous medium. A second important phenomenon is scattering, due to the various obstacles and inhomogeneities encountered by the wave. A third, more tenuous, phenomenon is absorption and dispersion of waves, due to dissipation processes.

The first and second phenomena need only a simple methodology: derivation of a wave equation and of a boundary condition from the linearized equations of the mechanics of continua for a homogeneous, steady, perfect simple fluid or elastic solid. That simple methodology allows us to solve a lot of problems, as other chapters of the book show. The third phenomenon, more subtle, needs many more conceptual tools, since it must call upon thermodynamics.

In a first step, we recall basic equations of the mechanics of continua for the rather general case of a thermo-viscous fluid or a thermo-elastic solid. One obtains all material necessary for derivation of the acoustic equations in complex situations.

In a second step we restrict ourselves to perfect simple fluids or elastic solids and linearize equations around an initial homogeneous steady state, leading to basic equations of elementary acoustics and elastic waves. Development of these simple cases are then proposed.

1.1 Review of Mechanics of Continua

Since acoustics is defined as small dynamic perturbations of a fluid or a solid, it is useful to make a quick recapitulation of the mechanics of continua in order to establish equations that are to be linearized. Those equations are conservation equations, the state equation and behaviour equations.

1.1.1. Conservation equations

Conservation equations describe conservation of mass (continuity equation), momentum (motion equation) and energy (from the first law of thermodynamics).

Let us consider a portion of material volume Ω filled with a (piecewise) continuous medium (for greater generality we suppose the existence of a discontinuity surface Σ – a shock wave or an interface – moving at velocity $\vec{V}$); the equations of conservation of mass, momentum, and energy are as follows.

Mass conservation equation (or continuity equation). The hypothesis of continuous medium allows us to introduce the notion of a (piecewise continuous) density function ρ, so that the total mass M of the material volume Ω is $M = \int_\Omega \rho \, d\Omega$; the mass conservation (or continuity) equation is written

$$\frac{d}{dt} \int_\Omega \rho \, d\Omega = 0 \tag{1.1}$$

where d/dt is the material time derivative of the volume integral.

Momentum conservation equation (or motion equation). Let $\vec{v}$ be the local velocity; the momentum of the material volume Ω is defined as $\int_\Omega \rho \vec{v} \, d\Omega$; and, if σ is the stress tensor and $\vec{F}$ the supply of body forces per unit volume (or volumic force source), the momentum balance equation for a volume Ω of boundary S with outward normal $\vec{n}$ is written

$$\frac{d}{dt} \int_\Omega \rho \vec{v} \, d\Omega = \int_S \sigma . \vec{n} \, dS + \int_\Omega \vec{F} \, d\Omega \tag{1.2}$$

Energy conservation equation (first law of thermodynamics). If e is the specific internal energy, the total energy of the material volume Ω is $\int_\Omega \rho(\vec{e} + \frac{1}{2}\vec{v}^2) \, d\Omega$; and, if $\vec{q}$ is the heat flux vector and r the heat supply per unit volume and unit time (or volumic heat source), the energy balance equation for a volume Ω of boundary S with outer normal $\vec{n}$ is written

$$\frac{d}{dt} \int_\Omega \rho(e + \tfrac{1}{2}\vec{v}^2) \, d\Omega = \int_S (\sigma \cdot \vec{v} - \vec{q}) \cdot \vec{n} \, dS + \int_\Omega (\vec{F} \cdot \vec{v} + r) \, d\Omega \tag{1.3}$$

Using the lemma on derivatives of integrals over a material volume Ω crossed by a discontinuity Σ of velocity $\vec{V}$:

$$\forall \phi; \quad \frac{d}{dt} \int_\Omega \phi \, d\Omega = \int_\Omega \left(\frac{\partial \phi}{\partial t} + \nabla \cdot (\phi \vec{v}) \right) d\Omega + \int_\Sigma [\phi(\vec{v} - \vec{V}) \cdot \vec{n}]_\Sigma \, d\Omega$$

where $[\Phi]_\Sigma$ designates the jump $\Phi^{(2)} - \Phi^{(1)}$ of the quantity Φ at the crossing of the discontinuity surface Σ; or

$$\frac{d}{dt}\int_\Omega \phi\, d\Omega = \int_\Omega \left(\frac{d\phi}{dt} + \phi\nabla\cdot\vec{v}\right) d\Omega + \int_\Sigma [\phi(\vec{v}-\vec{V})\cdot\vec{n}]_\Sigma\, d\sigma$$

with

$$\dot{\phi} \equiv \frac{d\phi}{dt} = \frac{\partial\phi}{\partial t} + \vec{v}\cdot\nabla\phi$$

the material time derivative of the function ϕ. Using the formula

$$\forall\vec{U}: \quad \int_S \vec{U}\cdot\vec{n}\, dS = \int_\Omega \nabla\cdot\vec{U}\, d\Omega + \int_\Sigma [\vec{U}\cdot\vec{n}]_\Sigma\, d\sigma$$

one obtains

$$\int_\Omega \left(\frac{d\rho}{dt} + \rho\nabla\cdot\vec{v}\right) d\Omega + \int_\Sigma [\rho(\vec{v}-\vec{V})\cdot\vec{n}]_\Sigma\, d\sigma = 0$$

$$\int_\Omega \left(\frac{d}{dt}(\rho\vec{v}) + \rho\vec{v}\nabla\cdot\vec{v}\right) d\Omega + \int_\Sigma [(\rho\vec{v}\otimes(\vec{v}-\vec{V}))\cdot\vec{n}]_\Sigma\, d\sigma$$
$$= \int_\Omega \nabla\cdot\sigma\, d\Omega + \int_\Sigma [\sigma\cdot\vec{n}]_\Sigma\, d\sigma + \int_\Omega \vec{F}\, d\Omega$$

$$\int_\Omega \left(\frac{d}{dt}\rho(e+\tfrac{1}{2}\vec{v}^2) + \rho(e+\tfrac{1}{2}\vec{v}^2)\nabla\cdot\vec{v}\right) d\Omega + \int_\Sigma [\rho(e+\tfrac{1}{2}\vec{v}^2)(\vec{v}-\vec{V})\cdot\vec{n}]_\Sigma\, d\sigma$$
$$= \int_\Omega \nabla\cdot(\sigma\cdot\vec{v}-\vec{q})\, d\Omega + \int_\Sigma [(\sigma\cdot\vec{v}-\vec{q})\cdot\vec{n}]_\Sigma\, d\sigma + \int_\Omega (\vec{F}\cdot\vec{v}+r)\, d\Omega$$

or

$$\int_\Omega \left(\frac{d\rho}{dt} + \rho\nabla\cdot\vec{v}\right) d\Omega + \int_\Sigma [\rho(\vec{v}-\vec{V})\cdot\vec{n}]_\Sigma\, d\sigma = 0$$

$$\int_\Omega \left(\frac{d}{dt}(\rho\vec{v}) + \rho\vec{v}\nabla\cdot\vec{v} - \nabla\cdot\sigma - \vec{F}\right) d\Omega + \int_\Sigma [(\rho\vec{v}\otimes(\vec{v}-\vec{V}) - \sigma)\cdot\vec{n}]_\Sigma\, d\sigma = 0$$

$$\int_\Omega \left(\frac{d}{dt}(\rho(e+\tfrac{1}{2}\vec{v}^2)) + \rho(e+\tfrac{1}{2}\vec{v}^2)\nabla\cdot\vec{v} - \nabla\cdot(\sigma\cdot\vec{v}-\vec{q}) - (\vec{F}\cdot\vec{v}+r)\right) d\Omega$$
$$+ \int_\Sigma [(\rho(e+\tfrac{1}{2}\vec{v}^2)(\vec{v}-\vec{V}) - (\sigma\cdot\vec{v}-\vec{q}))\cdot\vec{n}]_\Sigma\, d\sigma = 0$$

The continuity hypothesis states that all equations are true for any material volume Ω. So one finds the local forms of the conservation equations:

$$\frac{d\rho}{dt} + \rho\nabla \cdot \vec{v} = 0 \qquad [\rho(\vec{v} - \vec{V}) \cdot \vec{n}]_\Sigma = 0$$

$$\frac{d}{dt}(\rho\vec{v}) + \rho\vec{v}\nabla \cdot \vec{v} - \nabla \cdot \sigma - \vec{F} = 0 \qquad [(\rho\vec{v} \otimes (\vec{v} - \vec{V}) - \sigma) \cdot \vec{n}]_\Sigma = 0$$

$$\frac{d}{dt}(\rho(e + \tfrac{1}{2}\vec{v}^2)) + \rho(e + \tfrac{1}{2}\vec{v}^2)\nabla \cdot \vec{v} - \nabla \cdot (\sigma \cdot \vec{v} - \vec{q}) - (\vec{F} \cdot \vec{v} + r) = 0$$

$$[(\rho(e + \tfrac{1}{2}\vec{v}^2)(\vec{v} - \vec{V}) - (\sigma \cdot \vec{v} - \vec{q})) \cdot \vec{n}]_\Sigma = 0$$

So at the discontinuities, one obtains

$$\begin{cases} [\rho(\vec{N} - \vec{V}) \cdot \vec{n}]_\Sigma = 0 \\ [(\rho\vec{v} \otimes (\vec{v} - \vec{V}) - \sigma) \cdot \vec{n}]_\Sigma = 0 \\ [(\rho(e + \frac{1}{2}\vec{N}^2)(\vec{N} - \vec{V}) - (\sigma \cdot \vec{N} - \vec{q})) \cdot \vec{n}]_\Sigma = 0 \end{cases} \tag{1.4}$$

Away from the discontinuity surfaces combining the first three local forms of the conservation equations above, one obtains

$$\begin{cases} \dot{\rho} + \rho\nabla \cdot \vec{v} = 0 \\ \rho\vec{v} - \nabla \cdot \sigma = \vec{F} \\ \rho\dot{e} + \nabla \cdot \vec{q} = \sigma : D + r \end{cases} \tag{1.5}$$

or

$$\begin{cases} \dfrac{\partial\rho}{\partial t} + \nabla \cdot (\rho\vec{v}) = 0 \\ \rho\left(\dfrac{\partial\vec{v}}{\partial t} + \vec{v} \cdot \nabla\vec{v}\right) - \nabla \cdot \sigma = \vec{F} \\ \rho\left(\dfrac{\partial e}{\partial t} + \vec{v} \cdot \nabla e\right) + \nabla \cdot \vec{q} = \sigma : D + r \end{cases} \tag{1.6}$$

with $D = \frac{1}{2}(\nabla\vec{v} + {}^{\mathrm{T}}\nabla\vec{v})$ the strain rate tensor.

1.1.2. State equation

The state equation is the equation that describes the thermodynamic state of the material being considered. In particular, it describes whether it is a solid or a fluid.

A material admits a certain number of independent thermodynamic state variables. Given this number, one can choose any as primary ones, provided that they are independent. The others will become secondary state variables and are generally called state functions, since they become functions of the primary state variables. The state equation is then the equation that describes the dependance of the specific internal energy e upon the chosen primary state variables.

In acoustics, since motions are isentropic, it is judicious to choose as a first primary state variable the specific entropy s, whose existence is postulated by the second law of thermodynamics. For the choice of the others, one must distinguish between fluids and solids.

Simple fluid

A simple fluid is a medium that has the same response to loads from any two reference configurations having the same density. Thus the density ρ – or its inverse $v = \rho^{-1}$, the specific volume – constitutes another natural primary thermodynamic state variable.

The (caloric) state equation is the equation that describes the functional dependance of the specific internal energy upon the chosen pair of primary state variables (s, v):

$$e = e(s, v) \qquad \text{or} \qquad e = e(s, \rho)$$

Generally this equation *does not exist explicitly* (except for a perfect gas), but some measures allow us to reach some *local form* (gradient and Hessian) of the function around a given state point, sufficient to describe small perturbations, such as acoustic movements are, around a state. For example, the measurement of temperature T and pressure p at a given state point allows us to obtain the differential of the function $e(s, v)$ at this point:

$$de = T\,ds - p\,dv \tag{1.7}$$

This equation, called the Gibbs equation, defines the (thermodynamic) temperature $T = (\partial e/\partial s)_v$ and the (thermodynamic) pressure $p = -(\partial e/\partial v)_s$. A new general writing of the state equation is:

$$\begin{aligned} T &= T(s, v) \qquad \text{or} \qquad T = T(s, \rho) \\ p &= p(s, v) \qquad \text{or} \qquad p = p(s, \rho) \end{aligned}$$

The Gibbs equation also allows us to rewrite the energy conservation equation as:

$$T\rho\dot{s} = \tau : D - \nabla \cdot \vec{q} + r \tag{1.8}$$

having defined the *viscosity stress tensor*:

$$\tau = \sigma + pI \qquad (I \text{ is the unit tensor}) \tag{1.9}$$

Other measures allow us to obtain the second derivatives of the function $e(s, v)$, i.e. the first derivatives of the functions $T(s, v)$ and $p(s, v)$. With the chosen pair of primary state variables (s, v), they are the specific heat at constant volume, C_v, the isentropic (or adiabatic) compressibility χ_s, and the isentropic (or adiabatic) expansivity α_s. These can be written in the synthetic form:

$$\left\{ \begin{aligned} dT &= \frac{T}{C_v}\,ds - \frac{1}{\alpha_s v}\,dv \\ dp &= \frac{1}{\alpha_s v}\,ds - \frac{1}{\chi_s v}\,dv \end{aligned} \right. \qquad \text{or} \qquad \left\{ \begin{aligned} dT &= \frac{T}{C_v}\,ds + \frac{1}{\alpha_s \rho}\,d\rho \\ dp &= \frac{\rho}{\alpha_s}\,ds + \frac{1}{\chi_s \rho}\,d\rho \end{aligned} \right. \tag{1.10}$$

where

$$\frac{T}{C_v} = \left(\frac{\partial^2 e}{\partial s^2}\right)_v = \left(\frac{\partial T}{\partial s}\right)_v$$

defines the specific heat at constant volume, $C_v = T(\partial s/\partial T)_v$;

$$\frac{1}{\chi_s v} = \left(\frac{\partial^2 e}{\partial v^2}\right)_s = -\left(\frac{\partial p}{\partial v}\right)_s$$

defines the isentropic (or adiabatic) compressibility, $\chi_s = -(1/v)(\partial v/\partial p)_s$;

$$-\frac{1}{\alpha_s v} = \left(\frac{\partial^2 e}{\partial s\, \partial v}\right) = \left(\frac{\partial T}{\partial v}\right)_s = -\left(\frac{\partial p}{\partial s}\right)_v$$

defines the isentropic (or adiabatic) expansivity $\alpha_s = -(1/v)(\partial v/\partial T)_s$.

For acoustics, since one is concerned with very small fluctuations (typically 10^{-7}, always less than 10^{-3}), it will be sufficient to give these quantities at the chosen working state point $T_0,\ p_0, \ldots$.

Elastic solid. A solid exhibits different responses to loads from reference configurations involving deformation of the material in going from one to the other. So the strain tensor $\varepsilon = \frac{1}{2}(\nabla\vec{u} + {}^{\mathrm{T}}\nabla\vec{u})$ ($\vec{u}$ being the displacement), constitutes another natural primary variable, and the state equation takes the form, for an elastic solid (which is adequate to describe the small deformations involved in acoustics):

$$e = e(s, \varepsilon)$$

Here also, one knows no explicit formulation of such an equation and, for acoustics, one needs only the first and second derivatives of e.

The first derivative of the state equation can be written

$$de = T\, ds + \left(\frac{\sigma}{\rho}\right) : d\varepsilon \tag{1.11}$$

defining temperature $T = (\partial e/\partial s)_\varepsilon$ and stress tensor $\sigma_{ij} = \rho(\partial e/\partial \varepsilon_{ij})_{s,\varepsilon_{kl(kl \neq ij)}}$. This allows us to rewrite the energy conservation equation as

$$T\, \rho\dot{s} = -\nabla \cdot \vec{q} + r \tag{1.12}$$

Remark: One can decompose strain and stress tensors into spherical (index S) and deviatoric (index D) parts where tr D stands for the trace of the tensor D:

$$\varepsilon = \varepsilon^S + \varepsilon^D, \qquad \sigma = \sigma^S + \sigma^D, \qquad \text{with} \qquad \varepsilon^S = \tfrac{1}{3}(\mathrm{tr}\ \sigma)I, \qquad \sigma^S = \tfrac{1}{3}(\mathrm{tr}\ \sigma)\vec{I},$$

so that $\mathrm{tr}\ \varepsilon^S = \mathrm{tr}\ \varepsilon$, $\mathrm{tr}\ \varepsilon^D = 0$, $\mathrm{tr}\ \sigma^S = \mathrm{tr}\ \sigma$, $\mathrm{tr}\ \sigma^D = 0$.

The first differential becomes

$$de = T\,ds - P\,dv + \left(\frac{\sigma^D}{\rho}\right) : d\varepsilon^D$$

with $P \equiv -\frac{1}{3}\mathrm{tr}\ \sigma$ being the hydrostatic pressure and $dv/v = \mathrm{tr}(d\varepsilon)$ the dilatation.

The hydrostatic pressure is defined by an expression similar to that for the thermodynamic pressure for a fluid: $P = -(\partial e/\partial v)_{s,\varepsilon^D}$

Second derivatives of the state equation can be written in the synthetic form

$$\begin{cases} dT = \dfrac{T}{C_\varepsilon}\,ds + \displaystyle\sum_{ij} \beta_{ij}\,d\varepsilon_{ij} \\ d\left(\dfrac{\sigma_{ij}}{\rho}\right) = \beta_{ij}\,ds + \displaystyle\sum_{kl} \gamma_{ijkl}\,d\varepsilon_{kl} \qquad \text{or} \qquad d\sigma_{ij} = \rho\beta_{ij}\,ds + \displaystyle\sum_{kl} C_{ijkl}\,d\varepsilon_{kl} \end{cases}$$

with

$$\frac{T}{C_\varepsilon} = \left(\frac{\partial^2 e}{\partial s^2}\right)_\varepsilon = \left(\frac{\partial T}{\partial s}\right)_\varepsilon, \qquad \beta_{ij} = \left(\frac{\partial T}{\partial \varepsilon_{ij}}\right)_{s,\,\varepsilon_{kl(kl\neq ij)}} = \left(\frac{\partial^2 e}{\partial s\,\partial \varepsilon_{ij}}\right)_{s,\,\varepsilon_{kl(kl\neq ij)}},$$

$$\gamma_{ijkl} = \left(\frac{\partial(\sigma_{ij}/\rho)}{\partial \varepsilon_{kl}}\right)_s = \left(\frac{\partial^2 e}{\partial \varepsilon_{ij}\,\partial \varepsilon_{kl}}\right)_s$$

$C_{ijkl} = \rho\gamma_{ijkl} - \sigma_{ij}\delta_{kl}$ (with δ_{ij} the Kronecker symbol) or, in tensoral notation, $C = \rho\gamma - \sigma \otimes I$) defining the specific heat $C_\varepsilon = T(\partial s/\partial T)_\varepsilon$ and the stiffness tensor $C\ :\ C_{ijkl} = (\partial\sigma_{ij}/\partial\varepsilon_{kl})_s$

The equations for second derivatives can be rewritten, in tensorial notation,

$$\begin{cases} dT = \dfrac{T}{C_\varepsilon}\,ds + \beta : d\varepsilon \\ d\sigma = \rho\beta\,ds + \vec{C} : d\varepsilon \end{cases} \tag{1.13}$$

For an isotropic elastic solid the stiffness tensor will present only two independent components, and one can write $C_{ijkl} = \lambda\delta_{ij}\delta_{kl} + 2\mu\delta_{ik}\delta_{jl}$ with λ, μ being Lamé coefficients. Then

$$d\sigma = \rho\beta\,ds + \lambda\ \mathrm{tr}(d\varepsilon)I + 2\mu\,d\varepsilon$$

and, for isentropic deformations,

$$d\sigma = \lambda\ \mathrm{tr}(d\varepsilon)I + 2\mu\,d\varepsilon \tag{1.14}$$

These two examples of fluid and solid state equations are the simplest one can write, since they involve the minimum allowable number of independent state variables: specific entropy and specific volume for a simple fluid; specific entropy and strain tensor for an ideal elastic solid.

If one has to take into account other phenomena than purely thermo-mechanical phenomena, such as electrical or chemical effects, one has to introduce other independent state variables. For example, if a fluid is not a pure substance and is subjected to chemical reactions involving N species α, one has to introduce, in addition to specific entropy s and specific volume v, the $N-1$ independent concentrations c_α, $\alpha \in [1, N-1]$ (since $\sum_{\alpha=1}^{N} c_\alpha = 1$). Then the state equation is $e = e(s, v, c_1, c_2, ..., c_{N-1})$ and its first differential is $de = T\,ds - p\,dv + \sum_{\alpha=1}^{N-1} \mu_\alpha\, dc_\alpha$, defining temperature T, pressure p, and chemical potentials μ_α.

1.1.3. Constitutive equations

Constitutive equations give details of the behaviour of the material, specifying all the transport phenomena that occur in the medium; they are often called the phenomenological equations or the transport equations. Often the state equation is considered as one of them, simply more general than the others. The constitutive equations close the system of equations: with them one has the same number of equations as of unknowns.

A simple way to generate the constitutive equations is to use the concepts of generalized irreversible forces and fluxes and their proportionality in the framework of linear irreversible thermodynamics, which is widely sufficient for acoustics.

We begin by rewriting the energy conservation equation transformed by the state equation (i.e. with specific entropy instead of specific energy as state variable) in the shape of a balance equation:

$$\rho\dot{s} + \nabla \cdot \vec{J}_s = \phi_s \tag{1.15}$$

where $\vec{J}_s$ and ϕ_s are the input flux and source of entropy. This form of the result is not unique, but all formulations are equivalent. For example:

For a simple viscous fluid:

$$\rho\dot{s} + \nabla \cdot (T^{-1}\vec{q}) = \tau : (T^{-1}D) + \vec{q} \cdot \nabla(T^{-1}) + rT^{-1}$$

giving

$$\vec{J}_s = T^{-1}\vec{q} \qquad \text{and} \qquad \phi_s = \tau : (T^{-1}D) + \vec{q} \cdot \nabla(T^{-1}) + rT^{-1}$$

For a simple elastic solid:

$$\rho\dot{s} + \nabla \cdot (T^{-1}\vec{q}) = \vec{q} \cdot \nabla(T^{-1}) + rT^{-1}$$

giving

$$\vec{J}_s = T^{-1}\vec{q} \qquad \text{and} \qquad \phi_s = +\,\vec{q} \cdot \nabla(T^{-1}) + rT^{-1}$$

If now one decomposes the heat source r into its two components, an external one r^e and an internal one r^i (due to heat radiation or thermal relaxation processes for example), the entropy source ϕ_s is decomposed into two terms: one of external origin

$$\phi_s^e = r^e T^{-1}$$

and one of internal origin

$$\phi_s^i = \tau : (T^{-1}D) + \vec{q} \cdot \nabla(T^{-1}) + r^i T^{-1} \qquad \text{for a simple viscous fluid}$$
$$\phi_s^i = \vec{q} \cdot \nabla(T^{-1}) + r^i T^{-1} \qquad \text{for a simple elastic solid.}$$

The second law of thermodynamics postulates that, due to irreversible processes (dissipation phenomena), the internal entropy production rate ϕ_s^i is always positive or null (null when no irreversible processes occur or at equilibrium):

$$\phi_s^i \geqslant 0 \qquad \text{(Clausius–Duhem inequality)}$$

Within the framework of *linear irreversible thermodynamics*, which is sufficient for acoustics – i.e. assuming the system to be always in the vicinity of an equilibrium state ρ_e, T_e, etc. – one can linearize equations with respect to deviations from that reference state $\delta\rho$, δT, etc.:

$$\phi_s^i = (\tau) : (T^{-1}D) + (T^{-1}\vec{q}) \cdot (-T^{-1}\ \nabla T) + (T^{-1}r^i)(-T^{-1}\ \delta T)$$
$$\text{for a thermo-viscous fluid}$$
$$\phi_s^i = (T^{-1}\vec{q}) \cdot (-T^{-1}\ \nabla T) + (T^{-1}r^i)(-T^{-1}\ \delta T) \qquad \text{for a thermo-elastic solid}$$

Then noting that

$$X = (T^{-1}D, -T^{-1}\ \nabla T, -T^{-1}\ \delta T) \qquad \text{for a fluid}$$
$$X = (-T^{-1}\ \nabla T, -T^{-1}\ \delta T) \qquad \text{for a solid}$$

where the components X_j are called *generalized thermodynamic forces*, and

$$Y = (\tau, T^{-1}\vec{q}, T^{-1}r^i) \qquad \text{for a fluid}$$
$$Y = (T^{-1}\vec{q}, T^{-1}r^i) \qquad \text{for a solid}$$

where the components Y_i are called *generalized thermodynamic fluxes*, one can write the internal entropy production as a bilinear functional:

$$\phi_s^i = Y \cdot X \qquad \text{or} \qquad \phi_s^i = Y_i X_i.$$

Within the framework of linear irreversible thermodynamics, i.e. assuming we remain in the vicinity of an equilibrium state (where both fluxes and forces vanish), one assumes proportionality of fluxes with forces:

$$Y = L \cdot X \qquad \text{or} \qquad Y_i = L_{ij} X_j \tag{1.16}$$

so that generalized fluxes and forces, the former generally considered as consequences of the latter, vanish together at equilibrium.

The equations $Y_i = L_{ij}X_j$ are the desired *constitutive equations*, or *phenomenological equations*; the coefficients L_{ij} are called the *phenomenological coefficients*. These coefficients must satisfy certain laws:

- The Curie principle that states that there may exist no coupling between fluxes and forces having a different tensorial nature (a vector can be coupled only with vectors, etc.).
- The Onsager reciprocal relations: $L_{ij} = L_{ji}$.

Thus the phenomenological equations can be written, with suitable choice of parameters, as follows.

For a simple thermo-viscous fluid:

$$\tau = \lambda T I \ \mathrm{tr}(T^{-1} D) + 2\mu T(T^{-1} D), \qquad T^{-1}\vec{q} = k(-T^{-1}\ \nabla T),$$

$$T^{-1} r^i = \frac{\rho c}{\tau}(-T^{-1}\ \delta T)$$

leading to the classical Newtonian law of viscosity, Fourier's law of heat conduction, and Newton's law of cooling:

$$\tau = \lambda I \ \mathrm{tr}\ D + 2\mu \vec{D}, \qquad \vec{q} = -k\ \nabla T, \qquad r^i = -\frac{\rho C}{\tau}\ \delta T \tag{1.17}$$

For a simple thermoelastic solid, one obtain the same relations for heat conduction and heat radiation.

Here, since there is only one representative of each tensorial order in generalized forces and fluxes, there is no coupling. It would have been different if one has taken into account, for example, diffusion of species in a chemically reactive medium since species diffusion, like heat diffusion, results in a vector generalized flux, according to Fick's law of diffusion. So there would be coupling between heat diffusion and species diffusion: the so-called Dufour's and Soret's effects.

Remark: The framework of linear thermodynamics provides a domain of validity considerably wider than acoustics: it allows all classical linear constitutive relations. Only highly non-linear irreversible processes, strong departures from equilibrium and instabilities are excluded.

With conservation equations, state equations and constitutive equations, one now has as many equations as unknowns. We are ready for the linearization. Before that, we look at equations at discontinuities founded during the establishment of the conservation equations, since they play a prominent role in acoustics, for problems with interfaces and boundary problems.

1.1.4. Equations at discontinuities

We return to equations at discontinuities established when deriving local equations of conservation from the global ones for a material volume:

$$\begin{cases} [\rho(\vec{v} - \vec{V}) \cdot \vec{n}]_\Sigma = 0 \\ [(\rho\vec{v} \otimes (\vec{v} - \vec{V}) - \sigma) \cdot \vec{n}]_\Sigma = 0 \\ [(\rho(e + \frac{1}{2}\vec{v}^2)(\vec{v} - \vec{V}) - (\sigma \cdot \vec{v} - \vec{q})) \cdot \vec{n}]_\Sigma = 0 \end{cases} \tag{1.18}$$

where $[\Phi]_\Sigma$ designates the jump $\Phi^{(2)} - \Phi^{(1)}$ of the quantity Φ at the crossing of the discontinuity surface Σ.

Discontinuities can be encountered in two cases: shock waves and interfaces.

Shock waves
Such waves are characterized by the fact that matter effectively crosses the discontinuity surface. The earlier equations do not simplify.

Interfaces
In the case of an interface between two immiscible media (a perfect interface), matter does not cross the discontinuity surface and $\vec{v} \cdot \vec{n} = \vec{V} \cdot \vec{n}$ ($\vec{V}$ = velocity of the discontinuity surface Σ) in either direction.

Hence, $\vec{v}^{(1)} \cdot \vec{n} = \vec{v}^{(2)} \cdot \vec{n} = \vec{V} \cdot \vec{n}$, or $[\vec{v} \cdot \vec{n}]_\Sigma = 0$: there is continuity of the normal velocity of the medium. The second equation (1.18) becomes $[\sigma \cdot \vec{n}]_\Sigma = 0$: there is continuity of the normal stress. The third equation (1.18) becomes $[\vec{q} \cdot \vec{n}]_\Sigma = 0$: there is continuity of the normal heat flux. So at the crossing of a perfect interface, normal velocity, normal stress and normal heat flux are continuous.

Solid–solid interface. For an adhesive interface, $\vec{v}^{(1)} = \vec{V}$, $\vec{v}^{(2)} = \vec{V}$, so $[\vec{v}]_\Sigma = 0$: there is continuity of the velocity (and of the normal stress: $[\sigma \cdot \vec{n}]_\Sigma = 0$). For a non-adhesive interface (sliding interface), $\vec{v}^{(1)} \neq \vec{v}^{(2)}$, but $[\vec{v} \cdot \vec{n}]_\Sigma = 0$: there is only continuity of the normal velocity (and of the normal stress: $[\sigma \cdot \vec{n}]_\Sigma = 0$).

Fluid–solid interface. For a viscous fluid, there is adhesion and the conditions are the same as for adhesive bond between two solids: continuity of velocity and of normal stress, $[\vec{v}]_\Sigma = 0$, $[\sigma \cdot \vec{n}]_\Sigma = 0$. For a non-viscous fluid, there is sliding at the interface and only continuity of the normal velocity ($[\vec{v} \cdot \vec{n}]_\Sigma = 0$) and of normal stress, $[\sigma \cdot \vec{n}]_\Sigma = 0$; that is $-p^{(1)}\vec{n} = \sigma^{(2)} \cdot \vec{n}$.

Fluid–fluid interface. For two viscous fluids, there is adhesion at the interface, and the conditions are the same as for a bond between two solids: continuity of the velocity and of the normal stress, $[\vec{v}]_\Sigma = 0$, $[\sigma \cdot \vec{n}]_\Sigma = 0$. If one of the fluids is viscous and the other not, there is sliding, and the conditions are the same as for an interface between a non-viscous fluid and a solid: continuity of the normal velocity and of the normal stress, $[\vec{v} \cdot \vec{n}]_\Sigma = 0$, $[\sigma \cdot \vec{n}]_\Sigma = 0$. If the two fluids are perfect, i.e. non-viscous, there is also sliding and so continuity of the normal velocity only, $[\vec{v} \cdot \vec{n}]_\Sigma = 0$, and of the normal stress, $[\sigma \cdot \vec{n}]_\Sigma = 0$; that is $-p^{(1)}\vec{n} = -p^{(2)}\vec{n}$, i.e. $p^{(1)} = p^{(2)}$, or $[p]_\Sigma = 0$, i.e. continuity of the pressure.

These last continuity conditions are classical in acoustics: generally one considers *continuity of pressure* and *normal velocity* at interfaces.

1.2. Elementary Acoustics

Here we are interested in the simplest acoustics, linear acoustics (which describes low energy phenomena only) of a perfect simple fluid (subject to no dissipative phenomena) in an initial stationary homogeneous equilibrium. The main tool is linearization of equations, which reveals the main feature of sound: its wave nature. This wave propagation is the first phenomenon encountered in acoustics. The subsequent ones are wave refraction, reflection and diffraction, consecutive to interaction with interfaces and boundaries.

1.2.1. Linearization for a lossless homogeneous steady simple fluid

For a general (visco-thermal) fluid, the conservation equations are, from (1.6), (1.8) and (1.9)

$$\begin{cases} \dfrac{\partial \rho}{\partial t} + \nabla \cdot (\rho \vec{v}) = 0 \\ \rho \left(\dfrac{\partial \vec{v}}{\partial t} + \vec{v} \cdot \nabla \vec{v} \right) + \nabla p = \nabla \cdot \tau + \vec{F} \\ T\rho \left(\dfrac{\partial s}{\partial t} + \vec{v} \cdot \nabla s \right) = \tau : D - \nabla \cdot \vec{q} + r \end{cases} \tag{1.19}$$

where $\tau = \sigma + pI$ is the viscosity tensor and $D = \frac{1}{2}(\nabla \vec{v} + {}^{\mathrm{T}}\nabla \vec{v})$ strain rate tensor.

For a perfect fluid, one has $\tau = 0$ (no viscosity), $\vec{q} = 0$ (no heat conduction), $r^i = 0$ (no heat radiation) and so (omitting the index e for the external volumic heat source r^e):

$$\begin{cases} \dfrac{\partial \rho}{\partial t} + \nabla \cdot (\rho \vec{v}) = 0 \\ \rho \left(\dfrac{\partial \vec{v}}{\partial t} + \vec{v} \cdot \nabla \vec{v} \right) + \nabla p = \vec{F} \\ T\rho \left(\dfrac{\partial s}{\partial t} + \vec{v} \cdot \nabla s \right) = r \end{cases} \tag{1.20}$$

One can now linearize these equations around the homogeneous steady state $\rho^0 = ct$, $\vec{v}^0 = \vec{0}$, $p^0 = ct$, $T^0 = ct$, $s^0 = ct$, $\vec{F} = 0$, $r^0 = 0$.

Let ρ^1, $\vec{v}^1$, p^1, T^1, s^1 be the perturbations from that equilibrium induced by the source perturbations f^1, $\vec{F}^1$, r^1 (f^1 describes a small volumic source of mass): $\rho = \rho^0 + \rho^1$, $\vec{v} = \vec{v}^1$, $p = p^0 + p^1$, $T = T^0 + T^1$, $s = s^0 + s^1$. Identifying terms of the same order with respect to the perturbation, one obtains:

- At zeroth order, the equations governing the initial stationary homogeneous state, i.e. the tautology $0 = 0$.

- At first order, the first order equations governing perturbations, i.e. the equations of linear acoustics:

$$\begin{cases} \dfrac{\partial \rho^1}{\partial t} + \nabla \cdot (\rho^0 \vec{v}^1) = f^1 \\ \rho^0 \dfrac{\partial \vec{v}^1}{\partial t} + \nabla p^1 = \vec{F}^1 \\ T^0 \rho^0 \dfrac{\partial s^1}{\partial t} = r^1 \end{cases} \tag{1.21}$$

with, by a first order development of the state equations $T = T(s, \rho)$, $p = p(s, \rho)$ around (s_0, ρ_0), i.e. assimilating finite difference and differential in expressions of dT, dp in (1.10):

$$\begin{cases} T^1 = \dfrac{T^0}{C_v^0} s^1 + \dfrac{1}{\alpha_s^0 \rho^0} \rho^1 \\ p^1 = \dfrac{\rho^0}{\alpha_s^0} s^1 + \dfrac{1}{\chi_s^0 \rho^0} \rho^1 \end{cases} \tag{1.22}$$

1.2.2. Equations for entropy and vorticity: fundamental character of acoustic motion

The *linearized entropy balance equation* can be directly integrated:

$$s^1 = (\rho^0 T^0)^{-1} \int r^1 \, dt \tag{1.23}$$

This shows that s^1 and r^1 have the same spatial support, i.e. that s^1 vanishes outside the heat source r^1. One thus has the first property of acoustic movements (in perfect homogeneous fluids): *acoustic movements* (in perfect homogeneous fluids) *are isentropic* (or adiabatic) out of heat sources. This property is true only outside heat sources. In fact we will show later that acoustic entropy spreads out of heat sources (diffusion by heat conduction), so that there exists around heat sources a small zone of diffusion where entropy does not vanish. If there is no heat source, the isentropy property is valid everywhere, even at the location of other sources.

The *linearized momentum balance equation* gives, after applying the curl operator,

$$\rho^0 \frac{\partial}{\partial t} (\nabla \times \vec{v}^1) = \nabla \times \vec{F}^1$$

and so

$$\nabla \times \vec{v}^1 = (\rho^0)^{-1} \int \nabla \times \vec{F}^1 \, dt \tag{1.24}$$

Thus $\nabla \times \vec{v}^1$ vanishes outside curl sources $\nabla \times \vec{F}^1$, and one has the second property of acoustic movements: *acoustic movements* (in perfect homogeneous fluids) *are irrotational* out of velocity curl sources. As for entropy, one shows that in fact acoustic velocity curl spreads out of the curl sources (diffusion by viscosity), so that there exists around curl sources a small zone of diffusion where velocity curl does not vanish.

Consequence of isentropy. One has

$$\begin{cases} T^1 = \dfrac{1}{\alpha_s^0 \rho^0}\,\rho^1 + \dfrac{1}{C_v^0 \rho^0} \displaystyle\int r^1\,dt \\[2ex] p^1 = \dfrac{1}{\chi_s^0 \rho^0}\,\rho^1 + \dfrac{1}{\alpha_s^0 T^0} \displaystyle\int r^1\,dt \end{cases} \tag{1.25}$$

Also, by extracting ρ^1 from the second equation,

$$\rho^1 = \chi_s^0 \rho^0 p^1 - \frac{\chi_s^0 \rho^0}{\alpha_s^0 T^0} \int r^1\,dt \tag{1.26}$$

and substituting it in the first one,

$$T^1 = \frac{\chi_s^0}{\alpha_s^0}\,p^1 + \frac{1}{C_v^0 \rho^0}\left(1 - \frac{\rho^0 C_v^0 \chi_s^0}{(\alpha_s^0)^2 T^0}\right) \int r^1\,dt \tag{1.27}$$

Alternatively, introducing isobaric thermal expansivity $\alpha_p = -(1/v)(\partial v/\partial T)_p$, specific heat at constant pressure $C_p = T(\partial s/\partial T)_p$, and ratio of specific heats $\gamma = C_p/C_v$ (and noting that $\alpha_s = \alpha_p/(\gamma - 1)$):

$$\begin{cases} T^1 = \dfrac{\gamma^0 - 1}{\alpha_p^0 \rho^0}\,\rho^1 + \dfrac{\gamma^0}{C_p^0 \rho^0} \displaystyle\int r^1\,dt \\[2ex] \rho^1 = \chi_s^0 \rho^0 p^1 - \dfrac{\alpha_p^0}{C_p^0} \displaystyle\int r^1\,dt \qquad \left(\text{since } \dfrac{\rho^0 \chi_s^0}{\alpha_s^0 T^0} = \dfrac{\alpha_p^0}{C_p^0}\right) \\[2ex] T^1 = \dfrac{\alpha_p^0 T^0}{\rho^0 C_p^0}\,p^1 + \dfrac{1}{C_p^0 \rho^0} \displaystyle\int r^1\,dt \qquad \left(\text{since } \dfrac{\rho^0 C_p^0 \chi_s^0}{(\alpha_s^0)^2 T^0} = \gamma^0 - 1\right) \end{cases} \tag{1.28}$$

One sees that, out of heat sources, *all scalar components of the acoustic field are proportional to each other* (in a perfect fluid). The only exception is acoustic entropy, which moves alone, independently of the acoustic field, and vanishes out of heat sources.

1.2.3. Equations for pressure and other acoustic quantities: wave equations

Acoustic pressure is the main quantity that characterizes acoustic fields, since most acoustic gauges, such as the ear for example, are pressure sensitive. Hence one is first interested in that quantity.

By replacing in the linearized system (1.21) the acoustic density ρ^1 by its value taken from (1.28),

$$\rho^1 = \chi_s^0 \rho^0 p^1 - \frac{\alpha_p^0}{C_p^0} \int r^1 \, dt$$

one has

$$\begin{cases} \chi_s^0 \rho^0 \dfrac{\partial p^1}{\partial t} + \rho^0 \nabla \cdot \vec{v}^1 = f^1 + \dfrac{\alpha_p^0}{C_p^0} r^1 \\ \rho^0 \dfrac{\partial \vec{v}^1}{\partial t} + \nabla p^1 = \vec{F}^1 \end{cases} \tag{1.29}$$

By applying $-(\partial/\partial t)$(first equation) + $\nabla \cdot$ (second equation), one obtains

$$-\chi_s^0 \rho^0 \frac{\partial^2 p^1}{\partial t^2} + \nabla^2 p^1 = -\frac{\partial f^1}{\partial t} + \nabla \cdot \vec{F}^1 + \frac{\alpha_p^0}{C_p^0} \frac{\partial r^1}{\partial t}$$

Applying $-\chi_s^0(\partial/\partial t)$(second equation) + $(1/\rho^0)\nabla$(first equation), one obtains

$$-\chi_s^0 \rho^0 \frac{\partial^2 \vec{v}^1}{\partial t^2} + \nabla(\nabla \cdot \vec{v}^1) = \frac{1}{\rho^0} \nabla f^1 - \chi_s^0 \frac{\partial \vec{F}^1}{\partial t} + \frac{\alpha_p^0}{\rho^0 C_p^0} \nabla r^1$$

With $\chi_s^0 \rho^0 = 1/c_0^2$, where c_0 has the dimension of a velocity, one has the following.

For the acoustic pressure:

$$-\frac{1}{c_0^2} \frac{\partial^2 p^1}{\partial t^2} + \nabla^2 p^1 = -\frac{\partial f^1}{\partial t} + \nabla \cdot \vec{F}^1 + \frac{\alpha_p^0}{C_p^0} \frac{\partial r^1}{\partial t} \tag{1.30}$$

So acoustic pressure (and all other scalar components of the acoustic field) is governed by a wave equation (also named D'Alembert's equation), with wave velocity

$$c_0 = \frac{1}{\sqrt{\chi_s^0 \rho^0}} \tag{1.31}$$

This velocity is the velocity of acoustic waves, or the speed of sound in the case of sound waves.

The acoustic field is then a wave field and exhibits the phenomenology of wave fields: propagation, reflection/refraction, and diffraction.

For the acoustic velocity, using the identity: $\forall \vec{u}$, $\nabla^2 \vec{u} \equiv \nabla \cdot (\nabla \vec{u}) = \nabla(\nabla \vec{u}) - \nabla \times \nabla \times \vec{u}$:

$$-\frac{1}{c_0^2} \frac{\partial^2 \vec{v}^1}{\partial t^2} + \nabla^2 \vec{v}^1 = \frac{1}{\rho^0} \left(\nabla f^1 - \frac{1}{c_0^2} \frac{\partial \vec{F}^1}{\partial t} - \int \nabla \times \nabla \times \vec{F}^1 \, dt + \frac{\alpha_p^0}{C_p^0} \nabla r^1 \right) \tag{1.32}$$

For other acoustic quantities, one obtains similar wave equations, the only changing term being the source:

$$-\frac{1}{c_0^2}\frac{\partial^2 \rho^1}{\partial t^2}+\nabla^2\rho^1=\frac{1}{c_0^2}\left(-\frac{\partial f^1}{\partial t}+\nabla\cdot\vec{F}^1\right)-\frac{\alpha_p^0}{C_p^0}\int\nabla^2 r^1\,dt \tag{1.33}$$

$$-\frac{1}{c_0^2}\frac{\partial^2 T^1}{\partial t^2}+\nabla^2 T^1=\frac{\alpha_p^0 T^0}{\rho^0 C_p^0}\left(-\frac{\partial f^1}{\partial t}+\nabla\cdot\vec{F}^1\right)-\frac{1}{c_0^2\rho^0 C_v^0}\left(\frac{\partial r^1}{\partial t}-\frac{c_0^2}{\gamma^0}\int\nabla^2 r^1\,dt\right) \tag{1.34}$$

1.2.4. Velocity potential

Without restriction, one can always decompose a vector into its rotational component and its irrotational component: $\forall\vec{u},\ \vec{u}=\nabla\phi+\nabla\times\vec{\psi}$. So for the acoustic velocity, one can set

$$\vec{v}^1=\nabla\Phi+\nabla\times\vec{\Psi} \tag{1.35}$$

where Φ and Ψ are called the *scalar and vector potentials*. For the applied forces one can similarly set

$$:\vec{F}^1=\nabla G^1+\nabla\times\vec{H}^1 \tag{1.36}$$

Then the starting equations of acoustics (linearized conservation equations (1.21)) become

$$\begin{cases}\dfrac{\partial\rho^1}{\partial t}+\rho^0\nabla^2\Phi=f^1\\[2ex] \rho^0\nabla\dfrac{\partial\Phi}{\partial t}+\rho^0\nabla\times\dfrac{\partial\vec{\Psi}}{\partial t}+\nabla p^1=\nabla G^1+\nabla\times\vec{H}^1\\[2ex] T^0\rho^0\dfrac{\partial s^1}{\partial t}=r^1\end{cases}$$

or

$$\begin{cases}\dfrac{\partial\rho^1}{\partial t}+\rho^0\nabla^2\Phi=f^1\\[2ex] \rho^0\dfrac{\partial\Phi}{\partial t}+p^1=G^1\\[2ex] \rho^0\dfrac{\partial\vec{\Psi}}{\partial t}=\vec{H}^1\\[2ex] T^0\rho^0\dfrac{\partial s^1}{\partial t}=r^1\end{cases} \tag{1.37}$$

So

$$p^1 = -\rho^0 \frac{\partial \Phi}{\partial t} + G^1 \qquad \text{and} \qquad \vec{\Psi} = \frac{1}{\rho^0} \int \vec{H}^1 \, dt$$

and since from (1.28)

$$\rho^1 = \rho^0 \chi_s^0 p^1 - \frac{\alpha_p^0}{C_p^0} \int r^1 \, dt = -(\rho^0)^2 \chi_s^0 \frac{\partial \Phi}{\partial t} + \rho^0 \chi_s^0 G^1 - \frac{\alpha_p^0}{C_p^0} \int r^1 \, dt$$

the first equation of (1.37) becomes

$$-\rho^0 \chi_s^0 \frac{\partial^2 \Phi}{\partial t^2} + \nabla^2 \Phi = \frac{1}{\rho^0} \left(f^1 - \rho^0 \chi_s^0 \frac{\partial}{\partial t} G^1 + \frac{\alpha_p^0}{C_p^0} r^1 \right)$$

One sees that the acoustic field, in a perfect homogeneous fluid, is determined by only one quantity, termed the (scalar) *velocity potential.*

In the whole space one has

$$\begin{cases} \vec{v}^1 = \nabla \Phi + \dfrac{1}{\rho^0} \nabla \times \displaystyle\int \vec{H}^1 \, dt \\[2ex] p^1 = -\rho^0 \dfrac{\partial \Phi}{\partial t} + G^1 \\[2ex] \rho^1 = -\dfrac{\rho^0}{c_0^2} \dfrac{\partial \Phi}{\partial t} + \dfrac{1}{c_0^2} G^1 - \dfrac{\alpha_p^0}{C_p^0} \displaystyle\int r^1 \, dt \\[2ex] T^1 = -\dfrac{\alpha_p^0 T^0}{C_p^0} \dfrac{\partial \Phi}{\partial t} + \dfrac{\alpha_p^0 T^0}{\rho^0 C_p^0} G^1 + \dfrac{1}{C_p^0 \rho^0} \displaystyle\int r^1 \, dt \end{cases} \tag{1.38}$$

and

$$s^1 = (\rho^0 T^0)^{-1} \int r^1 \, dt \tag{1.39}$$

with a (scalar) potential governed by

$$-\frac{1}{c_0^2} \frac{\partial^2 \Phi}{\partial t^2} + \nabla^2 \Phi = \frac{1}{\rho^0} \left(f^1 - \frac{1}{c_0^2} \frac{\partial}{\partial t} G^1 + \frac{\alpha_p^0}{C_p^0} r^1 \right) \tag{1.40}$$

where $\vec{F}^1 = \nabla G^1 + \nabla \times \vec{H}^1$.

Domestic acoustics. With the usual sources of 'domestic' acoustics (usually electro-acoustic transducers), one can model sources as simple assemblages of volumic source of mass, so that one writes the simple equations:

$$\begin{cases} \vec{v}^1 = \nabla\Phi \\ p^1 = -\rho^0 \dfrac{\partial \Phi}{\partial t} \\ \rho^1 = -\dfrac{\rho^0}{c_0^2} \dfrac{\partial \Phi}{\partial t} \\ T^1 = -\dfrac{\alpha_p^0 T^0}{C_p^0} \dfrac{\partial \Phi}{\partial t}, \qquad s^1 = 0 \end{cases} \tag{1.41}$$

and

$$-\frac{1}{c_0^2}\frac{\partial^2 \Phi}{\partial t^2} + \nabla^2\Phi = \frac{1}{\rho^0} f^1 \tag{1.42}$$

1.2.5. Acoustic energy, acoustic intensity

Acoustic energy

Acoustic energy is defined as the variation of energy produced by the acoustic perturbation. The energy density is $w = \rho(e + \frac{1}{2}\vec{v}^2)$. In the general case (of a simple fluid), $w = w(\rho, s, \vec{v})$. Making a second order development of w:

$$\begin{aligned} \delta w = {} & \frac{\partial w}{\partial \rho}\,\delta\rho + \frac{\partial w}{\partial s}\,\delta s + \sum_{i=1}^{3} \frac{\partial w}{\partial v_i}\,\delta v_i \\ & + \frac{1}{2}\left[\frac{\partial^2 w}{\partial \rho^2}\,\delta\rho^2 + \frac{\partial^2 w}{\partial s^2}\,\delta s^2 + \sum_{i=1}^{3} \frac{\partial^2 w}{\partial v_i^2}\,\delta v_i^2\right] \\ & + \frac{\partial^2 w}{\partial \rho\,\partial s}\,\delta\rho\,\delta s + \sum_{i=1}^{3} \frac{\partial^2 w}{\partial \rho\,\partial v_i}\,\delta\rho\,\delta v_i + \sum_{i=1}^{3} \frac{\partial^2 w}{\partial s\,\partial v_i}\,\delta s\,\delta v_i \\ & + \sum_{i=1}^{3} \sum_{\substack{j=1 \\ j\neq i}}^{3} \frac{\partial^2 w}{\partial v_i\,\partial v_j}\,\delta v_i\,\delta v_j \end{aligned}$$

For a perfect homogeneous fluid at rest, $\delta s \equiv s^1 = 0$ (isentropicity of the acoustic perturbation outside heat sources) and $\vec{v}^0 = 0$, so

$$\delta w = \left[e + \rho\left(\frac{\partial e}{\partial \rho}\right)_s\right]\delta\rho + \frac{1}{2}\left[2\left(\frac{\partial e}{\partial \rho}\right)_s + \rho\left(\frac{\partial^2 e}{\partial \rho^2}\right)_s\right]\delta\rho^2 + \sum_{i=1}^{3} \tfrac{1}{2}\rho\,\delta v_i^2$$

and since

$$\left(\frac{\partial e}{\partial \rho}\right)_s = -\frac{1}{\rho^2}\left(\frac{\partial e}{\partial v}\right)_s = \frac{p}{\rho^2}$$

and

$$\left(\frac{\partial^2 e}{\partial \rho^2}\right)_s = \frac{\partial}{\partial \rho}\left(\frac{p}{\rho^2}\right)_s = \frac{1}{\rho^2}\left(\frac{\partial p}{\partial \rho}\right)_s - \frac{2p}{\rho^3} = \frac{c^2}{\rho^2} - \frac{2p}{\rho^3}$$

we have

$$\delta w = \left[e + \frac{p}{\rho}\right]\delta\rho + \frac{1}{2}\frac{c^2}{\rho}\delta\rho^2 + \sum_{i=1}^{3} \tfrac{1}{2}\rho\,\delta v_i^2$$

and since in the previous notations $\delta\rho \equiv \rho^1 = (1/c_0^2)p^1$, $\delta\vec{v} \equiv \vec{v}^1$,

$$\delta w = \left[e^0 + \frac{p^0}{\rho^0}\right]\frac{1}{c_0^2}p^1 + \frac{1}{2}\frac{1}{\rho^0 c_0^2}(p^1)^2 + \frac{1}{2}\rho^0(\vec{v}^1)^2 \tag{1.43}$$

Since acoustic perturbations are generally null mean-valued ($\langle p^1\rangle = 0$), the first term, proportional to the acoustic pressure, has a null mean value. It is eliminated by taking the mean value of the energy density perturbation: $w^A = \langle\delta w\rangle$.

Acoustic energy (density) is defined as the mean value of the energy density perturbation:

$$w^A = \langle\delta w\rangle = \frac{1}{2}\frac{1}{\rho^0 c_0^2}\langle(p^1)^2\rangle + \frac{1}{2}\rho^0\langle(\vec{v}^1)^2\rangle \tag{1.44}$$

Equation (1.43) is the expression of the acoustic energy (density) (for a perfect homogeneous fluid at rest).

Acoustic intensity

Acoustic intensity is defined as the variation of the energy flux produced by the acoustic perturbation. The energy flux density is $\vec{I} = -(\sigma\cdot\vec{v} - \vec{q})$. For a perfect fluid, since $\tau = \sigma + pI = 0$ and $\vec{q} = 0$, then $\vec{I} = p\vec{v}$.

The variation of the energy flux is to second order,

$$\delta\vec{I} = \delta p\vec{v} + p\,\delta\vec{v} + \delta p\,\delta\vec{v}$$

For a fluid initially at rest ($\vec{v}^0 = 0$) and noting $\delta p \equiv p^1$, and $\delta\vec{v} \equiv \vec{v}^1$,

$$\delta\vec{I} = p^0\vec{v}^1 + p^1\vec{v}^1 \tag{1.45}$$

As for acoustic energy, since acoustic perturbations are generally null mean-valued ($\langle\vec{v}\rangle = 0$), the first term, proportional to the acoustic velocity, has a null mean value. It is eliminated by taking the mean value of the energy flux: $\vec{I}^A = \langle\delta\vec{I}\rangle$.

Acoustic intensity is defined as the mean value of the energy flux perturbation:

$$\vec{I}^A = \langle \delta \vec{I} \rangle = \langle p^1 \vec{v}^1 \rangle \tag{1.46}$$

Equation (1.45) is the expression of the acoustic intensity (for a perfect homogeneous fluid at rest).

1.2.6. General solutions of the wave equation in free space

We consider the medium to be of infinite extent, in one or three dimensions. We are not interested in the two-dimensional case, which needs more mathematical tools (special functions) and does not introduce any further concepts.

One-dimensional case

The homogeneous wave equation (i.e. without source term) for a potential Φ is

$$-\frac{1}{c_0^2}\frac{\partial^2 \Phi}{\partial t^2} + \frac{\partial^2 \Phi}{\partial x^2} = 0 \tag{1.47}$$

One can show that it admits a general solution:

$$\Phi = f^+\left(t - \frac{x}{c_0}\right) + f^-\left(t + \frac{x}{c_0}\right) \tag{1.48}$$

The demonstration is very simple. Let us consider the variable change $(x, t) \rightarrow (\xi, n)$, where $\xi = t - (x/c_0)$ and $\eta = t + (x/c_0)$, and let $f(\xi, \eta) \equiv \Phi(x, t)$. The wave equation is transformed to $\partial^2 f/\partial \xi\, \partial \eta = 0$. This equation can be integrated:

$$\frac{\partial f}{\partial \eta} = F(\eta) \quad \Rightarrow \quad f = f^+(\xi) + \int F(\eta)\, d\eta = f^+(\xi) + f^-(\eta)$$

with $f^-(\eta) = \int F(\eta)\, d\eta$.

$f^+(t - (x/c_0))$ and $f^-(t + (x/c_0))$ are called respectively *progressive and regressive waves* (the first one propagates in the direction of positive x and the second one in the direction of negative x).

Green's function. With this result, one can show, using distribution theory, that the one-dimensional Green's function for an unbounded medium, i.e. the solution $G^{(1)}(x, x', t, t')$ of the inhomogeneous wave equation with a Dirac source at point x' and time t', $S(x, t) = \delta_{x'}(x)\, \delta_{t'}(t)$,

$$-\frac{1}{c_0^2}\frac{\partial^2 G^{(1)}}{\partial t^2} + \frac{\partial^2 G^{(1)}}{\partial x^2} = \delta_{x'}(x)\, \delta_{t'}(t) \tag{1.49}$$

is

$$G^{(1)}(x, x', t, t') = -\frac{c_0}{2} Y\left(t - t' - \frac{|x - x'|}{c_0}\right) \tag{1.50}$$

where $Y(t) = 0$ $(t < 0)$, $Y(t) = 1$ $(t > 0)$ is the Heaviside function.

Acoustic impedance. For a progressive wave, one has $\Phi = f^+(t - (x/c_0))$. So from (1.41), with $f^+(t) \equiv (\partial/\partial t)(f^+(t))$:

$$p^1 = -\rho^0 \frac{\partial \Phi}{\partial t} = -\rho^0 f^{+'}\left(t - \frac{x}{c_0}\right), \qquad v^1 = \frac{\partial \Phi}{\partial x} = -\frac{1}{c_0} f^+\left(t - \frac{x}{c_0}\right)$$

so that

$$v^1 = \frac{p^1}{\rho^0 c_0} \tag{1.51}$$

The quantity $Z = p^1/v^1$, homologous to an electric impedance in the electrical analogy of acoustics (where velocity and pressure are considered to be equivalent respectively to current and voltage), is called the *acoustic impedance of the wave.* The quantity $Z_0 = \rho^0 c_0$, characteristic of the propagation medium, is called the *characteristic (acoustic) impedance of the medium.* In the case of a progressive wave, the two quantities are equal: $Z = Z_0$.

For a harmonic progressive wave of angular frequency ω with time dependence $e^{-\iota\omega t}$ (classical choice in theoretical acoustics), one has $f^+(\xi) = A^+ e^{-\iota\omega t}\xi$, and so

$$\Phi = A^+ e^{-\iota\omega(t - x/c_0)} = A^+ e^{-\iota\omega t} e^{+\iota kx} \qquad \text{with } k = \omega/c_0 \text{ the } \textit{wavenumber}$$

Thus

$$p^1 = -\rho^0 \frac{\partial \Phi}{\partial t} = \iota\omega\rho^0 A^+ e^{-\iota\omega t} e^{+\iota kx} = \iota\omega\rho^0 \Phi \qquad \text{and}$$

$$v^1 = \frac{\partial \Phi}{\partial x} = \iota k A^+ e^{-\iota\omega t} e^{+\iota kx} = \iota k \Phi$$

For a time dependence in $e^{+\iota\omega t}$, one obtains (for a progressive wave)

$$\Phi = A^+ e^{+\iota\omega(t - x/c_0)} = A^+ e^{+\iota\omega t} e^{-\iota kx}$$

Three-dimensional case

Plane waves. One can again consider solutions of the same type as obtained in one dimension, i.e. waves propagating along a direction $\vec{n}_0$ (unit vector) and constant along directions normal to this direction. These are called *plane waves,* since wavefronts (planes of same field) are planes.

A plane wave propagating in the direction $\vec{n}_0$ can be described by the potential

$$\Phi = f^+\left(t - \frac{\vec{n}_0 \cdot \vec{x}}{c_0}\right)\left(f^+\left(t - \frac{r \cdot \vec{n}}{c}\right)\right) \tag{1.52}$$

One can verify that this solution actually satisfies the homogeneous wave equation

$$-\frac{1}{c_0^2}\frac{\partial^2 \Phi}{\partial t^2} + \nabla^2 \Phi = 0 \tag{1.53}$$

One then has from (1.41), with $f^{+'}(t) \equiv \partial(f^+(t))/\partial t$,

$$p^1 = -\rho^0 \frac{\partial \Phi}{\partial t} = -\rho^0 f^{+'}\left(t - \frac{\vec{n}_0 \cdot \vec{x}}{c_0}\right) \quad \text{and} \quad \vec{v}^1 = \nabla \Phi = -\frac{\vec{n}_0}{c_0} f^{+'}\left(t - \frac{\vec{n}_0 \cdot \vec{x}}{c_0}\right)$$

So

$$\vec{v}^1 = \frac{p^1}{\rho^0 c_0}\vec{n}_0 \tag{1.54}$$

Relation (1.54) shows on the one hand that plane waves are longitudinally polarized (the acoustic velocity is colinear with the propagation direction $\vec{n}_0$), and on the other that acoustic velocity and pressure are in phase and that their ratio $Z = p^1/v^1$, i.e. the acoustic impedance of the wave, equals the characteristic impedance of the medium $Z_0 = \rho^0 c_0$.

For a plane harmonic wave of angular frequency ω with time dependence $e^{-\iota\omega t}$ (classical choice in theoretical acoustics), one has $f^+(\xi) = A^+ e^{-\iota\omega\xi}$ and so

$$\Phi = A^+ e^{-\iota\omega(t - (\vec{n}_0 \cdot \vec{x}/c_0))} = A^+ e^{-\iota\omega t} e^{+\iota\vec{k} \cdot \vec{x}},$$

with $\vec{k} = (\omega/c_0)\vec{n}_0$ *wavevector*, the modulus of which is the wavenumber $k = \omega/c_0$. Then

$$p^1 = -\rho^0 \frac{\partial \Phi}{\partial t} = \iota\omega\rho^0 A^+ e^{-\iota\omega t} e^{+\iota\vec{k} \cdot \vec{x}} = \iota\omega\rho^0 \Phi \quad \text{and}$$

$$\vec{v}^1 = \nabla \Phi = \iota k A^+ e^{-\iota\omega t} e^{+\iota\vec{k} \cdot \vec{x}} \vec{n} = \iota k \Phi \vec{n}$$

For a time dependence $e^{+\iota\omega t}$, one obtains

$$\Phi = A^+ e^{+\iota\omega(t - (\vec{n} \cdot \vec{x}/c_0))} = A^+ e^{+\iota\omega t} e^{-\iota\vec{k} \cdot \vec{x}}$$

Spherical waves. One can also consider solutions of the homogeneous wave equation of the type $\Phi(\vec{x}, t) = \Phi(|\vec{x}|, t)$, i.e. depending only upon the distance $r = |\vec{x}|$. These are called *spherical waves*. They are solutions of the equation

$$-\frac{1}{c_0^2}\frac{\partial^2 \Phi}{\partial t^2} + \frac{1}{r^2}\frac{\partial \Phi}{\partial r}\left(r^2 \frac{\partial \Phi}{\partial r}\right) = 0 \tag{1.55}$$

By the function change $\Phi(r, t) = f(r, t)/r$, one obtains the classical one-dimensional wave equation

$$-\frac{1}{c_0^2}\frac{\partial^2 f}{\partial t^2}+\frac{\partial^2 f}{\partial r^2}=0 \tag{1.56}$$

the general solution of which is

$$f=f^+\left(t-\frac{r}{c_0}\right)+f^-\left(t+\frac{r}{c_0}\right)$$

i.e.

$$\Phi(r, t)=\frac{1}{r}f^+\left(t-\frac{r}{c_0}\right)+\frac{1}{r}f^-\left(t+\frac{r}{c_0}\right) \tag{1.57}$$

The solution $\Phi^+(r, t) = (1/r)f^+(t - (r/c_0))$ is called the *diverging* or *exploding* spherical wave; the solution $\Phi^-(r, t) = (1/r)f^-(t + (r/c_0))$ is called the *converging* or *imploding* spherical wave. Both have a dependence in $1/r$.

Green's function. With this result, one can show, using distribution theory, that the three-dimensional Green's function for an unbounded medium, i.e. the solution $G^3(\vec{x}, \vec{x}', t, t')$ of the inhomogeneous wave equation with a Dirac source at point $\vec{x}'$ and instant t', $S(\vec{x}, t) = \delta_{x'}(\vec{x})\ \delta_{t'}(t)$,

$$-\frac{1}{c_0^2}\frac{\partial^2 G^{(3)}}{\partial t^2}+\nabla^2 G^{(3)}=\delta_{x'}(\vec{x})\ \delta_{t'}(t) \tag{1.58}$$

satisfying a condition of no radiation coming from infinity (Sommerfeld condition) is

$$G^{(3)}(\vec{x}, \vec{x}', t, t')=\frac{\delta\left(t-t'-\dfrac{|\vec{x}-\vec{x}'|}{c_0}\right)}{4\pi\,|\vec{x}-\vec{x}'|} \tag{1.59}$$

Asymptotic behaviour of spherical waves. For a diverging spherical wave originating from $\vec{0}$, one has

$$\Phi(\vec{x}, t)=\frac{1}{|\vec{x}|}f^+\left(t-\frac{|\vec{x}|}{c_0}\right) \tag{1.60}$$

so, from (1.40), with $f^{+\prime}(t) \equiv \partial(f^+(t))/\partial t$,

$$p^1 = -\rho^0 \frac{\partial \Phi}{\partial t} = -\frac{\rho^0}{|\vec{x}|} f^{+\prime}\left(t - \frac{|\vec{x}|}{c_0}\right) \qquad \text{and}$$

$$\vec{v}^1 = \nabla \Phi = -\left[\frac{1}{|\vec{x}|^2} f^+\left(t - \frac{|\vec{x}|}{c_0}\right) + \frac{1}{c_0|\vec{x}|} f^{+\prime}\left(t - \frac{|\vec{x}|}{c_0}\right)\right]\vec{n}$$

where $\vec{n} = \vec{x}/|\vec{x}|$ is the unit radial vector, i.e. the normal to the wavefront.

Far from the source, i.e. for $|\vec{x}| f^{+\prime}/c_0 f^+ \gg 1$, one has

$$\vec{v}^1 \approx -\frac{1}{c_0|\vec{x}|} f^{+\prime}\left(t - \frac{|\vec{x}|}{c_0}\right)\vec{n}$$

and so

$$\vec{v}^1 \approx \frac{p^1}{\rho^0 c_0}\vec{n} \tag{1.61}$$

This is the same result as for a plane wave, replacing the unit direction vector of the plane wave by the unit radial vector of the spherical wave.

Relation (1.61) shows that, at large distance, on the one hand spherical diverging waves are longitudinally polarized (the acoustic velocity is colinear with the unit radial vector $\vec{n}$), and on the other that acoustic velocity and pressure are in phase and their ratio $Z = p^1/v^1$, i.e. the acoustic impedance of the wave, equals, as for plane waves, the characteristic impedance of the medium $Z_0 = \rho^0 c_0$.

At large distance a spherical wave behaves locally like a plane wave.

For a spherical diverging harmonic wave of angular frequency ω with time dependence $e^{-\iota\omega t}$ (classical choice in theoretical acoustics), one has $f^+(\xi) = A^+ e^{-\iota\omega\xi}$ and so

$$\Phi = \frac{A^+}{|\vec{x}|} e^{-\iota\omega(t - (|\vec{x}|/c_0))} = \frac{A^+}{|\vec{x}|} e^{-\iota\omega t} e^{+\iota k|\vec{x}|}, \qquad \text{with } k = \frac{\omega}{c_0} \text{ the wave number}$$

Thus,

$$p^1 = -\rho^0 \frac{\partial \Phi}{\partial t} = \iota\omega\rho^0 \frac{A^+}{|\vec{x}|} e^{-\iota\omega t} e^{+\iota k|\vec{x}|} = \iota\omega\rho^0 \Phi,$$

$$\vec{v}^1 = \nabla\Phi = \iota k \left[1 - \frac{1}{\iota k|\vec{x}|}\right] \frac{A^+}{|\vec{x}|} e^{-\iota\omega t} e^{+\iota k|\vec{x}|} \vec{n} = \iota k \left[1 - \frac{1}{\iota k|\vec{x}|}\right] \Phi \vec{n}$$

and the relation between acoustic pressure and acoustic velocity becomes:

$$\vec{v}^1 = \left[1 - \frac{1}{\iota k|\vec{x}|}\right] \frac{p^1}{\rho^0 c_0} \vec{n}$$

and the large distance condition becomes $k|\vec{x}| \gg 1$, or, introducing the wavelength $\lambda = 2\pi/k$, $|\vec{x}| \gg \lambda/2\pi$.

For a time dependence in $e^{+\iota\omega t}$, one obtains (for a diverging wave)

$$\Phi = \frac{A^+}{|\vec{x}|} e^{+\iota\omega(t - (|\vec{x}|/c_0))} = \frac{A^+}{|\vec{x}|} e^{+\iota\omega t} e^{-\iota k|\vec{x}|}$$

Energy and intensity of plane waves and of spherical diverging waves at large distance. In the two cases, one has seen that $\vec{v}^1 \approx (p^1/\rho^0 c_0)\vec{n}$, with $\vec{n}$ the unit vector normal to the wavefront. So the acoustic energy density is

$$w^A = \frac{1}{\rho^0 c_0^2} \langle (p^1)^2 \rangle = \rho^0 \langle (\vec{v}^1)^2 \rangle \tag{1.62}$$

and the acoustic intensity

$$\vec{I}^A = \frac{1}{\rho^0 c_0} \langle (p^1)^2 \rangle \vec{n} = \rho^0 c_0 \langle (\vec{v}^1)^2 \rangle \vec{n} \tag{1.63}$$

and the following relation between acoustic intensity and acoustic energy density:

$$\vec{I}^A = c_0 w^A \vec{n} \tag{1.64}$$

Thus, *acoustic intensity measures local energy density and local direction of propagation* of acoustic waves.

1.2.7. Harmonic waves

Let us consider pure sounds with angular frequency ω with time dependence $e^{-\iota\omega t}$ (classical choice in theoretical acoustics), i.e. sounds produced by sources of type $S(\vec{x}, t) = \tilde{S}_\omega(\vec{x})e^{-\iota\omega t}$, then solutions of the wave equation will be of the same type $\Phi(\vec{x}, t) = \tilde{\Phi}_\omega(\vec{x})e^{-\iota\omega t}$, $p^1(\vec{x}, t) = \tilde{p}^1_\omega(\vec{x})e^{-\iota\omega t}$, $\vec{v}^1(\vec{x}, t) = \tilde{\vec{v}}^1_\omega(\vec{x})e^{-\iota\omega t}$, and so on.

If one considers the simpler case of 'domestic' acoustics (simple volumic source of mass $f^1(\vec{x}, t) = \tilde{f}^1_\omega(\vec{x})e^{-\iota\omega t}$), one has from (1.41):

$$\begin{cases} \tilde{\vec{v}}^1_\omega(\vec{x}) = \nabla\tilde{\Phi}_\omega(\vec{x}) \\ \tilde{p}^1_\omega(\vec{x}) = \iota\omega\rho^0\tilde{\Phi}_\omega(\vec{x}) \\ \tilde{\rho}^1_\omega(\vec{x}) = \iota\omega \dfrac{\rho^0}{c_0^2} \tilde{\Phi}_\omega(\vec{x}) \\ \tilde{T}^1_\omega(\vec{x}) = \iota\omega \dfrac{\alpha_p^0 T^0}{C_p^0} \tilde{\Phi}_\omega(\vec{x}) \end{cases} \tag{1.65}$$

with, from (1.42),

$$\nabla^2\tilde{\Phi}_\omega(\vec{x}) + k^2\tilde{\Phi}_\omega(\vec{x}) = \frac{1}{\rho^0}\tilde{f}^1_\omega(\vec{x}), \qquad k = \frac{\omega}{c_0} \tag{1.66}$$

where k is the wavenumber.

Equation (1.66) is named a Helmholtz equation.

Major scattering problems are solved with this formalism, even for complex sounds. Indeed any time signal can be represented by the Fourier integral $x(t) = \int_{-\infty}^{+\infty} \hat{x}(\nu)e^{\iota 2\pi\nu t}\, d\nu$, with $\hat{x}(\nu) = \int_{-\infty}^{+\infty} \hat{x}(t)e^{-\iota 2\pi\nu t}\, dt$ the Fourier transform of $x(t)$, ν being the frequency. In the same way space-time functions such as the potential and other components of the field can be represented by the Fourier integrals $\Phi(\vec{x}, t) = \int_{-\infty}^{+\infty} \hat{\Phi}(\vec{x}, \nu)e^{\iota 2\pi\nu t}\, d\nu$ with $\hat{\Phi}(\vec{x}, \nu) = \int_{-\infty}^{+\infty} \Phi(\vec{x}, t)e^{-\iota 2\pi\nu t}\, dt$ the time Fourier transform of $\Phi(\vec{x}, t)$; $p^1(\vec{x}, t) = \int_{-\infty}^{+\infty} \hat{p}^1(\vec{x}, \nu)e^{\iota 2\pi\nu t}\, d\nu$ with $\hat{p}^1(\vec{x}, \nu) = \int_{-\infty}^{+\infty} p^1(\vec{x}, t)e^{-\iota 2\pi\nu t}\, dt$; $\vec{v}^1(\vec{x}, t) = \int_{-\infty}^{+\infty} \hat{\vec{v}}^1(\vec{x}, \nu)e^{\iota 2\pi\nu t}\, d\nu$ with $\hat{\vec{v}}^1(\vec{x}, \nu) = \int_{-\infty}^{+\infty} \vec{v}^1(\vec{x}, t)e^{-\iota 2\pi\nu t}\, dt$; and so on. Each component $\hat{\Phi}(\vec{x}, \nu)e^{\iota 2\pi\nu t}$, $\hat{p}^1(\vec{x}, \nu)e^{\iota 2\pi\nu t}$, $\hat{\vec{v}}^1(\vec{x}, \nu)e^{\iota 2\pi\nu t}$, etc., will behave like a harmonic signal $\tilde{\Phi}_\omega(\vec{x})e^{-\iota\omega t}$, $\tilde{p}^1_\omega(\vec{x})e^{-\iota\omega t}$, $\tilde{\vec{v}}^1_\omega(\vec{x})e^{-\iota\omega t}$, etc., with an angular frequency $\omega = -2\pi\nu$:

$$\tilde{\Phi}_\omega(\vec{x}) = \hat{\Phi}(\vec{x}, \nu),\ \tilde{p}^1_\omega(\vec{x}) = \hat{p}^1(\vec{x}, \nu),\ \tilde{\vec{v}}^1_\omega(\vec{x}) = \hat{\vec{v}}^1(\vec{x}, \nu),\ \ldots \qquad \text{with } \omega = -2\pi\nu$$

Since acoustic equations are linear, each frequency component of the field will satisfy (1.65) and (1.66), allowing the calculation of these frequency components and then, by Fourier synthesis, the space-time field.

Remark. One sees that the traditional convention of acousticians of choosing a harmonic dependence in $e^{-\iota\omega t}$ results in a negative frequency $\nu = -\omega/2\pi$ in the signal processing sense.

To solve scattering problems, it is often convenient to use the Green's functions of the Helmholtz equation:

$$\nabla^2 g_\omega(\vec{x}) + k^2 g_\omega(\vec{x}) = \delta_{x'}(\vec{x}), \qquad k = \frac{\omega}{c_0} \tag{1.67}$$

For an unbounded medium this is:

$$\text{one-dimensional:} \qquad g^{(1)}_\omega(x, x') = \frac{e^{+\iota k\,|\,x - x'\,|}}{2\iota k} \tag{1.68}$$

$$\text{three-dimensional:} \qquad g^{(3)}_\omega(\vec{x}, \vec{x}') = -\frac{e^{+\iota k\,|\,\vec{x} - \vec{x}'\,|}}{4\pi\,|\,\vec{x} - \vec{x}'\,|} \tag{1.69}$$

One must keep in mind, when returning to the time domain, that calculation with these Green's functions gives access to the $\nu = -\omega/2\pi$ component of frequency of the signal. Otherwise, exploding waves will be transformed into imploding waves and conversely.

1.2.8. Acoustic sources

In acoustics one is mainly concerned with acoustic pressure. Most usual detectors, beginning with the ear, are pressure sensitive, so we are interested here only in acoustic pressure sources, i.e. the source term of the equation that governs acoustic pressure.

From (1.30), but now taking into account non-linear terms describing acoustic emission by turbulence (Lighthill's theory), this source term is

$$S_{p^1} = -\frac{\partial f^1}{\partial t} + \nabla \cdot \vec{F}^1 + \frac{\alpha_p^0}{C_p^0}\frac{\partial r^1}{\partial t} - \nabla \cdot \nabla \cdot (\rho \vec{v} \otimes \vec{v}) \tag{1.70}$$

The first term $-\partial f^1/\partial t$ results from time fluctuations of f^1, the time rate input of mass density (i.e. time rate input of mass per unit volume, or mass injection per unit volume per unit time: dimension $\mathrm{ML}^{-3}\mathrm{T}^{-1}$ in mass M, length L, time T). The radiation condition of mass flow inputs is their non-stationarity. In this category one finds unsteady or transient flows and all usual sources of 'domestic' acoustics (vibrating surfaces acting as pistons injecting matter).

The second term $\nabla \cdot \vec{F}^1$ results from space fluctuations of $\vec{F}^1$, the time rate input of momentum density (or volumic density of supplied forces, i.e. supply of body forces per unit volume, dimension $\mathrm{ML}^{-2}\mathrm{T}^{-2}$). The radiation condition for supplied forces is their space non-uniformity. In this category one finds most aerodynamic forces on bodies moving through fluids.

The third term $(\alpha_p^0/C_p^0)(\partial r^1/\partial t)$ results from time fluctuations of r^1, the time rate input of heat density (i.e. time rate input of heat per unit volume, or heat injection per unit volume per unit time, dimension $\mathrm{ML}^{-1}\mathrm{T}^{-3}$). The radiation condition of heat flow inputs is their non-stationarity. The efficiency is proportional to the ratio of expansivity over specific heat, α_p^0/C_p^0. In this category one finds thermal shocks and impulsive or modulated laser beams.

The fourth term $-\nabla \cdot \nabla \cdot (\rho \vec{v} \otimes \vec{v})$ results from space fluctuations of the Reynolds tensor $(\rho \vec{v} \otimes \vec{v})$, i.e. from shear stresses within the fluid. The radiation condition for shear stresses is their space non-uniformity. This term explains acoustic emission by turbulence (jets, drags, wakes, boundary layers).

1.2.9. Boundary conditions

Generally media are homogeneous only over limited portions of space. On the other hand, one is often concerned with closed spaces (rooms, etc.) and various obstacles (walls, etc.). It is then necessary to distinguish interfaces between the fluid and an elastic object from interfaces between the fluid and a perfectly rigid obstacle (or so little penetrable that there is no need to consider what happens behind the interface).

Interface between two propagating media

We have found the general continuity conditions at an interface between two media (Section 1.1.4):

$$[\vec{v} \cdot \vec{n}]_\Sigma = 0 \qquad \text{continuity of normal velocity}$$
$$[\sigma \cdot \vec{n}]_\Sigma = 0 \qquad \text{continuity of normal stress}$$

For an interface between a perfect (i.e. non-viscous) fluid (1) and a solid (2), the second condition becomes

$$-p^{(1)}\vec{n} = \sigma^{(2)} \cdot \vec{n}$$

and for an interface between two perfect fluids

$$[p]_\Sigma = 0 \qquad \text{continuity of pressure}$$

By linearization, one obtains the acoustic continuity conditions:

$$\begin{aligned} &[\vec{v}^1 \cdot \vec{n}]_\Sigma = 0 \qquad \text{continuity of normal acoustic velocity} \\ &[\sigma^1 \cdot \vec{n}]_\Sigma = 0 \qquad \text{continuity of normal acoustic stress} \end{aligned} \tag{1.71}$$

For an interface between a perfect (i.e. non-viscous) fluid (1) and a solid (2), the second condition becomes

$$-p^{1^{(1)}}\vec{n} = \sigma^{1^{(2)}} \cdot \vec{n}$$

and for an interface between two perfect fluids

$$[p^1]_\Sigma = 0 \qquad \text{continuity of acoustic pressure.}$$

The general conditions of continuity for interfaces between perfect fluids are then

$$\begin{cases} [\vec{v}^1 \cdot \vec{n}]_\Sigma = 0 & \text{continuity of normal acoustic velocity} \\ [p^1]_\Sigma = 0 & \text{continuity of acoustic pressure} \end{cases} \tag{1.72}$$

In terms of the acoustic velocity potential,

$$\begin{cases} [\nabla\Phi \cdot \vec{n}]_\Sigma = 0 \\ [\rho^0 \Phi]_\Sigma = 0 \end{cases} \tag{1.73}$$

In terms of the acoustic pressure only,

$$\begin{cases} \left[\dfrac{1}{\rho^0} \nabla p^1 \cdot \vec{n}\right]_\Sigma = 0 \\ [p^1]_\Sigma = 0 \end{cases} \tag{1.74}$$

Plane interface between two perfect fluids. Consider a plane interface Σ between two perfect fluids denoted (1) (density $\rho^{(1)}$, wave velocity $c^{(1)}$) and (2) (density $\rho^{(2)}$, wave velocity $c^{(2)}$). An orthonormal frame (O, x, y, z) is chosen so that Σ lies in the plane (O, y, z) with Ox colinear with the normal $\vec{n}$ to the surface in the direction $(1) \to (2)$ $(\vec{n} = (1, 0, 0))$. Consider a plane harmonic wave impinging from (1) on the surface with velocity potential Φ_I of amplitude Φ_0, unit propagation vector $\vec{n}_I = (\cos\theta_I, \sin\theta_I, 0)$ (such that $(\vec{n}, \vec{n}_I) = \theta_I$), angular frequency ω (with time dependence $e^{-\iota\omega t}$), wavenumber $k^{(1)} = \omega/c^{(1)}$:

$$\Phi_I = \Phi_0 e^{-\iota\omega t} e^{+\iota k^{(1)} \vec{n}_I \cdot \vec{x}} = \Phi_0 e^{-\iota\omega t} e^{+\iota k^{(1)}(x \cos\theta_I + y \sin\theta_I)}$$

The two media being half-infinite, this wave gives rise to two plane waves:

- In medium (1) a reflected wave Φ_R with direction vector $\vec{n}_R = (\cos\theta_R, \sin\theta_R, 0)$ (such that $(\vec{n}, \vec{n}_R) = \theta_R$) and amplitude r_Φ:

$$\Phi_R = \Phi_0 e^{-\iota\omega t} e^{+\iota k^{(1)} \vec{n}_R \cdot \vec{x}} = r_\Phi \Phi_0 e^{-\iota\omega t} e^{+\iota k^{(1)}(x \cos\theta_R + y \sin\theta_R)}$$

- In medium (2) a reflected wave Φ_T with direction vector $\vec{n}_T = (\cos\theta_T, \sin\theta_T, 0)$ (such that $(\vec{n}, \vec{n}_T) = \theta_T$) and amplitude t_Φ:

$$\Phi_T = \Phi_0 e^{-\iota\omega t} e^{+\iota k^{(2)} \vec{n}_T \cdot \vec{x}} = t_\Phi \Phi_0 e^{-\iota\omega t} e^{+\iota k^{(2)}(x \cos\theta_T + y \sin\theta_T)}$$

r_Φ and t_Φ are called reflection and transmission coefficients for the velocity potential (of the interface Σ).

The velocity potential is:

$$\text{medium (1)}: \quad \Phi^{(1)} = \Phi_I + \Phi_R = \Phi_0 [e^{-\iota\omega t} e^{+\iota k^{(1)}(x \cos\theta_I + y \sin\theta_I)} + r_\Phi e^{-\iota\omega t} e^{+\iota k^{(1)}(x \cos\theta_R + y \sin\theta_R)}]$$

$$\text{medium (2)}: \quad \Phi^{(2)} = \Phi_T = t_\Phi \Phi_0 e^{-\iota\omega t} e^{+\iota k^{(2)}(x \cos\theta_T + y \sin\theta_T)}$$

The conditions of continuity at the interface:

$$\begin{cases} [\nabla\Phi \cdot \vec{n}]_\Sigma = 0 \\ [\rho^0 \Phi]_\Sigma = 0 \end{cases} \quad \text{i.e.} \quad \begin{cases} \left[\dfrac{\partial\Phi}{\partial x}\right]_\Sigma = 0 \\ [\rho^0 \Phi]_\Sigma = 0 \end{cases} \quad \text{or} \quad \begin{cases} \left(\dfrac{\partial\Phi^{(1)}}{\partial x}\right)_{x=0} = \left(\dfrac{\partial\Phi^{(2)}}{\partial x}\right)_{x=0} \\ (\rho^{(1)}\Phi^{(1)})_{x=0} = (\rho^{(2)}\Phi^{(2)})_{x=0} \end{cases}$$

give

$$k^{(1)}(\cos\theta_I e^{+\iota k^{(1)} y \sin\theta_I} + r_\Phi \cos\theta_R e^{+\iota k^{(1)} y \sin\theta_R}) = k^{(2)} t_\Phi \cos\theta_T e^{+\iota k^{(2)} y \sin\theta_T}$$

$$\rho^{(1)}(e^{+\iota k^{(1)} y \sin\theta_I} + r_\Phi e^{+\iota k^{(1)} y \sin\theta_R}) = \rho^{(2)} t_\Phi e^{+\iota k^{(2)} y \sin\theta_T}$$

Since these relations must be satisfied $\forall y$, one has, first,

$$k^{(1)} y \sin\theta_I = k^{(1)} y \sin\theta_R = k^{(2)} y \sin\theta_T$$

that is

$$\begin{cases} \sin\theta_I = \sin\theta_R, & \text{i.e. } \theta_R = \pi - \theta_I \\ k^{(1)} \sin\theta_I = k^{(2)} \sin\theta_T, & \text{i.e. } \left(\dfrac{\sin\theta_I}{c^{(1)}}\right) = \left(\dfrac{\sin\theta_T}{c^{(2)}}\right) \end{cases} \tag{1.75}$$

called the Snell–Descartes laws. Then

$$k^{(1)} \cos \theta_I(1 - r_\Phi) = k^{(2)} t_\Phi \cos \theta_T$$
$$\rho^{(1)}(1 + r_\Phi) = \rho^{(2)} t_\Phi$$

so that

$$t_\Phi = \frac{\rho^{(1)}}{\rho^{(2)}}(1 + r_\Phi) \quad \text{and} \quad \frac{1 + r_\Phi}{1 - r_\Phi} = \frac{z^{(2)}}{z^{(1)}}$$

with

$$z^{(1)} = \frac{Z^{(1)}}{\cos \theta_I} = \frac{\rho^{(1)} c^{(1)}}{\cos \theta_I} \quad \text{and} \quad z^{(2)} = \frac{Z^{(2)}}{\cos \theta_T} = \frac{\rho^{(2)} c^{(2)}}{\cos \theta_T}$$

giving

$$\begin{cases} r_\Phi = \dfrac{z^{(2)} - z^{(1)}}{z^{(2)} + z^{(1)}} \\ t_\Phi = 2 \dfrac{\rho^{(1)}}{\rho^{(2)}} \dfrac{z^{(2)}}{z^{(2)} + z^{(1)}} \end{cases} \tag{1.76}$$

In practice one is interested in *reflection* and *transmission* coefficients for the acoustic pressure p^1. Since $p^1 = -\rho^0 \, \partial\Phi/\partial t$, one has

$$p^{1^{(1)}} = \iota\omega\rho^{(1)}\Phi_0(\mathrm{e}^{-\iota\omega t}\mathrm{e}^{+\iota k^{(1)}(x \cos \theta_I + y \sin \theta_I)} + r_\Phi \mathrm{e}^{-\iota\omega t}\mathrm{e}^{+\iota k^{(1)}(x \cos \theta_R + y \sin \theta_R)})$$
$$p^{1^{(2)}} = \iota\omega\rho^{(2)} t_\Phi \Phi_0 \mathrm{e}^{-\iota\omega t}\mathrm{e}^{+\iota k^{(2)}(x \cos \theta_T + y \sin \theta_T)}$$

i.e.

$$p^{1^{(1)}} = P_0(\mathrm{e}^{-\iota\omega t}\mathrm{e}^{+\iota k^{(1)}(x \cos \theta_I + y \sin \theta_I)} + r_p \mathrm{e}^{-\iota\omega t}\mathrm{e}^{+\iota k^{(1)}(x \cos \theta_R + y \sin \theta_R)})$$
$$p^{1^{(2)}} = P_0 t_p \mathrm{e}^{-\iota\omega t}\mathrm{e}^{+\iota k^{(2)}(x \cos \theta_T + y \sin \theta_T)} \quad \text{with } P_0 = \iota\omega\rho^{(1)}\Phi_0$$

where r_p and t_p are defined as the reflection and transmission coefficients for the acoustic pressure (of the interface Σ).

Then one has

$$\begin{cases} r_p = r_\Phi = \dfrac{z^{(2)} - z^{(1)}}{z^{(2)} + z^{(1)}} \\ t_p = \dfrac{\rho^{(2)}}{\rho^{(1)}} t_\Phi = \dfrac{2z^{(2)}}{z^{(2)} + z^{(1)}} \end{cases} \tag{1.77}$$

One sees that the reflection coefficient for pressure and potential are the same and can be denoted r, but the transmission coefficients for pressure and potential are different.

Various cases

- If medium (2) has a smaller velocity than (1) ($c^2 < c^1$), then,

$$\sin\,\theta_T = \frac{c^{(2)}}{c^{(1)}}\sin\,\theta_I < \sin\,\theta_I, \qquad \text{so } \theta_T < \theta_I$$

There is always a transmitted wave.

- If medium (2) has a larger velocity than (1) ($c^2 > c^1$), then,

$$\sin\,\theta_T = \frac{c^{(2)}}{c^{(1)}}\sin\,\theta_I > \sin\,\theta_I, \qquad \text{so } \theta_T > \theta_I$$

There is a critical incidence angle $(\theta_I)_c$ beyond which there is no transmitted wave. This *critical angle* is reached when $\theta_T = \pi/2$, that is for $\sin\,(\theta_I)_c = c^{(1)}/c^{(2)}$, i.e.

$$(\theta_I)_c = \sin^{-1}\left(\frac{c^{(1)}}{c^{(2)}}\right) \tag{1.78}$$

Interface with a non-propagating medium: boundary conditions

Consider the most frequently met case of a *locally reacting surface*, i.e. a surface such that the acoustic field at a given point of the surface depends only on the properties of the surface at this point. These properties can be summarized in the ratio of pressure over normal velocity of the wave at the surface, termed the *normal impedance* of the surface

$$Z_n = \left(\frac{p^1}{\vec{v}^1\cdot\vec{n}_\Sigma}\right)_\Sigma$$

For a harmonic wave of velocity potential Φ with angular frequency ω and time dependence $e^{-\iota\omega t}$, one has from (1.65)

$$Z_n = \iota\omega\rho^0\left(\frac{\Phi}{\nabla\Phi\cdot\vec{n}_\Sigma}\right)_\Sigma = \iota\omega\rho^0\left(\frac{\Phi}{\partial_n\Phi}\right)_\Sigma = \iota\omega\rho^0\left(\frac{p^1}{\partial_n p^1}\right)_\Sigma$$

$$\text{with } \forall\phi,\ (\partial_n\phi)_\Sigma \equiv (\nabla\phi\cdot\vec{n}_\Sigma)_\Sigma$$

Then

$$\left(\frac{p^1}{\partial_n p^1}\right)_\Sigma = \left(\frac{\Phi}{\partial_n\Phi}\right)_\Sigma = \frac{Z_n}{\iota\omega\rho^0} = \frac{Z_n}{\iota k\rho^0 c_0} = \frac{\zeta}{\iota k} \tag{1.79}$$

with $\zeta = Z_n/Z_0$ the reduced (dimensionless) normal impedance of the surface or the *specific normal impedance* of the surface.

Let us now consider a plane harmonic wave (angular frequency ω, time dependence $e^{-\iota\omega t}$) impinging with incidence $\vec{n}_I = (\cos\,\theta_I,\ \sin\,\theta_I,\ 0)$ on an impenetrable interface of specific normal impedance ζ. As for the penetrable interface, the impinging plane wave gives rise to a reflected plane wave with direction $\vec{n}_R = (\cos\theta_R, \sin\theta_R, 0)$ with $\theta_R = \pi - \theta_I$, i.e. with direction $\vec{n}_R = (-\cos\theta_I, \sin\theta_I, 0)$. Then one has, using the same notation as for the penetrable interface,

$$\Phi = \Phi_0 e^{-\iota\omega t} e^{+\iota k y \sin\theta_I} [e^{+\iota k x \cos\theta_I} + r e^{-\iota k x \cos\theta_I}]$$

with r the reflection coefficient for pressure or potential. Thus

$$(\partial_n \Phi)_\Sigma \equiv \left(\frac{\partial\Phi}{\partial x}\right)_{x=0} = \iota k \cos\theta_I \Phi_0 e^{-\iota\omega t} e^{+\iota k y \sin\theta_I} (e^{+\iota k x \cos\theta_I} - r e^{-\iota k x \cos\theta_I})_{x=0}$$

$$= \iota k \cos\theta_I \Phi e^{-\iota\omega t} e^{+\iota k y \sin\theta_I} (1 - r)$$

and

$$\frac{\zeta}{\iota k} = \left(\frac{\Phi}{\partial_n \Phi}\right)_\Sigma = \frac{1}{\iota k \cos\theta_I} \frac{1+r}{1-r}$$

that is

$$\zeta = \frac{1}{\cos\theta_I} \frac{1+r}{1-r} \quad \text{or} \quad r = \frac{-1 + \zeta\cos\theta_I}{1 + \zeta\cos\theta} \tag{1.80}$$

Connection with the mathematical formulation of boundary conditions. The usual mathematical boundary conditions are as follows.

Dirichlet condition for a soft boundary:

$$(p^1)_\Sigma = 0, \qquad \text{i.e. } (\Phi)_\Sigma = 0 \tag{1.81}$$

Neumann condition for a rigid boundary:

$$(\vec{v}^1 \cdot \vec{n})_\Sigma = 0 \qquad \text{i.e. } (\partial_n \Phi)_\Sigma = 0 \tag{1.82}$$

Robin condition:

$$(a\Phi + b\partial_n \Phi)_\Sigma = 0 \qquad \text{i.e. } \left(\frac{\Phi}{\partial_n \Phi}\right)_\Sigma = -\frac{b}{a} \tag{1.83}$$

including Dirichlet ($b/a = 0$) and Neumann ($b/a = \infty$).

One sees that

$$\left(\frac{\Phi}{\partial_n \Phi}\right)_\Sigma = \frac{\zeta}{\iota k} = -\frac{b}{a}$$

that is

$$\zeta = -\iota k \frac{b}{a} \tag{1.84}$$

Thus the Dirichlet condition corresponds to an infinitely small specific normal impedance, i.e. to a surface with normal impedance infinitely smaller than the characteristic impedance of the propagation medium (sea surface in underwater acoustics for example). The Neumann condition corresponds to the inverse situation: a surface with normal impedance infinitely larger than the characteristic impedance of the propagation medium (rigid wall in aeroacoustics for example).

1.2.10. Units, orders of magnitude

The two main propagation media are air (aeroacoustics) and water (underwater acoustics). A third important medium, with properties close to those of water, is the human body, i.e. biological media (ultrasonography). The characteristics of the two main fluids, air and water, can also be considered as references for the two states of fluids: the gaseous and liquid states. These two states, the first characterized by strong compressibility and the second by weak compressibility, have homogeneous characteristics within a state and characteristics that differ greatly from one state to the other.

Characteristics are given at standard pressure and temperature: $P^0 = 1$ atm $= 10^5$ N m^{-2}, $T^0 = 20°\text{C} = 293$ K. *Air* behaves like a perfect gas with $R = p/\rho T \approx 286.9$: $\rho \approx 1.2$ kg m^{-3}, $\chi_s \approx 7.2 \times 10^{-6}$ m^2 N^{-1}, so that $c \approx 340$ m s^{-1} and $Z = \rho c \approx 408$ kg m^{-2} s^{-1} (408 rayls). For *water*, $\rho \approx 10^3$ kg m^{-3}, $\chi_s \approx 4.5 \times 10^{-10}$ m^2 N^{-1}, so that $c \approx 1482$ m s^{-1} and $Z = \rho c \approx 1.48 \times 10^6$ kg m^{-2} s^{-1} (1.48×10^6 rayls).

In **aeroacoustics**, the reference pressure level is $p_0^A = 2 \times 10^{-5}$ N m^{-2}, the approximate hearing threshold of the human ear. The reference intensity level is then $I_0^A = 10^{-12}$ W m^{-2}, the intensity of a plane wave of pressure $p_0^A = 2 \times 10^{-5}$ N m^{-2} propagating in an atmosphere at 20°C and normal pressure. Sound levels are given in decibels (dB) with respect to the reference intensity level and denoted IL (intensity level): IL (dB) $= 10 \log (I^A/I_0^A)$. Acoustic pressures are also given in decibels (dB) with respect to the reference pressure level and denoted SPL (sound pressure level): SPL (dB) $= 20 \log (p^A/p_0^A)$. For a plane wave, the two measures are identical: IL = SPL.

One may note that 20 dB is the noise level of a studio room, 40 dB the level of a normal conversation, 60 dB the level of an intense conversation, 90 dB the level of a symphony orchestra, 100 dB the noise level of a pneumatic drill at 2 m, 120 dB the noise level of a jet engine at 10 m, and 130–140 dB the pain threshold for the human ear.

A sound of 70 dB (mean level between hearing and pain thresholds) corresponds to an acoustic pressure of $p^A = 6 \times 10^{-2}$ N m^{-2}, i.e. to a relative pressure fluctuation (with respect to the atmospheric pressure) of $p^A/P_0 = 6 \times 10^{-7}$. All other relative fluctuations (density, temperature, etc.) are also of magnitude 10^{-7}. One sees that the linearization hypothesis is widely valid.

In **underwater acoustics**, the reference pressure level is $p_0^A = 1$ N m^{-2}. The reference intensity level corresponding to this pressure level is $I_0^A = 6.51 \times 10^{-7}$ W m^{-2}, the intensity level of a plane wave of reference pressure propagating in 3% salted sea water at 20°C and normal pressure (corresponding to water of the Pacific Ocean) ($\rho = 1.013 \times 10^3$ kg m^{-3}, $c = 1516$ m s^{-1}, $Z \equiv \rho c = 1.536 \times 10^{-6}$ rayls).

A 70 dB sonar pulse produces an acoustic pressure $p^A = 3.16 \times 10^3\ \mathrm{N\,m^{-2}}$, i.e. a relative pressure fluctuation $p^A/P_0 = 3.16 \times 10^{-2}$. All other relative fluctuations are of magnitude 10^{-6}–10^{-7}, widely justifying the linearization conditions.

1.2.11. Perfect gas

Most gases behave over a wide temperature and pressure range like a perfect gas with constant heat capacity. This is the case for air.

A perfect gas is a gas that obeys the laws $pv = f(T)$ and $e = g(T)$. This implies the well-known relation $pv = RT$, and also

$$de = C_v(T)\,dT, \qquad ds = C_v(T)\frac{dT}{T} + R\frac{dv}{v}, \qquad C_p(T) - C_v(T) = R$$

Then a perfect gas with constant heat capacity obeys the laws

$$pv = RT, \qquad e = C_vT, \qquad s - s_0 = C_v \log\left(\frac{p}{p_0}\left(\frac{v}{v_0}\right)^{\gamma}\right) \qquad \text{with } \gamma = \frac{C_p}{C_v}$$

Thus isentropic motion, peculiarly acoustic motion, satisfy

$$pv^{\gamma} = ct, \qquad Tv^{\gamma-1} = ct, \qquad Tp^{-(\gamma-1)/\gamma} = ct$$

Isentropic compressibility is then $\chi_s = 1/\gamma P$, and acoustic wave velocity is

$$c = \sqrt{\gamma\frac{P}{\rho}} = \sqrt{\gamma RT} \tag{1.85}$$

For air in particular one has

$$c \approx 20\sqrt{T} \tag{1.86}$$

a formula often used in aeroacoustics.

1.3. Elementary Acoustics of Solids: Elementary Elastic Waves

One is interested here in the elementary acoustics of solids, i.e. with small movements of perfect, homogeneous, isotropic elastic solids, with no dissipative phenomena.

1.3.1. Linearization for an isotropic, homogeneous, purely elastic solid

From (1.5) and (1.12), one has to linearize (if there is no heat source)

$$\begin{cases} \dot{\rho} + \rho\nabla\cdot\vec{v} = 0 \\ \rho\dot{\vec{v}} - \nabla\cdot\sigma = \vec{F} \\ T\rho\dot{s} = 0 \end{cases} \tag{1.87}$$

that is

$$\begin{cases} \dfrac{\partial \rho}{\partial t} + \nabla \cdot (\rho \vec{v}) = 0 \\ \rho\left(\dfrac{\partial \vec{v}}{\partial t} + \vec{v} \cdot \nabla \vec{v}\right) - \nabla \cdot \sigma = \vec{F} \\ T\rho\left(\dfrac{\partial s}{\partial t} + \vec{v} \cdot \nabla s\right) = 0 \end{cases} \tag{1.88}$$

The result is

$$\begin{cases} \dfrac{\partial \rho^1}{\partial t} + \rho^0 \nabla \cdot \vec{v}^1 = 0 \\ \rho^0 \dfrac{\partial \vec{v}^1}{\partial t} - \nabla \cdot \sigma^1 = \vec{F}^1 \\ s^1 = ct = 0 \end{cases} \tag{1.89}$$

As for perfect fluids, small perfectly elastic movements are isentropic.

As a first consequence, assimilating finite difference and differential (1.14), $\sigma^1 = \lambda^0 \operatorname{tr}(\varepsilon^1) I + 2\mu^0 \varepsilon^1$, and if one chooses as variable the elementary displacement $\vec{u}^1$:

$$\vec{v}^1 = \frac{\partial \vec{u}^1}{\partial t}, \qquad \varepsilon^1 = \tfrac{1}{2}(\nabla \vec{u}^1 + {}^{\mathrm{T}}\nabla \vec{u}^1) \qquad \text{and} \qquad \operatorname{tr}(\varepsilon^1) = \nabla \vec{u}^1$$

Then

$$\begin{aligned} \nabla \cdot \sigma^1 &= \lambda^0 \nabla(\nabla \cdot \vec{u}^1) + \mu^0(\nabla \cdot (\nabla \vec{u}^1) + \nabla \cdot ({}^{\mathrm{T}}\nabla \vec{u}^1)) \\ &= (\lambda^0 + \mu^0)\nabla(\nabla \cdot \vec{u}^1) + \mu^0 \nabla \cdot (\nabla \vec{u}^1) \\ &= (\lambda^0 + 2\mu^0)\nabla(\nabla \cdot \vec{u}^1) - \mu^0 \nabla \times \nabla \times \vec{u}^1, \end{aligned}$$

$$\text{since } \nabla \cdot (\nabla \vec{u}) = \nabla(\nabla \cdot \vec{u}) - \nabla \times \nabla \times \vec{u}$$

and the linearized equation of motion becomes

$$\rho^0 \frac{\partial^2 \vec{u}^1}{\partial t^2} - (\lambda^0 + 2\mu^0)\nabla(\nabla \cdot \vec{u}^1) + \mu^0 \nabla \times \nabla \times \vec{u}^1 = \vec{F}^1 \tag{1.90}$$

1.3.2. Compression/expansion waves and shear/distorsion waves

Without restriction, one can decompose the displacement vector into its rotational component and its irrotational component: $\vec{u}^1 = \vec{u}_P^1 + \vec{u}_S^1$, with $\vec{u}_P^1 = \nabla \phi$ and

$\vec{u}_S^1 = \nabla \times \vec{\psi}$, i.e. with $\nabla \times \vec{u}_P^1 = 0$ and $\nabla \cdot \vec{u}_S^1 = 0$, where ϕ and $\vec{\psi}$ are the *scalar* and *vector potentials*. Similarly for the applied forces: $\vec{F}^1 = \nabla G^1 + \nabla \times \vec{H}^1$.

Then the linearized motion equation is decomposed into

$$\rho^0 \frac{\partial^2 \vec{u}_P^1}{\partial t^2} - (\lambda^0 + 2\mu^0)\nabla(\nabla \cdot \vec{u}_P^1) = \nabla G^1$$

$$\rho^0 \frac{\partial^2 \vec{u}_S^1}{\partial t^2} + \mu^0 \nabla \times \nabla \times \vec{u}_S^1 = \nabla \times \vec{H}^1$$

or, since $\nabla^2 \vec{u} \equiv \nabla \cdot (\nabla \vec{u}) = \nabla(\nabla \cdot \vec{u}) - \nabla \times \nabla \times \vec{u}$, $\nabla \times \vec{u}_P^1 = 0$ and $\nabla \cdot \vec{u}_S^1 = 0$,

$$\begin{cases} \rho^0 \dfrac{\partial^2 \vec{u}_P^1}{\partial t^2} - (\lambda^0 + 2\mu^0)\nabla^2 \vec{u}_P^1 = \nabla G^1 \\ \rho^0 \dfrac{\partial^2 \vec{u}_S^1}{\partial t^2} - \mu^0 \nabla^2 \vec{u}_S^1 = \nabla \times \vec{H}^1 \end{cases} \tag{1.91}$$

or

$$\begin{cases} -\dfrac{1}{c_P^2} \dfrac{\partial^2 \vec{u}_P^1}{\partial t^2} + \nabla^2 \vec{u}_P^1 = -\dfrac{1}{\lambda^0 + 2\mu^0} \nabla G^1 \\ -\dfrac{1}{c_S^2} \dfrac{\partial^2 \vec{u}_S^1}{\partial t^2} + \nabla^2 \vec{u}_S^1 = -\dfrac{1}{\mu^0} \nabla \times \vec{H}^1 \end{cases} \tag{1.92}$$

with

$$c_P = \sqrt{\frac{\lambda^0 + 2\mu^0}{\rho^0}}; \qquad c_s = \sqrt{\frac{\mu^0}{\rho^0}} \tag{1.93}$$

For the potentials, one has

$$\begin{cases} -\dfrac{1}{c_P^2} \dfrac{\partial^2 \phi}{\partial t^2} + \nabla^2 \phi = -\dfrac{1}{\lambda^0 + 2\mu^0} G^1 \\ -\dfrac{1}{c_S^2} \dfrac{\partial^2 \vec{\psi}}{\partial t^2} + \nabla^2 \psi = -\dfrac{1}{\mu^0} \vec{H}^1 \end{cases} \tag{1.94}$$

Equations (1.92) and (1.94) are wave equations. They show the existence of two types of waves, propagating at different wave velocities:

- Compression/expansion, i.e. pressure waves $\vec{u}_P^1$, derived from a scalar potential ϕ by $\vec{u}_P^1 = \nabla \phi$, characterized by $\nabla \times \vec{u}_P^1 = 0$ and $\nabla \cdot \vec{u}_P^1 \neq 0$, and propagating at velocity c_P.
- Shear/distorsion waves $\vec{u}_S^1$, derived from a vector potential $\vec{\psi}$ by $\vec{u}_S^1 = \nabla \times \vec{\psi}$, characterized by $\nabla \cdot \vec{u}_S^1 = 0$ and $\nabla \times \vec{u}_S^1 \neq 0$, and propagating at velocity $c_S < c_P$.

In seismology, pressure waves are also called primary waves (P-waves) and shear waves secondary waves (S-waves), since the latter propagate at lower speed and so arrive later than the former.

1.3.3. Plane waves: Longitudinal and transverse waves

Consider a plane pressure wave propagating in direction $\vec{n}$ with velocity c_P, $\phi = \phi^+(t - \vec{n}\cdot\vec{x}/c_P)$. One can verify that this is a solution of equation (1.94) without source term. Then

$$\vec{u}_P^1 = \nabla\phi = -\frac{\vec{n}}{c_P}\,\phi^{+'}\left(t - \frac{\vec{n}\cdot\vec{x}}{c_P}\right)$$

$\vec{u}_P^1$ is collinear with $\vec{n}$. The vibration is parallel to the direction of propagation: one says that the wave is *longitudinal.*

Consider now a plane shear wave propagating in direction $\vec{n}$ with velocity c_S and polarization of potential $\vec{s}_\psi$, $\vec{\psi} = \vec{s}_\psi\psi^+(t - \vec{n}\cdot\vec{x}/c_S)$. One can verify that this is a solution of equation (1.94) for the vector potential. Then

$$\vec{u}_S^1 = \nabla\times\vec{\psi} = -\frac{\vec{n}}{c_S}\times\vec{s}_\psi\psi^{+'}\left(t - \frac{\vec{n}\cdot\vec{x}}{c_S}\right)$$

$\vec{u}_S^1$ is perpendicular to $\vec{n}$. The vibration is perpendicular to the direction of propagation: one says that the wave is a *transverse.*

1.3.4. Orders of magnitude

Wave velocities

$$c_P = \sqrt{\frac{\lambda^0 + 2\mu^0}{\rho^0}}, \qquad c_S = \sqrt{\frac{\mu^0}{\rho^0}}$$

can also be written as functions of Young's modulus E^0 and Poisson's ratio ν^0:

$$E^0 = \mu^0\,\frac{3\lambda^0 + 2\mu^0}{\lambda^0 + \mu^0} \quad \text{and} \quad \nu^0 = \frac{\lambda^0}{2(\lambda^0 + \mu^0)}$$

i.e.

$$\lambda^0 = \frac{E^0\nu^0}{(1 + \nu^0)(1 - 2\nu^0)} \quad \text{and} \quad \mu^0 = \frac{E^0}{2(1 + \nu^0)}$$

and

$$c_P = \sqrt{\frac{E^0(1 - \nu^0)}{\rho^0(1 + \nu^0)(1 - 2\nu^0)}}, \qquad c_S = \sqrt{\frac{E^0}{2\rho^0(1 + \nu^0)}} \tag{1.95}$$

Generally one has roughly:

$$c_S \sim \frac{c_P}{2} \tag{1.96}$$

Typical values at 20°C are:

- Aluminium: $c_P = 6300\ \text{m s}^{-1}$, $c_S = 3080\ \text{m s}^{-1}$, $\rho = 2.7 \times 10^3\ \text{kg m}^{-3}$, $Z_P \equiv \rho c_P = 1.7 \times 10^7$ rayls, $Z_S \equiv \rho c_S = 0.8 \times 10^7$ rayls.
- Copper: $c_P = 4700\ \text{m s}^{-1}$, $c_S = 2260\ \text{m s}^{-1}$, $\rho = 8.9 \times 10^3\ \text{kg m}^{-3}$, $Z_P \equiv \rho c_P = 4.2 \times 10^7$ rayls, $Z_S \equiv \rho c_S = 1.6 \times 10^7$ rayls.
- Steel: $c_P = 5900\ \text{m s}^{-1}$, $c_S = 3230\ \text{m s}^{-1}$, $\rho = 7.7 \times 10^3\ \text{kg m}^{-3}$, $Z_P \equiv \rho c_P = 4.5 \times 10^7$ rayls, $Z_S \equiv \rho c_S = 2.5 \times 10^7$ rayls.
- Plexiglas: $c_P = 2730\ \text{m s}^{-1}$, $c_S = 1430\ \text{m s}^{-1}$, $\rho = 1.18 \times 10^3\ \text{kg m}^{-3}$, $Z_P \equiv \rho c_P = 3.2 \times 10^7$ rayls, $Z_S \equiv \rho c_S = 1.7 \times 10^7$ rayls.

1.3.5. General behaviour

Elastic waves in solids behave like acoustic waves in fluids: they are subjected to propagation, refraction, reflection and scattering phenomena. Simply, they are polarized waves (vector waves). One has merely to consider independently pressure waves and shear waves and make them interact at discontinuities.

1.4. Conclusion

We have given the material necessary to derive acoustic equations in rather general cases but have treated only the simplest case of perfect simple fluid or elastic solid initially homogeneous and at rest. This leads to classical wave equations and constitutes elementary acoustic and elastic wave theory.

More complex situations can also be treated on the basis of given equations of mechanics: non-homogeneous media, non-steady initial states, dissipative media. The recipe is simple – linearization – but developments are numerous and have given rise to an abundant bibliography that it is not possible to summarize within the context of this book.

Bibliography

This chapter requires only basic knowledge in mechanics of continua and thermodynamics. One can find it in [1]–[3].

[1] EHRINGEN, A.C., 1967. *Mechanics of continua.* John Wiley & Sons, New York.
[2] MALVERN, L.E., 1969. *Introduction to the mechanics of continuous media.* Prentice-Hall, Englewood Cliffs, NJ.
[3] LAVENDA, B.H., 1993. *Thermodynamics of irreversible processes.* Dover Publications, New York.

Derivation of acoustic equations can be found in [4]–[6]:

[4] LANDAU, L.D. and LIFSCHITZ, E.M., 1986. *Fluid mechanics.* Course of Theoretical Physics, vol. 6. Pergamon Press, Oxford.
[5] LANDAU, L.D. and LIFSCHITZ, E.M., 1986. *Theory of elasticity.* Course of Theoretical Physics, vol. 7. Pergamon Press, Oxford.
[6] BREKHOVSKIKH, L.M. and GONCHAROV, V. 1985. *Mechanics of continua and wave dynamics.* Springer series in wave phenomena Vol. 1, Springer Verlag, New York.

One cannot discuss general acoustics without quoting the classic book:

[7] MORSE, P.M. and INGARD, K.U., 1968. *Theoretical acoustics.* McGraw-Hill, New York. This book also gives developments on acoustics of inhomogeneous media, moving media and dissipative media.

More developments on acoustics of dissipative media can be found in [8]–[10].

[8] BHATHIA, A.B., 1967. *Ultrasonic absorption.* Oxford University Press, London.
[9] HERZFELD, K.F. and LITOVITZ, T.A., 1959. *Absorption and dispersion of ultrasonic waves.* Academic Press, New York.
[10] MASON, W.P., 1971. *Physical Acoustics*, vol.II, part A: Properties of gases, liquids and solutions. Academic Press, New York.

CHAPTER 2

Acoustics of Enclosures

Paul J.T. Filippi

Introduction

This chapter deals with a few methods of common use for the study of the sound field inside an enclosure: factory halls, concert halls, theatres, airplane cabins, cars, trucks ... .

In the first section, the equations which govern the phenomena are presented and the conditions for existence and uniqueness of the solution are stated: the notions of *resonance frequencies* and *resonance modes* (free oscillations), as well as *eigenfrequencies* and *eigenmodes* are introduced. The following section is devoted to the simple example of a parallelepipedic enclosure: in particular, it is shown that the response of the room to a harmonic excitation can be expanded into a series of eigenmodes. In the third section, the response of an enclosure to a transient signal (sound establishment and sound decay) is calculated: it is expressed as a series of resonance modes which, if there is energy absorption by the walls, for instance, are different from the eigenmodes. In Section 2.4, the interest is focused on a very academic problem: a two-dimensional circular enclosure. It provides the opportunity to introduce the rather general method of *separation of variables*, and to compare the corresponding solution with the eigenmodes series. Section 2.5 introduces briefly the *method of images*: it is an approximation which comes from geometrical optics, which is mainly valid at high frequency and for polyhedral boundaries only (for simplicity, convex polyhedra will be considered, only). In the last section, the Green's representation of the sound field inside an enclosure is introduced: this is one of the most general mathematical representations of the solution in which the acoustic pressure reflected by the boundaries is expressed as the radiation of fictitious sources located on the enclosure boundaries.

The classical mathematical proofs concerning the theory of partial differential equations and boundary value problems are not given in this manual. The reader can find them in specialized textbooks such as those which are mentioned in the short bibliographic list at the end of the chapter.

2.1. General Statement of the Problem

Let us consider a domain Ω, with a regular boundary σ. By regular boundary we mean, for example, a piecewise indefinitely differentiable surface (or curve in $\mathbb{R}^2$). This allows us to define almost everywhere a unit vector $\vec{n}$, normal to σ and pointing out to the exterior of Ω. This bounded domain is filled with a homogeneous isotropic perfect fluid, characterized by a density ρ_0 and a sound velocity c_0. Acoustic sources are present: they are described by a function (or, more generally, a distribution) denoted $F(M, t)$, depending on the space variable $M(x, y, z)$ and the time variable t.

2.1.1. The wave equation

The acoustic pressure $\psi(M, t)$ generated by the sources satisfies the following wave equation:

$$\left(\Delta - \frac{1}{c_0^2}\frac{\partial^2}{\partial t^2}\right)\psi(M, t) = F(M, t) \qquad M \in \Omega, \quad t \in]-\infty, +\infty[$$
$$\psi(M, t) = \partial_t\psi(M, t) = 0 \qquad t < t_0 \tag{2.1}$$

where t_0 is the time at which the sources $F(M, t)$ start. In general, the sources stop after a bounded duration. But this is not a reason for the fluid motion to stop, too. In fact, the energy conservation equations used to describe the phenomenon show that, after the sources have stopped, the acoustic energy inside the enclosure decreases more or less exponentially. Indeed, when a wave front arrives on an obstacle, it gives it a certain amount of its energy, the remaining part being reflected; the motion of the fluid lasts indefinitely with an amplitude which is reduced each time the wave encounters an obstacle. In practice, the sound level decreases rapidly under the hearing threshold.

The description of the influence of the boundary σ must be added to the set of equations (2.1). Two simple cases can be considered which correspond to a conservative (no energy loss) physical system (for such an ideal system, the sound field keeps a constant level when the sources have been turned off).

The Neumann boundary condition. Assume that the boundary σ is made of a perfectly rigid solid which forces the normal particle velocity $\vec{n} \cdot \vec{V}(M, t)$ to be zero. The momentum equation

$$\rho_0 \frac{\partial \vec{V}}{\partial t} + \nabla\psi = 0$$

shows that the acoustic pressure $\psi(M, t)$ must satisfy the boundary condition

$$\partial_n\psi(M, t) = 0, \qquad M \in \sigma, \quad \forall t \tag{2.2}$$

where the normal derivative $\partial_n\psi(M, t)$ is defined by the scalar product

$$\partial_n\psi(M, t) = \vec{n}(M) \cdot \nabla\psi(M, t)$$

The corresponding boundary value problem is called the Neumann problem.

The Dirichlet boundary condition. Assume that the fluid which the acoustic wave is propagating in is a liquid which is in contact with a gas along σ (an example is the surface of the sea). An excellent approximation of the influence of such a boundary on the sound field is to neglect the transmission of acoustic energy into the gas. Thus, the gas exerts on the liquid boundary a constant pressure (the atmospheric pressure in the example of the sea) and, as a consequence, the acoustic pressure is zero, that is:

$$\psi(M, t) = 0, \qquad M \in \sigma, \quad \forall t \tag{2.3}$$

The corresponding boundary value problem is called the Dirichlet problem.

It is shown that equation (2.1) together with one of the two conditions (2.2) or (2.3) has one and only one solution. The same result is valid if the Neumann condition is imposed on a part σ' of σ and the Dirichlet condition is assumed on the other part σ''. The proof is a classical result of the theory of boundary value problems.

For a boundary which absorbs energy, the expression of the boundary condition is not so easy to establish. To be totally rigorous, it is necessary to mathematically describe how the acoustic energy is transmitted to the boundaries of the propagation domain: this implies solving a problem which involves a coupling between a fluid and an elastic (possibly viscous) solid. This is, for instance, the case in building acoustics: rooms are often bounded by thin elastic structures (light walls, doors, windows ...) which allow sound energy transmission. In many real life situations, the signals involved have a narrow frequency spectrum (no more than an octave): a reasonable approximation is to consider that the ratio of the acoustic pressure to the normal particle velocity (impedance) is a constant, which, of course, depends on the central frequency of the signal. More realistic signals can be considered as a linear combination of narrow frequency spectrum components which satisfy different boundary conditions.

2.1.2. The Helmholtz equation

Many realistic considerations have led scientists to focus their efforts on the response of physical systems to periodic (and mainly harmonic) excitations. A physical time function can, in general, be expressed as an inverse Fourier transform (the validity conditions are not constraining in physics): this means that almost every deterministic acoustic excitation can be considered as the linear combination – in fact, an integral – of elementary harmonic components. Many real life acoustic phenomena have a periodic (or quasi periodic) time dependency, that is they are linear combinations of harmonic components, the number of which can be considered as finite. Typical examples include electric converters, machine tools, thermal or electric motors. Under certain conditions, the sound emitted by a musical instrument can also be considered as a quasi periodic excitation. In room acoustics engineering, the first step for the designer is to investigate the room response to a sine excitation and to try to make it as independent as possible of the frequency; then, the response to transient excitations is examined, sometimes with much less precise methods, the attention being mainly focused on the propagation of the wave fronts.

Assume that the source $F(M, t)$ has a harmonic time dependency and is represented by a complex function $f(M)e^{-\iota\omega t}$, where ι is $\sqrt{-1}$. The physical quantity is given by

$$F(M, t) = \Re[f(M)e^{-\iota\omega t}] \tag{2.4}$$

It is shown that the acoustic pressure can equally be represented by a complex function $p(M)e^{-\iota\omega t}$ which is related to the physical pressure by

$$\psi(M, t) = \Re[p(M)e^{-\iota\omega t}] \tag{2.5}$$

The functions $f(M)$ and $p(M)$ are respectively called the *complex amplitudes* of the source and of the sound pressure. It must be kept in mind that a microphone or the ear is sensitive to the function $\psi(M, t)$ and not to its complex representation. Finally, the particle velocity $\vec{V}(M, t)$ is associated to a complex function $\vec{v}(M)e^{-\iota\omega t}$ by

$$\vec{V}(M, t) = \Re[\vec{v}(M)e^{-\iota\omega t}] \tag{2.6}$$

Expressions (2.4) and (2.5) are introduced into the wave equation (2.1) to get

$$\left(\Delta - \frac{1}{c_0^2}\frac{\partial^2}{\partial t^2}\right)\Re[p(M)e^{-\iota\omega t}] = \Re[f(M)e^{-\iota\omega t}]$$

Elementary considerations show that this equation is satisfied if and only if the complex pressure amplitude is a solution of

$$(\Delta + k^2)p(M) = f(M), \qquad M \in \Omega, \quad k^2 = \frac{\omega}{c_0^2} \tag{2.7}$$

This equation is called the *Helmholtz equation.*

2.1.3. Boundary condition for harmonic regimes

A perfectly rigid boundary remains characterized by the Neumann condition; the liquid–gas interface is still described by a Dirichlet condition.

In many practical situations of an absorbing boundary, it is possible to adopt the *Robin boundary condition* which we write:

$$\frac{\partial p(M)}{\partial \vec{n}(M)} - \frac{\iota k}{\zeta(M)}p(M) = 0, \qquad M \in \sigma \tag{2.8}$$

This relationship states that, at any point $M \in \sigma$, the normal component of the acoustic pressure gradient (which is proportional to the normal component of the particle velocity) is proportional to the acoustic pressure itself. The quantity $\zeta(M)$, which can vary from point to point, is called the *specific normal impedance* of the boundary; its inverse is called the *specific normal admittance*. The Robin boundary condition is, like the Neumann and the Dirichlet ones, a local condition. It describes accurately the physical phenomenon as far as the acoustic wave is rapidly attenuated inside the boundary material in the tangential direction. This is, in particular, the behaviour of the porous materials commonly used as acoustic absorbers.

The specific normal impedance is a complex quantity the real part of which is necessarily positive. Indeed, let us calculate the energy flux δE which flows across a boundary element $d\sigma$ during one period T. It is given by

$$\delta E = d\sigma \int_0^T \Re(p e^{-\iota\omega t}) \Re(\vec{v} \cdot \vec{n} e^{-\iota\omega t})\, dt \tag{2.9}$$

The momentum equation expresses the particle velocity in terms of the pressure gradient:

$$-\iota\omega\rho_0\vec{v} + \nabla p = 0$$

Let $\dot{p}$ and $\dot{\zeta}$ (resp. $\hat{p}$ and $\hat{\zeta}$) be the real parts (resp. the imaginary parts) of the acoustic pressure p and of the impedance ζ. One obtains

$$\vec{n} \cdot \nabla p = \frac{\iota\omega}{c_0 \mid \zeta \mid^2} [(\dot{p}\dot{\zeta} + \hat{p}\hat{\zeta}) + \iota(\hat{p}\dot{\zeta} - \dot{p}\hat{\zeta})]$$

or equivalently

$$\vec{n} \cdot \vec{v} = \frac{1}{\rho_0 c_0 \mid \zeta \mid^2} [(\dot{p}\dot{\zeta} + \hat{p}\hat{\zeta}) + \iota(\hat{p}\dot{\zeta} - \dot{p}\hat{\zeta})]$$

Using this last equality, the energy flux is written

$$\begin{aligned}\delta E = \frac{d\sigma}{\rho_0 c_0 \mid \zeta \mid^2} \int_0^T & \{\dot{p}(\dot{p}\dot{\zeta} + \hat{p}\hat{\zeta}) \cos^2 \omega t + \hat{p}(\hat{p}\dot{\zeta} - \dot{p}\hat{\zeta}) \sin^2 \omega t \\ & + [2\dot{p}\hat{p}\dot{\zeta} + (\hat{p}^2 - \dot{p}^2)\hat{\zeta}] \sin \omega t \cos \omega t\}\, dt\end{aligned} \tag{2.10}$$

The integration interval being one period, the last term has a zero contribution. The first two terms lead to

$$\delta E = \frac{T\, d\sigma}{2\rho_0 c_0 \mid \zeta \mid^2} \mid p \mid^2 \dot{\zeta} \tag{2.11}$$

If the boundary element $d\sigma$ absorbs energy, the flux δE which flows across it must be positive. This implies that the real part $\dot{\zeta}$ of the specific normal impedance is positive. The quantity $\dot{\zeta}$ is called the *acoustic resistance*, while the imaginary part of the normal specific impedance is called the *acoustic reactance*.

The specific normal impedance is a function of frequency. It is useful to have an idea of its variations. As a rough general rule, at low frequencies, every material is perfectly reflecting ($|\zeta| \to \infty$) while, at high frequencies, every material is perfectly absorbing ($\dot{\zeta} \to 1, \hat{\zeta} \to 0$). The surface of a porous material can be accurately characterized by such a local boundary condition. Delany and Bazley have proposed a simple model of the specific normal impedance of a porous medium which is expressed as a function of the frequency by

$$\zeta = 1. + 9.08\left(\frac{s}{f}\right)^{0.75} + \iota 11.9\left(\frac{s}{f}\right)^{0.73}$$

where the parameter s, called the *flow resistance*, characterizes the porosity of the medium. Figure 2.1 corresponds to the value $s = 300$ and represents the impedance curve of standard materials used as ceiling covering. This simple model of impedance is quite satisfactory in many practical situations providing the thickness of material used is large enough. There are, of course, many other models which provide a more accurate description of acoustic porous materials by taking into account more details: thickness of the material layer, variations of the porosity, vibrations and damping of the solid structure, etc.

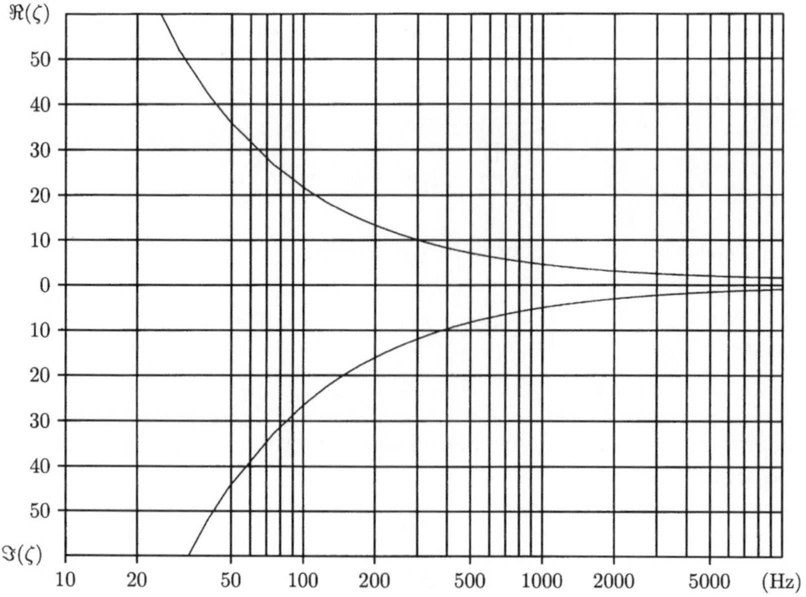

Fig. 2.1. Delany and Bazley's model of impedance: real and imaginary parts of ζ as a function of frequency ($s = 300$).

The three types of boundary conditions can be gathered in a unique expression:

$$\alpha \partial_n p(M) + \beta p(M) = 0, \qquad M \in \sigma \tag{2.12}$$

with

$$\alpha = 1, \beta = 0 \longrightarrow \text{Neumann boundary condition}$$
$$\alpha = 0, \beta = 1 \longrightarrow \text{Dirichlet boundary condition}$$
$$\alpha = 1, \beta \neq 0 \longrightarrow \text{Robin boundary condition}$$

When the boundary is composed of several panels with different homogeneous materials, the coefficients α and β are piecewise constant functions. It is also possible to consider a boundary with continuously varying properties: then α and β are continuous functions. The most general case is to consider piecewise continuous functions α and β.

2.1.4. Eigenmodes and eigenfrequencies, condition for existence and uniqueness of the solution

Let us consider the following boundary value problem:

$$(\Delta + k^2)p(M) = f(M), \qquad M \in \Omega \tag{2.7}$$
$$\alpha \partial_n p(M) + \beta p(M) = 0, \qquad M \in \sigma \tag{2.12}$$

where σ, the boundary of the domain Ω is assumed to be a piecewise indefinitely differentiable surface (or curve in $\mathbb{R}^2$). The following theorem stands:

Theorem 2.1 (Existence and uniqueness of the solution)

(a) There exists a countable sequence $k_n (n = 1, 2, \ldots, \infty)$ of wavenumbers such that the homogeneous boundary value problem (2.7, 2.12) has non-zero solutions. To each such wavenumber, called an *eigenwavenumber*, a frequency f_n, called an *eigenfrequency* is associated.
(b) For each eigenwavenumber k_n, the homogeneous boundary value problem has a finite number N_n of non-zero solutions $\Psi_{nm} (n = 1, 2, \ldots, \infty;$ $m = 1, 2, \ldots, N_n < \infty)$, which are linearly independent. These solutions are called *eigenmodes* of the Laplace operator for the boundary condition (2.12). The number N_n of independent solutions corresponding to the wavenumber k_n is called its *order of multiplicity*.
(c) The eigenwavenumbers k_n are real if the ratio α/β is real; if $\Im(\alpha/\beta) \neq 0$, then $\Im k_n \neq 0$.
(d) If k is equal to any eigenwavenumber, then the non-homogeneous boundary value problem (2.7, 2.12) has no solution. If k is not equal to any eigenwavenumber, then the solution $p(M)$ exists and is unique for any source term $f(M)$.

Remark. It is equally possible to consider a boundary value problem with a homogeneous Helmholtz equation ($f(M) \equiv 0$) and a non-homogeneous boundary condition:

$$\alpha \partial_n p(M) + \beta p(M) = g(M), \qquad M \in \sigma$$

Such a boundary condition appears when the energy is furnished to the enclosure by a motion of the boundary (or a part of the boundary). As an example, assume that a part σ_1 of σ is the external surface of a rotating machine. This surface is mechanically excited by the vibrations induced by the rotating parts of the machine. It thus has a non-zero normal velocity which is transmitted to the surrounding fluid. This energy transmission is expressed by the boundary condition

$$\partial_n p(M) = \frac{u(M)}{\iota \omega \rho_0}, \qquad \forall M \in \sigma_1$$

On the remaining part of the boundary, a homogeneous condition is assumed. It can be shown that the former theorem is still valid: the solution exists and is unique for any source term $u(M)$ if and only if $k \neq k_n$.

2.2. Sound Field inside a Parallelepipedic Enclosure: Free Oscillations and Eigenmodes

In an enclosure, the fluid oscillations which can occur without any source contribution are called *free oscillations*, or *free regimes*, or *resonance modes*. Such non-causal phenomena can, of course, appear as a purely intellectual concept; but, in fact, they correspond to a certain aspect of the physical reality, as will be explained. It is well known that if someone sings a given note close to a piano, when he stops he can hear the corresponding string, which continues to produce sound. Though the mechanical energy given by the singer to the piano string is very small, the resulting sound is quite easy to hear. The physical phenomenon which occurs in this experiment is that the infinitely small amount of energy absorbed by the string is quite sufficient to excite its free oscillation mode or resonance mode, which has a very low damping constant.

For an enclosure, the eigenmodes are purely mathematical functions which, in general, are not simply related to the resonance modes which, in a certain way, describe the physical properties of the system. In mathematics, an eigenfunction (or eigenmode) of an operator A is a non-trivial solution of the equation $AU = \Lambda U$, where the constant Λ is called an eigenvalue. For an acoustics problem, the phenomenon is governed by the combined Laplace operator and boundary condition operator. It is obvious that, in the general case, this operator is frequency dependent because of the frequency dependence of the boundary conditions. This implies that the eigenfunctions (usually called eigenmodes by acousticians) and eigenvalues are frequency dependent, too.

This section starts with the Neumann problem: the eigenfrequencies and eigenmodes, as defined in Section 2.1.4, are calculated. Because the boundary condition is independent of the frequency, these modes also describe the free oscillations of the fluid contained in the enclosure: in this case, eigenmodes and resonance modes are identical. Then, the Robin problem is considered. The free oscillations regimes (resonance modes) are looked for: it is necessary to consider that the impedance can vary with frequency, which implies that the different resonance modes satisfy different boundary conditions.

The next step is to show that the response of the enclosure to a harmonic excitation can be expanded into a series of eigenmodes, that is into a series of functions which satisfy the same boundary condition: an explicit expression of the coefficients of the series is obtained. For the Robin problem, these modes are different from the resonance modes.

It must be noticed that, for the Dirichlet problem, the resonance modes are identical to the eigenmodes. The calculations, which are essentially the same as those developed for the Neumann problem, are left to the reader.

The parallelepipedic domain Ω which is considered here is defined by

$$-\frac{a}{2} < x < +\frac{a}{2}, \qquad -\frac{b}{2} < y < +\frac{b}{2}, \qquad -\frac{c}{2} < z < +\frac{c}{2}$$

2.2.1. Eigenfrequencies and eigenmodes for the Neumann problem

Let us look for the existence of wavenumbers k such that the following homogeneous boundary value problem has a non-zero solution:

$$(\Delta + k^2)p(x,y,z) = 0, \quad x \in \left]-\frac{a}{2}, +\frac{a}{2}\right[, \quad y \in \left]-\frac{b}{2}, +\frac{b}{2}\right[, \quad z \in \left]-\frac{c}{2}, +\frac{c}{2}\right[$$

$$\begin{aligned} &\partial_x p(-a/2, y, z) = 0, \qquad \partial_x p(+a/2, y, z) = 0 \\ &\partial_y p(x, -b/2, z) = 0, \qquad \partial_y p(x, +b/2, z) = 0 \\ &\partial_z p(x, y, -c/2) = 0, \qquad \partial_z p(x, y, +c/2) = 0 \end{aligned} \tag{2.13}$$

Let us consider a function $R(x)$ solution of

$$\begin{aligned} &\left(\frac{d^2}{dx^2} + k^2\right) R(x) = 0, \qquad x \in \left]-\frac{a}{2}, +\frac{a}{2}\right[\\ &d_x R(-a/2) = 0, \qquad d_x R(+a/2) = 0 \end{aligned} \tag{2.14}$$

It is obvious that such a function, if it exists, satisfies system (2.13) too. This suggests seeking a solution of this system in a *separate variables form*:

$$p(x,y,z) = R(x)S(y)T(z) \tag{2.15}$$

Introducing this expression into the Helmholtz equation leads to

$$\frac{R''(x)}{R(x)} + \frac{S''(y)}{S(y)} + \frac{T''(z)}{T(z)} + k^2 = 0 \tag{2.16}$$

Because k^2 is a constant, this equation can be satisfied if and only if each term R''/R, S''/S and T''/T is independent of the variables (x, y, z). The homogeneous boundary value problem (2.13) is thus reduced to three one-dimensional boundary value problems:

$$\frac{R''(x)}{R(x)} + \alpha^2 = 0, \qquad R'(-a/2) = 0, \qquad R'(+a/2) = 0 \tag{2.17}$$

$$\frac{S''(y)}{S(y)} + \beta^2 = 0, \qquad S'(-b/2) = 0, \qquad S'(+b/2) = 0 \tag{2.17'}$$

$$\frac{T''(z)}{T(z)} + \gamma^2 = 0, \qquad T'(-c/2) = 0, \qquad T'(+c/2) = 0 \tag{2.17''}$$

Then, the function p given by expression (2.15) is a solution of (2.13) with

$$k^2 = \alpha^2 + \beta^2 + \gamma^2 \tag{2.18}$$

The general solution of the differential equation (2.17) is

$$R(x) = A\mathrm{e}^{-\iota\alpha x} + B\mathrm{e}^{+\iota\alpha x} \tag{2.19}$$

in which expression the coefficients A and B must satisfy the following system of homogeneous equations:

$$A\mathrm{e}^{+\iota\alpha a/2} - B\mathrm{e}^{-\iota\alpha a/2} = 0$$

$$A\mathrm{e}^{-\iota\alpha a/2} - B\mathrm{e}^{+\iota\alpha a/2} = 0$$

This system has a non-zero solution if and only if its determinant is zero; this occurs if α satisfies

$$\sin \alpha a = 0$$

that is if α belongs to the following sequence:

$$\alpha_r = r\frac{\pi}{a}, \qquad r = 0, 1, \ldots, \infty \tag{2.20}$$

The corresponding solution is written

$$R_r = 2B\mathrm{e}^{-\iota\alpha_r a/2} \cos \alpha_r(x - a/2) \tag{2.21}$$

where the coefficient B is arbitrary. One of the most useful choices is to impose that the function $R_r(x)$ has an $\mathbf{L}^2$-norm equal to 1:

$$\int_{-a/2}^{+a/2} | R_r(x) |^2 \, dx = 4 \, | B |^2 \int_{-a/2}^{+a/2} \cos^2 \alpha_r(x - a/2) \, dx = 1$$

Choosing $B = e^{\iota\alpha_r a/2}/\sqrt{2a})$, one gets:

$$R_r(x) = \frac{1}{\sqrt{2a}} \cos \frac{r\pi(x - a/2)}{a}, \qquad r = 0, 1, \ldots, \infty \tag{2.22}$$

In the same way, the functions $S(y)$ and $T(z)$ are given by

$$\begin{aligned} S_s(x) &= \frac{1}{\sqrt{2b}} \cos \frac{s\pi(y - b/2)}{b}, \qquad s = 0, 1, \ldots, \infty \\ T_t(x) &= \frac{1}{\sqrt{2c}} \cos \frac{t\pi(z - c/2)}{c}, \qquad t = 0, 1, \ldots, \infty \end{aligned} \tag{2.23}$$

So, a countable sequence of solutions of the homogeneous boundary value problem (2.13) can be defined by

$$P_{rst}(x, y, z) = \frac{1}{\sqrt{8abc}} \cos \frac{r\pi(x - a/2)}{a} \cos \frac{s\pi(y - b/2)}{b} \cos \frac{t\pi(z - c/2)}{c} \tag{2.24}$$

$$r, s, t \text{ integers} \geqslant 0$$

The corresponding wavenumbers are written

$$k_{rst} = \pi \sqrt{\frac{r^2}{a^2} + \frac{s^2}{b^2} + \frac{t^2}{c^2}} \tag{2.25}$$

Let us recall that the eigenmodes are defined up to an arbitrary amplitude factor: with the choice here adopted, they have an $\mathbf{L}^2$-norm equal to unity.

Remark. If b is a multiple of a, say $b = 3a$ for instance, then the eigenmodes

$$p_{r\,0\,0}(x, y, z) = \frac{1}{\sqrt{2a}} \cos \frac{r\pi(x - a/2)}{a} \quad \text{and}$$

$$p_{0\,3r\,0}(x, y, z) = \frac{1}{\sqrt{2b}} \cos \frac{3r\pi(y - b/2)}{b}$$

correspond to the same wavenumber $r\pi/a$ which is said to have a multiplicity order equal to 2. But for this wavenumber, the homogeneous Neumann problem has two linearly independent solutions. It is also possible to have wavenumbers with a multiplicity order of 3, to which three linearly independent eigenmodes are associated.

2.2.2. Resonance frequencies and resonance modes for the Robin problem

Let us consider the following boundary value problem:

$$(\Delta + k^2)p(x,y,z) = 0, \quad x \in \left]-\frac{a}{2}, +\frac{a}{2}\right[, \quad y \in \left]-\frac{b}{2}, +\frac{b}{2}\right[, \quad z \in \left]-\frac{c}{2}, +\frac{c}{2}\right[$$

$$\begin{aligned}
\partial_x p(x,y,z) - \iota k\alpha p(x,y,z) &= 0 \quad \text{at} \quad x = +a/2 \\
-\partial_x p(x,y,z) - \iota k\alpha p(x,y,z) &= 0 \quad \text{at} \quad x = -a/2 \\
\partial_y p(x,y,z) - \iota k\beta p(x,y,z) &= 0 \quad \text{at} \quad y = +b/2 \\
-\partial_y p(x,y,z) - \iota k\beta p(x,y,z) &= 0 \quad \text{at} \quad y = -b/2 \\
\partial_z p(x,y,z) - \iota k\gamma p(x,y,z) &= 0 \quad \text{at} \quad z = +c/2 \\
-\partial_z p(x,y,z) - \iota k\gamma p(x,y,z) &= 0 \quad \text{at} \quad z = -c/2
\end{aligned} \qquad (2.26)$$

A solution of system (2.26), if it exists, describes the free oscillations of the fluid which fills the domain Ω. Such an oscillation is called a *resonance mode*; the corresponding wavenumber (resp. frequency) is called a *resonance wavenumber* (resp. *resonance frequency*). The role of these resonance modes will appear clearly in the calculation of transient regimes. It must be noticed that, if two resonance modes correspond to two different resonance frequencies, they satisfy different boundary conditions: indeed, the coefficients $\iota k\alpha$, $\iota k\beta$ and $\iota k\gamma$ are generally frequency dependent (through the factor k but also because the impedance of a material depends on the frequency). To simplify the following calculations, it is assumed that the three admittances α, β and γ are constants independent of the frequency.

The technique of separation of variables used for the Neumann problem is again applied: the solution is sought as a product $R(x)S(y)T(z)$ of three functions, each of them depending on one variable only. Thus, the function $R(x)$ must satisfy the following boundary value problem:

$$\begin{aligned}
R''(x) + \xi^2 R(x) &= 0, \quad x \in]-a/2, +a/2[\\
\partial_x R(x) - \iota k\alpha R(x) &= 0 \quad \text{at} \quad x = +a/2 \\
-\partial_x R(x) - \iota k\alpha R(x) &= 0 \quad \text{at} \quad x = -a/2
\end{aligned} \qquad (2.27)$$

Its general expression is

$$R(x) = A\mathrm{e}^{\iota\xi x} + B\mathrm{e}^{-\iota\xi x}$$

and system (2.27) has a non-zero solution if and only if ξ is one of the roots of the following equation

$$\mathrm{e}^{2\iota\xi a} = \left(\frac{\xi + k\alpha}{\xi - k\alpha}\right)^2 \qquad (2.28)$$

It must be noticed that expression (2.28) depends on the parameter k, and so do the solutions of this equation.

In the same way, it is possible to define $S(y)$ and $T(z)$ by

$$S(y) = C\mathrm{e}^{\iota\eta y} + D\mathrm{e}^{-\iota\eta y}$$
$$T(z) = E\mathrm{e}^{\iota\zeta z} + F\mathrm{e}^{-\iota\zeta z} \tag{2.29}$$

where η and ζ must be solutions of the following two equations:

$$\mathrm{e}^{2\iota\eta b} = \left(\frac{\eta + k\beta}{\eta - k\beta}\right)^2 \tag{2.30}$$

$$\mathrm{e}^{2\iota\zeta c} = \left(\frac{\zeta + k\gamma}{\zeta - k\gamma}\right)^2 \tag{2.31}$$

Any function of the form $R(x)S(y)T(z)$ satisfies a homogeneous Helmholtz equation, the wavenumber of which is

$$k^2 = \xi^2 + \eta^2 + \gamma^2 \tag{2.32}$$

Thus, the parameters ξ, η and ζ are defined by the four equations (2.28, 2.30, 2.31, 2.32). It does not seem possible to obtain an analytic expression of the solution of this system, even under the hypothesis of reduced admittances α, β and γ independent of the frequency. The proof of the existence of a countable sequence of solutions is not at all an elementary task. The existence of such a sequence (ξ_r, η_s, ζ_t) is stated without proof. To each solution, a wavenumber k_{rst} is associated by the definition

$$k_{rst}^2 = \xi_r^2 + \eta_s^2 + \zeta_t^2$$
$$k_{rst} > 0 \text{ if } k_{rst}^2 > 0, \qquad \Im(k_{rst}) > 0 \text{ else}$$
$$r, s, t, \text{ integers}$$

The corresponding resonance mode is written

$$P_{rst}(x, y, z) = A_{rst}\left[\mathrm{e}^{\iota\xi_r(x - a/2)} + \frac{\xi_r - k_{rst}\alpha}{\xi_r + k_{rst}\alpha}\mathrm{e}^{-\iota\xi_r(x - a/2)}\right]$$
$$\times \left[\mathrm{e}^{\iota\eta_s(y - b/2)} + \frac{\eta_s - k_{rst}\beta}{\eta_s + k_{rst}\beta}\mathrm{e}^{-\iota\eta_s(y - b/2)}\right]$$
$$\times \left[\mathrm{e}^{\iota\zeta_t(z - c/2)} + \frac{\zeta_t - k_{rst}\gamma}{\zeta_t + k_{rst}\gamma}\mathrm{e}^{-\iota\zeta_t(z - c/2)}\right] \tag{2.33}$$

2.2.3. Forced regime for the Neumann problem: Eigenmodes series expansion of the solution

Let us now try to establish an expression for the solution of the non-homogeneous equation

$$(\Delta + k^2)p(x, y, z) = f(x, y, z), \qquad M = (x, y, z) \in \Omega \tag{2.34}$$

which satisfies a homogeneous Neumann boundary condition. It can be proved that the sequence of functions $P_{rst}(x, y, z)$ as given by (2.24) is a basis of the Hilbert space which $p(x, y, z)$ belongs to: the space of functions which are square integrable in Ω together with their first derivatives and which satisfy a homogeneous Neumann boundary condition on σ, the boundary of Ω. If the source term $f(x, y, z)$ is a square integrable function, it can be expanded into a series of the eigenmodes:

$$\begin{aligned} f(x, y, z) &= \sum_{r,s,t=0}^{\infty} f_{rst} P_{rst}(x, y, z) \\ f_{rst} &= \int_{\Omega} f(x, y, z) P_{rst}(x, y, z)\, dx\, dy\, dz \end{aligned} \tag{2.35}$$

The acoustic pressure $p(x, y, z)$ is sought as a series expansion of the same functions:

$$p(x, y, z) = \sum_{r,s,t=0}^{\infty} \Psi_{rst} P_{rst}(x, y, z) \tag{2.36}$$

Expression (2.36) is introduced into (2.34), leading to

$$\sum_{r,s,t=0}^{\infty} \Psi_{rst}(k^2 - k_{rst}^2) P_{rst}(x, y, z) = \sum_{r,s,t=0}^{\infty} f_{rst} P_{rst}(x, y, z) \tag{2.37}$$

Intuitively, this equation is satisfied everywhere in Ω if and only if the coefficients of the terms $P_{rst}(x, y, z)$ on both sides of the equality are equal, that is if

$$\Psi_{rst}(k^2 - k_{rst}^2) P_{rst}(x, y, z) = f_{rst}\,, \qquad r, s, t = 0, 1, 2, \ldots, \infty$$

If k is different from any eigenwavenumber k_{rst}, one gets:

$$\Psi_{rst} = \frac{f_{rst}}{k^2 - k_{rst}^2}, \qquad r, s, t = 0, 1, 2, \ldots, \infty \tag{2.38}$$

and, thus, the series which represents the response $p(x, y, z)$ of the fluid to the excitation $f(x, y, z)$ is uniquely determined. *A priori*, this series does not converge to the solution at each point but it converges in the sense of the $\mathbf{L}^2$-norm. It can be shown that there is a point convergence outside the support of the source term.

The hypothesis that $f(x, y, z)$ is a square integrable function can be relaxed and replaced by the less restrictive one that f is a distribution with compact support: for example, an isotropic small source located around a point (u, v, w) can be very accurately described as a Dirac measure

$$f(x, y, z) = \delta_u(x) \otimes \delta_v(y) \otimes \delta_w(z)$$

Rigorously, equation (2.37) must be solved in the weak sense. This means that it must be replaced by

$$\int_\Omega \sum_{r,s,t=0}^{\infty} \Psi_{rst}(k^2 - k_{rst}^2)P_{rst}(x, y, z)\Phi(x, y, z)\, dx\, dy\, dz$$

$$= \int_\Omega \sum_{r,s,t=0}^{\infty} f_{rst}P_{rst}(x, y, z)\Phi(x, y, z)\, dx\, dy\, dz$$

where $\Phi(x, y, z)$ is any function of the Hilbert space which $p(x, y, z)$ belongs to. If we chose the sequence of eigenmodes $P_{r's't'}(x, y, z)$, the result already given is straightforward. If f is a distribution, the integrals in the former equality are replaced by the duality products and $\Phi(x, y, z)$ is any indefinitely differentiable function with compact support in Ω: in particular, the eigenmodes are such functions. Then, for a Dirac measure the following result is easily established:

$$f_{rst} = P_{rst}(u, v, w)$$

Expressions of the coefficients Ψ_{rst} and of the series (2.36) remain unchanged. Nevertheless, the series which represents the acoustic pressure converges in the distribution sense. Practically, there is a point convergence outside the source term support. Such an elementary result is a direct application of the basic concepts of the theory of distributions developed in the book by L. Schwartz [10] referenced at the end of this chapter.

2.2.4. Forced regime for the Robin problem: Expansion into a series of the eigenmodes of the Laplace operator

Let us consider now the non-homogeneous Helmholtz equation

$$(\Delta + k^2)p(x, y, z) = f(x, y, z), \qquad M = (x, y, z) \in \Omega \tag{2.34}$$

associated to the Robin boundary conditions (2.26). It is not possible – at least in a straightforward way – to expand the solution into a series of the resonance modes presented in subsection 2.2.2 since these functions satisfy different boundary conditions as, for example,

$$\partial_x p_{rst}(x, y, z) - \iota k_{rst}\alpha p_{rst}(x, y, z) = 0 \qquad \text{for} \qquad x = a/2$$

It is much simpler to introduce the eigenmodes of the boundary value problem defined by the Laplace operator and the Robin conditions for a fixed wavenumber k.

They are defined by the homogeneous set of equations

$$(\Delta + \chi^2)\Phi(x, y, z) = 0, \quad x \in \left]-\frac{a}{2}, +\frac{a}{2}\right[, \quad y \in \left]-\frac{b}{2}, +\frac{b}{2}\right[, \quad z \in \left]-\frac{c}{2}, +\frac{c}{2}\right[$$

$$\begin{aligned}
\partial_x\Phi(x, y, z) - \iota k\alpha\Phi(x, y, z) &= 0 \quad \text{at} \quad x = +a/2 \\
-\partial_x\Phi(x, y, z) - \iota k\alpha\Phi(x, y, z) &= 0 \quad \text{at} \quad x = -a/2 \\
\partial_y\Phi(x, y, z) - \iota k\beta\Phi(x, y, z) &= 0 \quad \text{at} \quad y = +b/2 \\
-\partial_y\Phi(x, y, z) - \iota k\beta\Phi(x, y, z) &= 0 \quad \text{at} \quad y = -b/2 \\
\partial_z\Phi(x, y, z) - \iota k\gamma\Phi(x, y, z) &= 0 \quad \text{at} \quad z = +c/2 \\
-\partial_z\Phi(x, y, z) - \iota k\gamma\Phi(x, y, z) &= 0 \quad \text{at} \quad z = -c/2
\end{aligned} \tag{2.39}$$

It will be assumed that there exists a countable sequence of solutions Φ_{rst}, called *eigenmodes* corresponding to a sequence of wavenumbers χ_{rst} called *eigenwavenumbers*, which depend on the wavenumber k. The method of separation of variables can again be used and $\Phi_{rst}(x, y, z;\, k)$ is sought as a product of three functions each of which depends on one variable only:

$$\Phi_{rst}(x, y, z;\, k) = \tilde{R}_r(x;\, k)\tilde{S}_s(y;\, k)\tilde{T}_t(z;\, k)$$

Each function is sought as a linear combination of two exponential functions. For example, we have

$$\tilde{R}_r(x;\, k) = A_r \mathrm{e}^{\iota\tilde{\xi}_r x} + B_r \mathrm{e}^{-\iota\tilde{\xi}_r x} \tag{2.40}$$

in which expression $\tilde{\xi}_r$ is a function of k, solution of

$$\mathrm{e}^{\iota\tilde{\xi}a}(\tilde{\xi} - k\alpha)^2 - \mathrm{e}^{-\iota\tilde{\xi}a}(\tilde{\xi} + k\alpha)^2 = 0 \tag{2.41}$$

This equation cannot be solved analytically. We assume, without a proof, that it has a countable sequence of solutions which depend analytically on the parameter k. The coefficients A_r and B_r are related by

$$B_r = A_r \mathrm{e}^{\iota\tilde{\xi}_r a}\,\frac{\tilde{\xi}_r - k\alpha}{\tilde{\xi}_r + k\alpha}$$

Finally, the function $\tilde{R}_r$ is written

$$\tilde{R}_r(x;\, k) = A_r\left[\mathrm{e}^{\iota\tilde{\xi}_r x} + \frac{\tilde{\xi}_r - k\alpha}{\tilde{\xi}_r + k\alpha}\,\mathrm{e}^{-\iota\tilde{\xi}_r x}\right] \tag{2.42}$$

Let us now prove that these functions are orthogonal to each other. To this aim, use is made of the obvious equalities

$$\int_{-a/2}^{+a/2}\left\{\left[\frac{d^2\tilde{R}_r}{dx^2} + \tilde{\xi}_r^2\tilde{R}_r\right]\tilde{R}_q - \tilde{R}_r\left[\frac{d^2\tilde{R}_q}{dx^2} + \tilde{\xi}_q^2\tilde{R}_q\right]\right\} dx = 0 \tag{2.43}$$

$$\frac{d^2\tilde{R}_r}{dx^2}\tilde{R}_q = \frac{d}{dx}\left[\frac{d\tilde{R}_r}{dx}\tilde{R}_q\right] - \frac{d\tilde{R}_r}{dx}\frac{d\tilde{R}_q}{dx}$$

$$\frac{d^2\tilde{R}_q}{dx^2}\tilde{R}_r = \frac{d}{dx}\left[\frac{d\tilde{R}_q}{dx}\tilde{R}_r\right] - \frac{d\tilde{R}_q}{dx}\frac{d\tilde{R}_r}{dx}$$

Expression (2.43) is equivalent to

$$\int_{-a/2}^{+a/2}\frac{d}{dx}\left[\frac{d\tilde{R}_r}{dx}\tilde{R}_q - \tilde{R}_r\frac{d\tilde{R}_q}{dx}\right]dx + (\tilde{\xi}_r^2 - \tilde{\xi}_q^2)\int_{-a/2}^{+a/2}\tilde{R}_r\tilde{R}_q\,dx = 0$$

The first integral is easily shown to be zero: an integration by parts leaves boundary terms only which cancel out because the two functions $\tilde{R}_r$ and $\tilde{R}_q$ satisfy the same boundary condition. This implies that the second term is zero too. As a consequence, if $r \neq q$ the second integral is zero, and the functions $\tilde{R}_r$ and $\tilde{R}_q$ are orthogonal. The amplitude coefficients A_r are commonly chosen so that the functions $\tilde{R}_r$ satisfy

$$\int_{-a/2}^{+a/2}\tilde{R}_r^2\,dx = 1$$

The functions $\tilde{S}_s(y;\,k)$ and $\tilde{T}_t(z;\,k)$ are determined in the same way. The eigenmodes are thus given by

$$\Phi_{rst}(x, y, z;\,k) = \tilde{R}_r(x;\,k)\tilde{S}_s(y;\,k)\tilde{T}_t(z;\,k)$$

and satisfy a homogeneous Helmholtz equation with a wavenumber $\chi_{rst}(k)$ defined by

$$\chi_{rst}^2(k) = \tilde{\xi}_r^2(k) + \tilde{\eta}_s^2(k) + \tilde{\zeta}_t^2(k)$$
$$\Re[\chi_{rst}(k)] \geqslant 0$$

Let $\tilde{f}_{rst}$ be the coefficients of the expansion of the source function f in terms of the Φ_{rst}

$$\tilde{f}_{rst} = \int_\Omega f(x, y, z)\Phi_{rst}(x, y, z;\,k)\,dx\,dy\,dz \tag{2.44}$$

The solution of equation (2.34), which satisfies the Robin boundary conditions (2.39), is given by the series

$$p(x, y, z) = \sum_{r,s,t=0}^{\infty}\frac{\tilde{f}_{rst}}{k^2 - \chi_{rst}^2}\Phi_{rst}(x, y, z;\,k) \tag{2.45}$$

This points out a new difficulty which appears with the Robin problem: for each wavenumber, it is necessary to determine a set of eigenwavenumbers; being the solution of a transcendental equation, they cannot be determined analytically like those of the Neumann and the Dirichlet problems. It must be recalled that the eigenmodes $\Phi_{rst}(x, y, z; k)$ depend on the wavenumber k even if the impedance of the boundary of the domain does not depend on this parameter.

2.3. Transient Phenomena – Reverberation Time

This section deals with the solution of the wave equation inside an enclosure, for two particular transient sources: the source produces a harmonic signal with angular frequency ω_0 which starts at $t = 0$ (sound establishment) or which stops at $t = 0$ (sound stopping). It is assumed that the boundary of the propagation domain absorbs energy. It is shown that, during sound establishment, the acoustic pressure tends asymptotically to the forced harmonic response of the enclosure, following roughly an exponential law. When the source is stopped, it goes asymptotically to zero, following a very similar exponential law. This will lead to the introduction of the notion of *reverberation time* and the Sabine formula: this formula enables us to approximately relate the time rate of energy decrease to a mean absorption coefficient of the walls.

2.3.1. A simple one-dimensional example: Statement of the problem

Let us consider a waveguide with constant cross section, extending over the domain $0 < x < \infty$. The sub-domain $0 < x < L$ – which will be considered as the enclosure – contains a fluid with density ρ and sound velocity c. The remaining part $L < x < \infty$ is occupied by a fluid with density ρ' and sound velocity c'. It is assumed that $\zeta = \rho c/\rho' c' > 1$ (this assumption is not essential). The boundary $x = 0$ is perfectly rigid. A sound source is located at $x = s < L$ and emits plane longitudinal harmonic waves, with angular frequency ω_0, starting at $t = 0$; it is modelled by the distribution $\delta_s \cos \omega_0 t = \delta_s \Re(\mathrm{e}^{-\iota\omega_0 t})$. The equations to solve are:

$$
\begin{gathered}
\left[\frac{\partial^2}{\partial x^2} - \frac{1}{c^2}\frac{\partial^2}{\partial t^2}\right]\tilde{P}(x, t) = Y(t)\Re(\mathrm{e}^{-\iota\omega_0 t})\delta_s, \qquad x \in]0, L[, \qquad t \in]0, \infty[\\
\left[\frac{\partial^2}{\partial x^2} - \frac{1}{c'^2}\frac{\partial^2}{\partial t^2}\right]\tilde{P}'(x, t) = 0, \qquad x \in]L, \infty[, \qquad t \in]0, \infty[\\
\partial_x \tilde{P}(x, t) = 0 \qquad \text{at } x = 0, \forall t \\
\tilde{P}(x, t) = \tilde{P}'(x, t) \qquad \text{at } x = L, \forall t \\
\tilde{V}(x, t) = \tilde{V}'(x, t) \qquad \text{at } x = L, \forall t \\
\text{outgoing waves conditions at } x \to \infty, \forall t
\end{gathered}
\tag{2.46}
$$

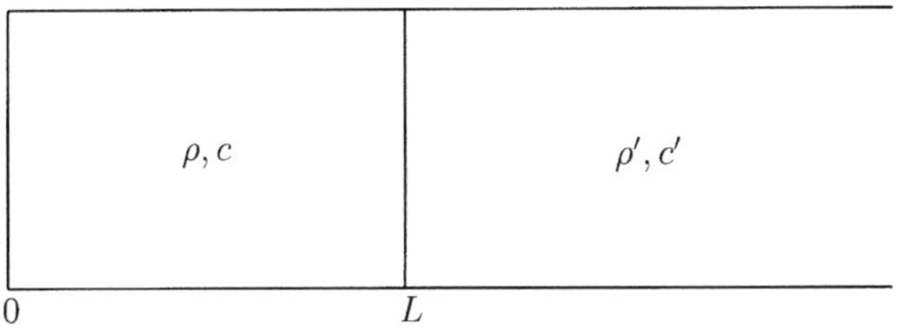

Fig. 2.2. Scheme of the one-dimensional enclosure.

where $\tilde{P}(x, t)$ (resp. $\tilde{P}'(x, t)$) stands for the acoustic pressure in the first (resp. second) fluid; $\tilde{V}(x, t)$ and $\tilde{V}'(x, t)$ are the corresponding particle velocities; $Y(t)$ is the Heaviside step function ($=0$ for $t<0$, $=1$ for $t>0$). Because the aim of this academic problem is to point out the main phenomena which occur in room acoustics, attention will be paid to the function $\tilde{P}(x, t)$ only.

It is useful to associate to $\tilde{P}(x, t)$ (resp. $\tilde{P}'(x, t)$) the complex pressure $P(x, t)$ (resp. $P'(x, t)$) and to $\tilde{V}(x, t)$ (resp. $\tilde{V}'(x, t)$) the complex particle velocity $V(x, t)$ (resp. $V'(x, t)$) which are solutions of the following system:

$$\left[\frac{\partial^2}{\partial x^2} - \frac{1}{c^2}\frac{\partial^2}{\partial t^2}\right] P(x, t) = Y(t)\mathrm{e}^{-\iota\omega_0 t}\delta_s, \qquad x \in]0, L[, \qquad t \in]0, \infty[$$

$$\left[\frac{\partial^2}{\partial x^2} - \frac{1}{c^2}\frac{\partial^2}{\partial t^2}\right] P'(x, t) = 0, \qquad x \in]L, \infty[, \qquad t \in]0, \infty[$$

$$\partial_x P(x, t) = 0 \qquad \text{at } x = 0, \forall t \tag{2.46a}$$

$$P(x, t) = P'(x, t) \qquad \text{at } x = L, \forall t$$

$$V(x, t) = V'(x, t) \qquad \text{at } x = L, \forall t$$

$$\text{outgoing waves conditions at } x \rightarrow \infty, \forall t$$

This system is somewhat simpler to solve, and the real part of its solution is the physical acoustic pressure and acoustic particle velocity.

To solve equations (2.46a), the Laplace transform is used. In a first step the transformed equations are solved and, in a second step, the inverse Laplace transform of the obtained solution is calculated. Let us recall that the direct and inverse Laplace transforms of a function are defined by:

$$f(q) = \int_0^\infty F(t)\mathrm{e}^{-qt}\, dt$$

$$F(t) = \frac{1}{2\iota\pi}\int_{\varepsilon - \iota\infty}^{\varepsilon + \iota\infty} f(q)\mathrm{e}^{qt}\, dq$$

The transformed system of equations is:

$$\left[\frac{d^2}{dx^2} - \frac{q^2}{c^2}\right] p(x, q) = \frac{1}{q + \iota\omega_0}\, \delta_s, \qquad x \in]0, L[\tag{2.47}$$

$$\left[\frac{d^2}{dx^2} - \frac{q^2}{c'2}\right] p'(x, q) = 0, \qquad x \in]L, \infty[\tag{2.47a}$$

$$d_x p(x, q) = 0 \qquad \text{at } x = 0 \tag{2.47b}$$

$$p(x, q) = p'(x, q) \qquad \text{at } x = L \tag{2.47c}$$

$$\frac{1}{\rho}\, d_x p(x, q) = \frac{1}{\rho'}\, d_x p'(x, q) \qquad \text{at } x = L \tag{2.47d}$$

Equality (2.47d) has been obtained by using the momentum equation which relates the acoustic pressure to the particle velocity. The 'outgoing waves condition', which expresses that the energy entering the domain $L < x < \infty$ propagates toward the increasing abscissae only, implies that the function p' must remain finite. This is achieved by choosing the following form:

$$p'(x, q) = A\, \frac{e^{-qx/c'}}{2qc'}$$

This results in the following relationship between p' and its derivative with respect to x:

$$\frac{dp'(x, q)}{dx} = -\frac{q}{c'}\, p'(x, q)$$

The continuity conditions (2.47b, 2.47c) enable us then to write a Robin boundary condition for the pressure p:

$$\frac{dp(x, q)}{dx} + \frac{q}{\zeta c}\, p(x, q) = 0, \qquad \text{at } x = L, \qquad \text{with } \zeta = \frac{\rho' c'}{\rho c} \tag{2.48}$$

We are thus left with the boundary value problem governed by equations (2.47, 2.48).

2.3.2. Eigenmodes expansion of the solution

Following the method used in subsection 2.2.4, the eigenmodes of the one-dimensional Robin problem are defined by:

$$\tilde{p}_m(x, q) = A_m \left[\frac{e^{\iota\tilde{K}_m(L - x)/c}}{2\iota\tilde{K}_m/c} + \frac{e^{\iota\tilde{K}_m(L + x)/c}}{2\iota\tilde{K}_m/c}\right] \tag{2.49}$$

where $\tilde{K}_m$ is the solution of the transcendental equation

$$e^{2\iota\tilde{K}_m L/c}(q+\iota\tilde{K}_m\zeta)+(q-\iota\tilde{K}_m\zeta)=0 \tag{2.50}$$

The choice of the coefficients A_m is made by the second of the following relationships:

$$\int_0^L \tilde{p}_m(x,q)\tilde{p}_n(x,q)\,dx = 0 \qquad \text{for } m \neq n$$
$$= 1 \qquad \text{for } m = n$$

The solution $p(x, q)$ is given by the series

$$p(x,q) = -\frac{c^2}{q+\iota\omega_0}\sum_m \frac{\tilde{p}_m(s,q)\tilde{p}_m(x,q)}{q^2+\tilde{K}_m^2(q)} \tag{2.51}$$

To get the time response of the system, it is now necessary to calculate the inverse Laplace transform of each term of this series, that is to calculate the following integrals:

$$P_m(x,t) = -\frac{c^2}{2\iota\pi}\int_{\varepsilon-\iota\infty}^{\varepsilon+\iota\infty}\frac{\tilde{p}_m(s,q)\tilde{p}_m(x,q)}{(q+\iota\omega_0)(q^2+\tilde{K}_m^2)}\,e^{qt}dq \tag{2.52}$$

The residues theorem is a suitable tool. The integration contours, which consist of a half-circle and its diameter, are shown on Fig. 2.3: the path γ^- is adopted for $t<0$ and the path γ^+ is adopted for $t>0$. Thus, the term e^{qt} decreases exponentially along the half-circle as its radius R tends to infinity. The physical phenomenon must be causal, that is the pressure signal cannot start before the source. To prove that the mathematical solution satisfies this condition, it must be shown that the functions to be integrated have their poles located in the half-plane $\Re(q)<0$, that is the solutions of

$$q^2+\tilde{K}_m^2(q)=0 \tag{2.53}$$

have a negative imaginary part. This equation leads us to look for solutions of equation (2.50) which satisfy one of the following two equalities:

$$\iota\tilde{K}_m(q)=q \qquad \text{or} \qquad \iota\tilde{K}_m(q)=-q$$

These two relationships lead, in fact, to a unique equation

$$e^{2\iota\tilde{K}_m L/c}(1+\zeta)+(1-\zeta)=0 \tag{2.54}$$

which is deduced from (2.50) with $q=\iota\tilde{K}_m$. The solutions will be denoted by $K_m=-\Omega_m+\iota\tau_m$. It is easily shown that one has

$$\tau_m = \frac{c}{2L}\ln\left|\frac{1+\zeta}{1-\zeta}\right|$$
$$\Omega_m = \frac{m\pi c}{L}, \qquad m=-\infty,\ldots,-2,-1,0,1,2,\ldots,+\infty \tag{2.55}$$

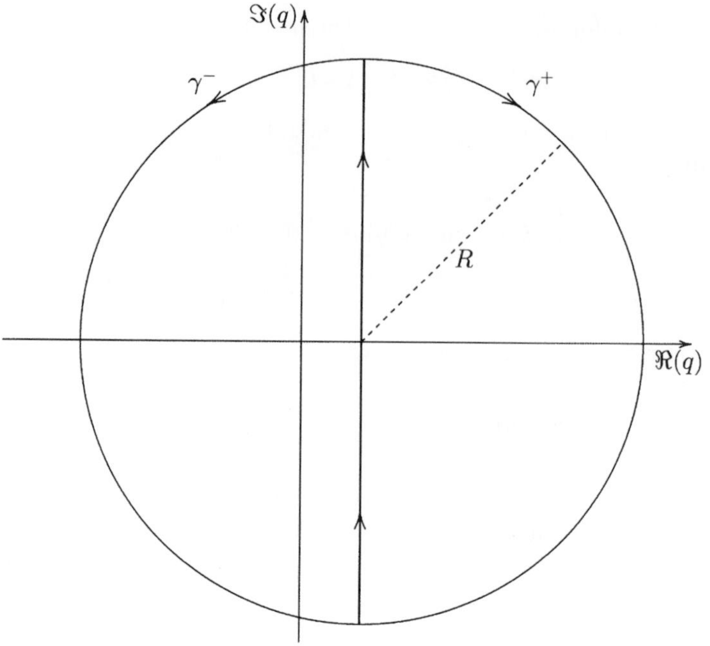

Fig. 2.3. Integration contours for the inverse Laplace transforms.

It is clear that these solutions are located in the correct half of the complex plane. Two other remarks must be made: the damping factors τ_m are positive and have the same value denoted by τ; the resonance wavenumbers can be gathered by pairs, Ω_m and Ω_{-m}, which are symmetrical with respect to the imaginary axis. The acoustic pressure $P(x, t)$ is readily shown to be given by:

$$\tilde{P}(x, t) = c^2 Y(t)\Re\left\{ \sum_{m=-\infty}^{+\infty} \left[\frac{\tilde{p}_m(x, \iota K_m)\tilde{p}_m(s, \iota K_m)\mathrm{e}^{-\iota\Omega_m t}\mathrm{e}^{-\tau t}}{2(\omega_0 - \Omega_m + \iota\tau)(-\Omega_m + \iota\tau)(1 + \iota\tilde{K}_m'(-\Omega_m + \iota\tau))} \right.\right.$$
$$\left.\left. + \frac{\tilde{p}_m(x, -\iota\omega_0)\tilde{p}_m(s, -\iota\omega_0)}{\omega_0^2 - \tilde{K}_m^2(-\iota\omega_0)}\mathrm{e}^{-\iota\omega_0 t} \right] \right\}$$
$$\text{with} \quad \tilde{K}_m'(q) = \frac{d\tilde{K}_m(q)}{dq} \tag{2.56}$$

The first series of terms contains the resonance modes – or free oscillation regimes – of the cavity; it tends exponentially to zero as $t \to \infty$. The second series of terms is the response of the cavity to a harmonic excitation with angular frequency ω_0: it is expressed in terms of the eigenmodes associated to the Robin boundary condition.

Remark. In the book by Ph. Morse and U. Ingard [8] cited in the bibliography of this chapter, a similar calculation is presented, but the term $\tilde{K}_m'(-\Omega_m + \iota\tau)$ which appears in the denominators of the first series is missing.

2.3.3. The 'boundary sources' method

The boundary value problem (2.47, 2.48) can be solved by a method which provides a good image of the physical reality: the effect of the boundary $x = L$ is modelled by a point source $A\delta_L$ the amplitude of which is chosen so that the Robin boundary condition is satisfied. The perfectly rigid boundary at $x = 0$ is accounted for by the fields corresponding to the image sources δ_{-s} and $A\delta_{-L}$, located at the points $x = -s$ and $x = -L$ respectively.

The function $p(x, q)$ takes the form:

$$p(x,q) = -\frac{1}{q+\iota\omega_0}\left\{\frac{\mathrm{e}^{-q|s-x|/c}}{2q/c} + \frac{\mathrm{e}^{-q(s+x)/c}}{2q/c}\right\} - \frac{A}{q+\iota\omega_0}\left\{\frac{\mathrm{e}^{-q(L-x)/c}}{2q/c} + \frac{\mathrm{e}^{-q(L+x)/c}}{2q/c}\right\} \tag{2.57}$$

and the coefficient A is given by

$$A = \frac{\zeta - 1}{(\zeta+1) - (\zeta-1)\mathrm{e}^{-2qL/c}}\left[\mathrm{e}^{-q(L-s)/c} + \mathrm{e}^{-q(L+s)/c}\right] \tag{2.58}$$

This leads to the final result

$$p(x,q) = -\frac{1}{q+\iota\omega_0}\left\{\frac{\mathrm{e}^{-q|s-x|/c}}{2q/c} + \frac{\mathrm{e}^{-q(s+x)/c}}{2q/c}\right\} - \frac{1}{q+\iota\omega_0}\frac{\zeta-1}{(\zeta+1)-(\zeta-1)\mathrm{e}^{-2qL/c}} \times\left\{\frac{\mathrm{e}^{-q(2L-s-x)/c}}{2q/c} + \frac{\mathrm{e}^{-q(2L+s-x)/c}}{2q/c} + \frac{\mathrm{e}^{-q(2L-s+x)/c}}{2q/c} + \frac{\mathrm{e}^{-q(2L+s+x)/c}}{2q/c}\right\} \tag{2.59}$$

The successive terms of equality (2.59) have the following interpretation (see Fig. 2.4):

1. direct wave;
2. wave reflected at $x = 0$;
3. wave reflected at $x = L$;
4. wave reflected at $x = 0$ and then at $x = L$;
5. wave reflected at $x = L$ and then at $x = 0$;
6. wave reflected at $x = 0$, then at $x = L$ and again at $x = 0$.

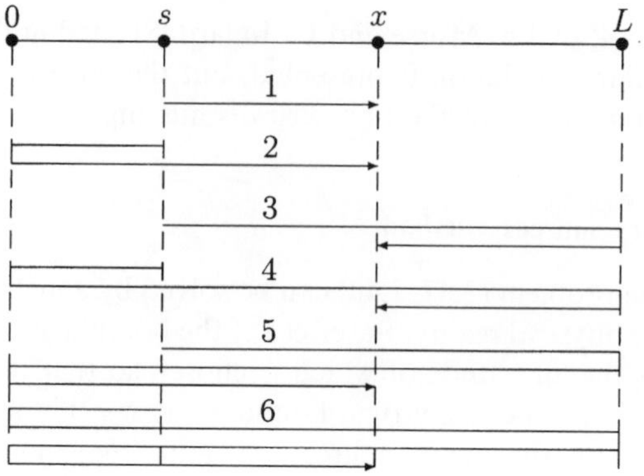

Fig. 2.4. Scheme of the successive wavefronts.

This decomposition is, of course, not unique. For example, it is possible to point out an infinite number of successive reflections at each end of the guide. This simple expression shows the first and most important steps of sound establishment in a room.

The inverse Laplace transform of the first two terms is readily obtained using the residue theorem. The result is

$$-\frac{1}{2\iota\pi}\int_{\varepsilon-\iota\infty}^{\varepsilon+\iota\infty}\frac{e^{-q|s\pm x|/c}}{(q+\iota\omega_0)q/c}\,e^{qt}\,dq$$

$$=Y(t-|s\pm x|/c)\frac{c}{\iota\omega_0}+Y(t-|s\pm x|/c)\frac{c e^{\iota\omega_0|s\pm x|/c}}{2\iota\omega_0}\,e^{-\iota\omega_0 t} \tag{2.60}$$

In a similar way, the residue theorem enables us to obtain the inverse Laplace transform of the other terms:

$$-\frac{1}{2\iota\pi}\int_{\varepsilon-\iota\infty}^{\varepsilon+\iota\infty}\frac{\zeta-1}{(\zeta+1)-(\zeta-1)e^{-2qL/c}}\,\frac{e^{-q(2L\pm s\pm x)/c}}{2q/c}\,\frac{e^{qt}}{q+\iota\omega_0}\,dt$$

$$=-Y(t-(2L\pm s\pm x)/c)\left\{-\frac{c}{4\iota\omega_0}(\zeta-1)\right.$$

$$+\frac{c^2}{4L}\sum_m\frac{e^{-(\iota\Omega_m+\tau)(2L\pm s\pm x)/c}}{(\iota\Omega_m+\tau)[\iota(\omega_0-\Omega_m)-\tau]}\,e^{-\iota\Omega_m t}e^{-\tau t}$$

$$\left.+c\,\frac{\zeta-1}{(\zeta+1)-(\zeta-1)e^{2\iota\omega_0 L/c}}\,\frac{e^{\iota\omega_0(2L\pm s\pm x)/c}}{2\iota\omega_0}\,e^{-\iota\omega_0 t}\right\} \tag{2.61}$$

Gathering equalities (2.60) and (2.61) and keeping the real parts only, the following result is obtained:

$$
\begin{aligned}
\tilde{P}(x,t) = {} & cY(t - |s-x|/c)\Re\left\{\frac{e^{\iota\omega_0|s-x|/c}}{2\iota\omega_0}\,e^{-\iota\omega_0 t}\right\} \\
& + cY(t-(s+x)/c)\Re\left\{\frac{e^{\iota\omega_0(s+x)/c}}{2\iota\omega_0}\,e^{-\iota\omega_0 t}\right\} \\
& + c\Re\left\{\frac{\zeta-1}{(\zeta+1)-(\zeta-1)e^{2\iota\omega_0 L/c}}\left[Y(t-(2L-s-x)/c)\,\frac{e^{\iota\omega_0(2L-s-x)/c}}{2\iota\omega_0}\right.\right. \\
& + Y(t-(2L+s-x)/c)\,\frac{e^{\iota\omega_0(2L+s-x)/c}}{2\iota\omega_0} \\
& + Y(t-(2L-s+x)/c)\,\frac{e^{\iota\omega_0(2L-s+x)/c}}{2\iota\omega_0} \\
& \left.\left. + Y(t-(2L+s+x)/c)\,\frac{e^{\iota\omega_0(2L+s+x)/c}}{2\iota\omega_0}\right]e^{-\iota\omega_0 t}\right\} \\
& + \frac{c^2}{4L}\Re\left\{Y(t-(2L-s-x)/c)\sum_{m=-\infty}^{+\infty}\frac{e^{(\iota\Omega_m+\tau)(2L-s-x)/c}}{(\iota\Omega_m+\tau)[\iota(\omega_0-\Omega_m)-\tau]}\,e^{-\iota\Omega_m t}\right. \\
& + Y(t-(2L+s-x)/c)\sum_{m=-\infty}^{+\infty}\frac{e^{(\iota\Omega_m+\tau)(2L+s-x)/c}}{(\iota\Omega_m+\tau)[\iota(\omega_0-\Omega_m)-\tau]}\,e^{-\iota\Omega_m t} \\
& + Y(t-(2L-s+x)/c)\sum_{m=-\infty}^{+\infty}\frac{e^{(\iota\Omega_m+\tau)(2L-s+x)/c}}{(\iota\Omega_m+\tau)[\iota(\omega_0-\Omega_m)-\tau]}\,e^{-\iota\Omega_m t} \\
& \left. + Y(t-(2L+s+x)/c)\sum_{m=-\infty}^{+\infty}\frac{e^{(\iota\Omega_m+\tau)(2L+s+x)/c}}{(\iota\Omega_m+\tau)[\iota(\omega_0-\Omega_m)-\tau]}\,e^{-\iota\Omega_m t}\right\}e^{-\tau t}
\end{aligned}
\tag{2.62}
$$

The sum of the first three groups of terms (with the factor $e^{-\iota\omega_0 t}$) represents the permanent regime which is set up for $t > (2L+s+x)/c$. The series of terms with $e^{-\tau t}$ as common factor represents the transient part of the signal: mathematically, it lasts indefinitely; but, from a physical point of view, because it decreases exponentially, it becomes more or less rapidly undetectable and can be considered as having a finite duration (it can be considered as zero when its level is about 60 dB less than that of the permanent signal).

This representation of the transient response of an enclosure shows some important features of sound establishment. The main point to be noticed is that sound establishment is not a continuous phenomenon: the energy level increases by steps corresponding to the arrival of the direct sound and of the waves successively reflected at each end of the guide; expression (2.62) points out the first five reflected waves. In a three-dimensional room, the phenomenon is essentially the same: the first signal which arrives on a receiver is the direct wavefront; then this wavefront impinges on a boundary and a reflected attenuated secondary wave appears, which provides an additional signal on the receiver; the process goes on indefinitely, but, in practice, a permanent regime is rapidly reached.

It is possible to expand the solution described by formula (2.62) into a series of successive reflected waves only, but the permanent regime will not appear. This is the topic of the next subsection.

2.3.4. Description of sound establishment by a series of successive reflections

The coefficient of the second group of terms in the Laplace transform $p(x, q)$ given by expression (2.59) can be expanded into a formal series:

$$\frac{\zeta - 1}{(\zeta + 1) - (\zeta - 1)e^{-2qL/c}} = \frac{\zeta - 1}{\zeta + 1} \sum_{n=0}^{\infty} \left(\frac{\zeta - 1}{\zeta + 1}\right)^n e^{-2nqL/c}$$

This series contains two factors:

- $e^{-2qL/c}$ which is less than 1 as soon as $\Re q$ is positive;
- $(\zeta - 1)/(\zeta + 1)$ the modulus of which is always less than 1 (it represents the reflection coefficient of the boundary at $x = L$).

Thus, the series is convergent in the half-plane $\Re q > 0$. The inverse Laplace transform can be calculated by integrating each term of its series representation. The following result is obtained:

$$\tilde{P}(x, t) =$$

$$-cY(t - |s - x|/c)\Re\left[\frac{e^{\iota\omega_0 |s - x|/c}}{2\iota\omega_0} e^{-\iota\omega_0 t}\right]$$

$$- cY(t - (s + x)/c)\Re\left[\frac{e^{\iota\omega_0 (s + x)/c}}{2\iota\omega_0} e^{-\iota\omega_0 t}\right]$$

$$+ c\sum_{n=0}^{\infty}\left\{ Y[t - (2(n + 1)L - s - x)/c]\Re\left[\left(\frac{\zeta - 1}{\zeta + 1}\right)^{n+1} \frac{e^{\iota\omega_0[(2(n + 1)L - s - x)/c]}}{\iota\omega_0} e^{-\iota\omega_0 t}\right]\right.$$

$$+ Y[t-(2(n+1)L+s-x)/c]\Re\left[\left(\frac{\zeta-1}{\zeta+1}\right)^{n+1}\frac{e^{\iota\omega_0[(2(n+1)L+s-x)/c]}}{\iota\omega_0}e^{-\iota\omega_0 t}\right]$$

$$+ Y[t-(2(n+1)L-s+x)/c]\Re\left[\left(\frac{\zeta-1}{\zeta+1}\right)^{n+1}\frac{e^{\iota\omega_0[(2(n+1)L-s+x)/c]}}{\iota\omega_0}e^{-\iota\omega_0 t}\right]$$

$$+ Y[t-(2(n+1)L+s+x)/c]\Re\left[\left(\frac{\zeta-1}{\zeta+1}\right)^{n+1}\frac{e^{\iota\omega_0[(2(n+1)L+s+x)/c]}}{\iota\omega_0}e^{-\iota\omega_0 t}\right]\Bigg\} \tag{2.63}$$

This shows clearly that the sound is set up after an infinite number of steps which correspond to the successive reflections of the initial wavefront at each end of the waveguide. After m reflections at the boundary $x = L$, the amplitude of the wave is $|(\zeta-1)/(\zeta+1)|^m$ and decreases to 0 as $m \to \infty$. This series representation has two disadvantages: first, the permanent regime which is reached asymptotically does not appear; second, the fundamental role of the resonance modes has totally disappeared.

As a conclusion of these last two subsections, it is obvious that the mathematical representation of the physical phenomenon which must be adopted depends on the properties the physicist is interested in.

2.3.5. Sound decay – reverberation time

This is the complementary phenomenon of sound establishment. Let us assume that a sound source, emitting a harmonic signal, is stopped at $t = 0$. The first equation in (2.46) is replaced by

$$\left[\frac{\partial^2}{\partial x^2}-\frac{1}{c^2}\frac{\partial^2}{\partial t^2}\right]\tilde{P}(x,t) = [1 - Y(t)]\Re(e^{-\iota\omega_0 t})\delta_s, \qquad x \in]0, L[, \qquad t \in]0, \infty[\tag{2.64}$$

The other equations remain unchanged. It is obvious that the solution of this problem is obtained by subtracting the expression of the sound establishment from that of the permanent regime. Using representation (2.56), one gets:

$$\tilde{P}(x,t) = -c^2 Y(t)\Re\left\{\sum_{m=-\infty}^{+\infty}\frac{\tilde{p}_m(x,\iota K_m)\tilde{p}_m(s,\iota K_m)e^{-\iota\Omega_m t}}{2(\omega_0-\Omega_m+\iota\tau)(-\Omega_m+\iota\tau)(1+\iota\tilde{K}_m'(-\Omega_m+\iota\tau))}\right\}e^{-\tau t} \tag{2.65}$$

The time signal is a series of harmonic damped signals which, in the present example, have the same damping factor τ. Instead of this factor, the enclosure is more conveniently characterized by a *reverberation time* T_r. It is defined as the time delay which is required for the amplitude level of the acoustic pressure to decay 60 dB, that is:

$$60 = 20 \log e^{\tau T_r} \qquad \text{or equivalently} \qquad T_r = \frac{3}{\tau \log e} = \frac{6.91}{\tau} \tag{2.66}$$

In this simple example, the reverberation time is a perfectly defined quantity which is related to the physical constants which define the system by exact relationships. In an actual room, the various resonance modes each have their own damping factors: as a consequence, when a harmonic source is cut off, the various resonance components of the signal decrease with different rates and it is not possible to define a unique reverberation time. Nevertheless, some restrictive but realistic assumptions can be made which allow us to define a reverberation time.

2.3.6. Reverberation time in a room

For any enclosure, the existence of a sequence of resonance frequencies and resonance modes can be established: this is a difficult mathematical task which is beyond the scope of this course. From the point of view of the physicist, it is just necessary to know that the results established on the former one-dimensional example are valid for two- and three-dimensional domains. Let us consider a domain Ω in $\mathbb{R}^3$. Its boundary σ has (almost everywhere) an external unit normal vector $\vec{n}$. We look for the values of ω such that the homogeneous boundary value problem

$$\begin{aligned} \left(\Delta + \frac{\omega^2}{c^2}\right) p(M, \omega) = 0, \qquad & M \in \Omega \\ \ell(p(M), \partial_n p(M)) = 0, \qquad & M \in \sigma \end{aligned} \tag{2.67}$$

has a non-zero solution. In equation (2.67), $\ell(p(M), \partial_n p(M))$ is any linear relationship between the pressure $p(M)$ and its normal gradient, which expresses that there is an energy loss through σ. A linear combination of these two functions (with piecewise continuous coefficients) is a local boundary condition; a relationship involving an integral operator is necessary for non-local boundary conditions. Furthermore, the energy loss is generally frequency dependent. The existence of a sequence of resonance angular frequencies can be proved. To each such frequency

$$\omega_m = \Omega_m + \iota \tau_m, \qquad \tau_m > 0 \tag{2.68}$$

corresponds a resonance mode $P_m(M)$.

Let us assume that the room was initially excited by a harmonic source with angular frequency ω_0 which stops at $t = 0$. It can be shown that the acoustic pressure can then be expressed as a series of terms which are the product of a harmonic function and an exponentially decreasing amplitude:

$$P(M,t) = Y(t) \sum_m \frac{a_m P_m(M)}{\iota(\omega_0 - \Omega_m) - \tau_m} \mathrm{e}^{-\iota\Omega_m t} \mathrm{e}^{-\tau_m t} \tag{2.69}$$

where the coefficients a_m depend on the function which represents the source. It is obvious that such an expression does not allow us to define a reverberation time as was done in the former subsection. Nevertheless, this concept can be used in most real life situations. Assume that the boundary σ of the room absorbs a rather low rate of energy. This implies that the values of τ_m are small. If the driving angular frequency ω_0 is equal to the real part Ω_q of a given resonance angular frequency, an excellent approximation of the acoustic pressure is given by

$$P(M,t) \simeq -Y(t) \frac{a_q P_q(M)}{\tau_q} \mathrm{e}^{-(\iota\Omega_q + \tau_q)t} \tag{2.70}$$

the other terms having a denominator $\iota(\omega_0 - \Omega_m) - \tau_m$ which has a modulus large compared with τ_q. In such a case, the acoustic pressure decrease follows a law very close to an exponential one. The notion of reverberation time is quite meaningful.

Assume now that ω_0 satisfies the double inequality $\Omega_q < \omega_0 < \Omega_{q+1}$ with Ω_{q+1} close to Ω_q, then $P(M, t)$ is accurately approximated by

$$P(M,t) \simeq Y(t) \left\{ \frac{a_q P_q(M) \mathrm{e}^{-(\iota\Omega_q + \tau_q)t}}{\iota(\omega_0 - \Omega_q) - \tau_q} + \frac{a_q P_{q+1}(M) \mathrm{e}^{-(\iota\Omega_{q+1} + \tau_{q+1})t}}{\iota(\omega_0 - \Omega_{q+1}) - \tau_{q+1}} \right\} \tag{2.71}$$

In addition, suppose that the following inequalities are satisfied

$$\left| \frac{a_q P_q(M)}{\iota(\omega_0 - \Omega_q) - \tau_q} \right| > \left| \frac{a_{q+1} P_{q+1}(M)}{\iota(\omega_0 - \Omega_{q+1}) - \tau_{q+1}} \right|, \qquad \tau_q > \tau_{q+1}$$

Thus, in a first step, the acoustic pressure is mainly governed by the first term and a first reverberation time corresponding to τ_q can be defined. Then, the second term becomes the most important one and a second reverberation corresponding to τ_{q+1} is pointed out. The model can obviously be improved by defining three, four, ... reverberation times. But the physical meaning of such a complicated model is no longer satisfying. In general, to describe the sound pressure level decrease, acousticians use laws having no more than two different slopes: this simple model is quite sufficient in practice.

2.3.7. The formula of Sabine

When experiment shows that the sound level decrease in a room can be approximated by a linear function of time (exponential decrease of the pressure amplitude), the slope of this curve is easily related to the mean absorption of the room walls. We will give the main steps to establish this relationship, which was obtained by Sabine and is called the formula of Sabine.

Let us consider a room with volume V. The total area of its walls is S. The walls are assumed to have roughly constant acoustic properties; they can be characterized by a mean energy absorption coefficient $\overline{\alpha}$. Using an optical analogy, it is possible to adopt acoustic rays for modelling the sound field emitted by a source: each time a ray impinges on the room boundary, it is reflected as by a mirror and its energy is reduced by the factor $(1 - \overline{\alpha})$. Between two successive reflections, a ray travels on a straight line of variable length. But it is possible to define a mean free path with length ℓ_m,

$$\ell_m = 4V/S \tag{2.72}$$

After one time unit, the length of the ray travel is equal to c_0, the value of the sound velocity. This corresponds to a total number of reflections equal to c_0/ℓ_m. Then, after one second, the energy of the ray is reduced by the factor $(1 - \overline{\alpha})^{c_0/\ell_m}$. It is reasonable to assume that statistically all the rays which reach a given point follow the same energy loss law. Thus, the law which governs the energy decrease in a room after the source is stopped is given by the expression

$$E = E_0(1 - \overline{\alpha})^{c_0 t/\ell_m} \tag{2.73}$$

where E_0 is the energy initially contained in the room. This leads to the following law for the sound pressure level N:

$$N = N_0 + 10\,\frac{c_0 t}{\ell_m}\,\log_{10}(1 - \overline{\alpha})$$

The sound pressure level decay is 60 dB after a time T_r, which is called *reverberation time*, and given by

$$T_r = -\frac{24V}{c_0 S \log_{10}(1 - \overline{\alpha})} \tag{2.74}$$

This formula is known as the *formula of Eyring*.

Conversely, from the measurement of the reverberation time, a *Sabine absorption coefficient* $\overline{\alpha}_S$ can be defined: it represents the mean energy absorption coefficient of the room walls, and is given by:

$$\overline{\alpha}_S = \frac{24V}{c_0 S T_r}\ln 10 \tag{2.75}$$

This expression is obtained under the assumption that the mean energy absorption coefficient $\overline{\alpha}$ of the walls is small, so leading to

$$-\log_{10}(1 - \overline{\alpha}) \simeq \frac{\overline{\alpha} + \mathcal{O}(\overline{\alpha}^2)}{\ln 10}$$

and expression (2.74) reduces to

$$T_r \simeq \frac{24V}{c_0 S\overline{\alpha}_S} = 0.16\,\frac{V}{S\overline{\alpha}} \tag{2.76}$$

This approximation of the formula of Eyring is known as the *formula of Sabine*. The Sabine absorption coefficient, which is equal to $-\ln(1-\overline{\alpha})$, is always greater than $\overline{\alpha}$. It can be larger than 1: this is *a priori* in contradiction with the notion of absorption coefficient. This shows that the concept of a reverberation time does not correspond to a rigorous modelling of the physical phenomenon: various hypotheses are necessary to justify its use as a good approximation of physical reality. Though they are generally valid, it is necessary to keep in mind that we are dealing with an approximation which is quite useful in room acoustics for concert halls or conference room design and characterization, for noise control in factory halls or offices as well as any acoustic problem in an enclosure.

2.4. Acoustic Field inside a Circular Enclosure: Introduction to the Method of Separation of Variables

Let us consider the following two-dimensional boundary value problem: the domain of propagation Ω is a disc with radius ρ_0; its boundary σ has an exterior normal unit vector $\vec{n}$ (see Fig. 2.5). We seek a function $p(M)$ solution of the Neumann problem:

$$\begin{aligned} (\Delta + k^2)p(M) &= f(M), \qquad M \in \Omega \\ \partial_n p(M) &= 0, \qquad M \in \sigma \end{aligned} \tag{2.77}$$

In what follows, use will be made of cylindrical coordinates with origin the centre of Ω: the coordinates of a point M inside the propagation domain are denoted by (ρ, θ) while those of a point P on σ are denoted by (ρ_0, θ_0). Two analytic methods to solve this problem are proposed: the eigenmodes are calculated and the solution is expanded into a series of these functions; the solution of the problem is sought as a series of functions of the form $R_n(\rho)\Psi_n(\theta)$. These two representations are very much similar. Nevertheless, it will be shown that, for a given accuracy, the second one – called *variables separation representation* – requires many fewer terms than the first one (the first series is less rapidly convergent that the second one).

2.4.1. Determination of the eigenmodes of the problem

Because of the geometry of the domain, it is interesting to write the Helmholtz equation in cylindrical coordinates with origin at the centre of the domain Ω. It is a classical result that the Laplace operator is written as follows:

$$\Delta = \frac{1}{\rho}\frac{\partial}{\partial \rho}\left(\rho\,\frac{\partial}{\partial \rho}\right) + \frac{1}{\rho^2}\frac{\partial^2}{\partial \theta^2} \tag{2.78}$$

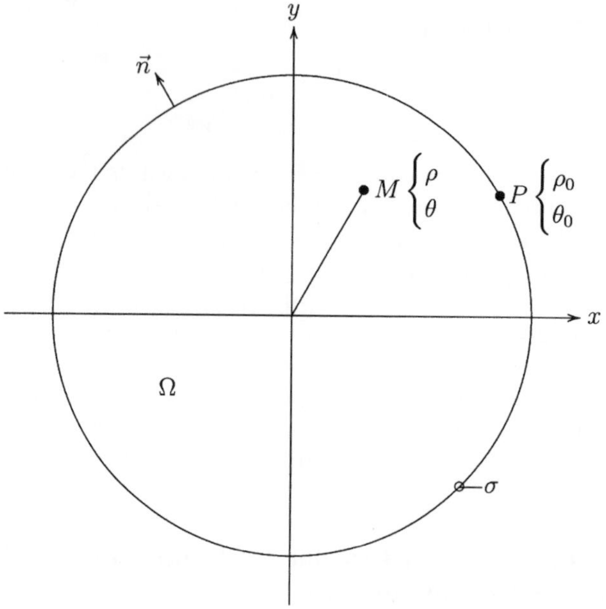

Fig. 2.5. Geometry of the circular domain.

Let us look for the existence of solutions of the homogeneous Neumann problem of the form $\Phi(\rho, \theta) = R(\rho)\Psi(\theta)$. The two functions $R(\rho)$ and $\Psi(\theta)$ must first satisfy the homogeneous partial differential equation

$$\frac{1}{\rho}\frac{d}{d\rho}\left(\rho\frac{dR(\rho)}{d\rho}\right)\Psi(\theta) + R(\rho)\frac{1}{\rho^2}\frac{d^2\Psi(\theta)}{d\theta^2} + k^2 R(\rho)\Psi(\theta) = 0$$

Dividing by $R\Psi/\rho^2$ one gets

$$\left\{\frac{\rho}{R(\rho)}\frac{d}{d\rho}\left(\rho\frac{dR(\rho)}{d\rho}\right) + k^2\rho^2\right\} + \frac{1}{\Psi(\theta)}\frac{d^2\Psi(\theta)}{d\theta^2} = 0 \tag{2.79}$$

The first term in (2.79) is independent of θ, while the second one does not depend on ρ. Thus, this equation can be satisfied if and only if each term is equal to the same constant, that is if R and Ψ are solutions of the equations

$$\begin{aligned} &\frac{\rho}{R(\rho)}\frac{d}{d\rho}\left(\rho\frac{dR(\rho)}{d\rho}\right) + k^2\rho^2 - \alpha^2 = 0 \\ &\frac{1}{\Psi(\theta)}\frac{d^2\Psi(\theta)}{d\theta^2} + \alpha^2 = 0 \end{aligned} \tag{2.80}$$

where α^2 is a separation constant to be determined. Any solution of a homogeneous partial differential equation with analytical coefficients must be an analytical function. This implies that $\Phi(\rho, \theta)$ and, as a consequence, $\Psi(\theta)$ must be periodic functions of the angular variable. The only possible solutions for the second equation (2.80) are given by

$$\alpha = n, \qquad \Psi(\theta) = e^{\iota n\theta}$$
$$n = -\infty, \ldots, -2, -1, 0, +1, +2, \ldots, +\infty \tag{2.81}$$

Then the first equation (2.80) becomes:

$$\frac{d^2 R(z)}{dz^2} + \frac{1}{z}\frac{dR(z)}{dz} + \left(1 - \frac{n^2}{z^2}\right) R(z) = 0, \qquad \text{with } z = k\rho \tag{2.82}$$

This equation is known as the Bessel equation (the theory of Bessel functions is presented in many textbooks, for example the book by Ph.M. Morse and K.U. Ingard cited in the bibliography). It has a solution which is regular at $z = 0$ and which is denoted by $J_n(z)$ and called the *Bessel function of order n*; among all its properties, we just recall the equality $J_{-n}(z) = (-1)^n J_n(z)$. The boundary condition will be satisfied if

$$J_n'(k\rho_0) = 0 \tag{2.83}$$

It is shown that for any n there exists a countable sequence k_{nm} $(m = 1, 2, \ldots, \infty)$ such that the former equality is satisfied and which provides a countable sequence of functions

$$R_{nm}(\rho) = J_n(k_{nm}\rho) \tag{2.84}$$

From (2.81) and (2.84), the eigenmodes of the problem are built up

$$\Phi_{nm}(\rho, \theta) = A_{nm} J_n(k_{nm}\rho) e^{\iota n\theta}$$
$$n = -\infty, \ldots, -2, -1, 0, +1, +2, \ldots, +\infty \tag{2.85}$$
$$m = 1, 2, \ldots, \infty$$

the corresponding eigenwavenumbers being k_{nm}. The coefficients A_{nm} are chosen so that each eigenmode has an $\mathbf{L}^2$-norm equal to 1.

Orthogonality of the eigenmodes. Let Φ_{nm} and Φ_{pq} be two different eigenfunctions. If $n \neq p$, then the orthogonality of the functions $e^{\iota n\theta}$ and $e^{\iota p\theta}$ implies that of the two eigenfunctions. Let us now calculate the scalar product of Φ_{nm} and Φ_{nq}; we have:

$$I = (k_{nm}^2 - k_{np}^2) \int_\Omega \Phi_{nm} \Phi_{nq}^* \, d\Omega$$
$$= 2\pi A_{nm} A_{nq}^* (k_{nm}^2 - k_{np}^2) \int_0^{\rho_0} R_{nm}(\rho) R_{nq}(\rho) \rho \, d\rho \tag{2.86}$$

Equation (2.82) implies the following equalities:

$$k_{nm}^2 R_{nm}(\rho) = \frac{n^2}{\rho^2} R_{nm}(\rho) - \frac{1}{\rho}\frac{d}{d\rho}\left(\rho \frac{dR_{nm}(\rho)}{d\rho}\right)$$

$$k_{nq}^2 R_{nq}(\rho) = \frac{n^2}{\rho^2} R_{nq}(\rho) - \frac{1}{\rho}\frac{d}{d\rho}\left(\rho \frac{dR_{nq}(\rho)}{d\rho}\right)$$

which lead to:

$$I = 2\pi A_{nm} A_{nq}^* \int_0^{\rho_0} \left\{ \frac{1}{\rho}\frac{d}{d\rho}\left(\rho \frac{dR_{nq}(\rho)}{d\rho}\right) R_{nm}(\rho) - \frac{1}{\rho}\frac{d}{d\rho}\left(\rho \frac{dR_{nm}(\rho)}{d\rho}\right) R_{nq}(\rho) \right\} \rho \, d\rho \tag{2.87}$$

This expression is integrated by parts:

$$\begin{aligned} I = {} & 2\pi A_{nm} A_{nq}^* \left[\rho \frac{dR_{nq}(\rho)}{d\rho} R_{nm}(\rho) - \rho \frac{dR_{nm}(\rho)}{d\rho} R_{nq}(\rho) \right]_0^{\rho_0} \\ & - 2\pi A_{nm} A_{nq}^* \int_0^{\rho_0} \left\{ \frac{dR_{nq}(\rho)}{d\rho}\frac{dR_{nm}(\rho)}{d\rho} - \frac{dR_{nm}(\rho)}{d\rho}\frac{dR_{nq}(\rho)}{d\rho} \right\} \rho \, d\rho \end{aligned} \tag{2.88}$$

The integrated term is zero because: (1) the coefficient ρ is zero at the first end of the integration interval; (2) the derivatives of $R_{nm}(\rho)$ and $R_{nq}(\rho)$ are zero at the other end. In the remaining integral, the term to be integrated is zero. Thus the eigenfunctions $\Phi_{nm}(\rho, \theta)$ are orthogonal to each other.

Normalization coefficients. The set of eigenmodes is a basis for the functional space which any solution of the Neumann problem belongs to. It is generally simpler to deal with a basis of functions having a unit norm. To this end, the value of the coefficients A_{nm} must be

$$A_{nm} = \frac{1}{\sqrt{2\pi}} \left\{ \int_0^{\rho_0} [J_n(k_{nm}\rho)]^2 \rho \, d\rho \right\}^{-1/2} \tag{2.89}$$

2.4.2. Representation of the solution of equation (2.77) as a series of the eigenmodes

Let us first expand the second member of equation (2.77) into a series of the eigenmodes Φ_{nm}:

$$\begin{aligned} f(M) &= \sum_{n=-\infty}^{+\infty} \sum_{m=1}^{+\infty} f_{nm} A_{nm} J_n(k_{nm}\rho) e^{\iota n\theta} \\ f_{nm} &= A_{nm} \int_0^{2\pi} e^{-\iota n\theta} \, d\theta \int_0^{\rho_0} f(\rho, \theta) J_n(k_{nm}\rho) \rho \, d\rho \end{aligned} \tag{2.90}$$

The function $p(M)$ is sought as a similar series expansion:

$$p(m) = \sum_{n=-\infty}^{+\infty} \sum_{m=1}^{+\infty} p_{nm} A_{nm} J_n(k_{nm}\rho) \mathrm{e}^{\iota n\theta} \tag{2.91}$$

and it is easily proved that the coefficients p_{nm} are given by:

$$p_{nm} = \frac{f_{nm}}{k^2 - k_{nm}^2} \tag{2.92}$$

These coefficients, and, as a consequence, the solution $p(M)$, are all defined if and only if the wavenumber k is different from all the eigenwavenumbers k_{nm} of the Neumann problem.

In the simple situation where the source is a point source located at point $S(\rho_S, \theta_S)$ (Dirac measure located at S), the coefficients f_{nm} are

$$f_{nm} = A_{nm} J_n(k_{nm}\rho_S) \mathrm{e}^{-\iota n\theta_S}$$

and series (2.91) takes the form

$$p(M) = \sum_{n=-\infty}^{+\infty} \sum_{m=1}^{+\infty} \frac{A_{nm}^2 J_n(k_{nm}\rho_S) J_n(k_{nm}\rho)}{k^2 - k_{nm}^2} \mathrm{e}^{\iota n(\theta - \theta_S)} \tag{2.93}$$

It must be remarked that $p(M)$ has a logarithmic singularity at the source location, that is at the point $(\rho = \rho_S, \theta = \theta_S)$: this singularity is that of the sound pressure that a point source radiates in free space. For this reason, the convergence rate of the series decreases as the point M draws nearer to S: the number of terms required to evaluate the pressure with a given accuracy rapidly becomes large in the vicinity of the point source.

2.4.3. The method of separation of variables

Let us pay attention to the simple example of the response of the system to a point isotropic source of unit amplitude located at a point S. The solution is sought in the following form:

$$\begin{aligned} p(M) &= p_0(M) + \psi(M) \\ p_0(M) &= -\frac{\iota}{4} H_0^{(1)}(k \mid MS \mid) \end{aligned} \tag{2.94}$$

where $H_0^{(1)}(k \mid MS \mid)$ is the Hankel function of the first kind and $\mid MS \mid$ is the distance between the two points M and S (for details on the Hankel functions, the reader can refer to the book by Ph.M. Morse and K.U. Ingard); $p_0(M)$ represents the radiation of a point isotropic source in free space. The function $\psi(M)$, which represents the sound field reflected by the boundary σ of the domain Ω, satisfies a non-homogeneous Neumann boundary value problem:

$$\begin{aligned} (\Delta + k^2)\psi(M) &= 0, \qquad M \in \Omega \\ \partial_\rho \psi(M) &= -\partial_\rho p_0(M), \qquad M \in \sigma \end{aligned} \tag{2.95}$$

The reflected field $\psi(M)$ is sought as a series of products of functions depending on one variable only:

$$\psi(M) = \sum_{n=-\infty}^{+\infty} a_n R_n(\rho)\phi_n(\theta)$$

Following the same method as in subsection 2.4.1, it is easily seen that the functions $R_n(\rho)$ must satisfy the first equation (2.80) while the functions $\phi_n(\theta)$ must be solutions of the second one. The necessary regularity of $\psi(M)$ (solution of a homogeneous partial differential equation with analytic coefficients) implies that this function, and, as a consequence, the $\phi_n(\theta)$, must be 2π-periodic, that is

$$\phi_n(\theta) = \mathrm{e}^{\iota n\theta}, \qquad n = -\infty, \ldots, -2, -1, 0, 1, 2, +\infty$$

To each of these functions, there corresponds a solution of the Bessel equation which must be regular at $\rho = 0$; we can take:

$$R_n(\rho) = -\frac{\iota}{4} J_n(k\rho)$$

Thus, the reflected field can be sought as a series expansion of the form

$$\psi(M) = \frac{\iota}{4} \sum_{n=-\infty}^{+\infty} a_n J_n(k\rho)\mathrm{e}^{\iota n\theta} \tag{2.96}$$

The coefficients a_n are determined by the boundary condition. To this end, use is made of the following classical result from the Bessel functions theory:

$$\begin{aligned} H_0^{(1)}(k \mid MS \mid) &= \sum_{n=-\infty}^{+\infty} H_n^{(1)}(k\rho_S) J_n(k\rho)\mathrm{e}^{\iota n(\theta - \theta_S)} \quad \text{for} \quad \rho < \rho_S \\ &= \sum_{n=-\infty}^{+\infty} H_n^{(1)}(k\rho) J_n(k\rho_S)\mathrm{e}^{\iota n(\theta - \theta_S)} \quad \text{for} \quad \rho < \rho_S \end{aligned} \tag{2.97}$$

The Neumann boundary condition can thus be written

$$\sum_{n=-\infty}^{+\infty} a_n J_n'(k\rho_0)\mathrm{e}^{\iota n\theta} = \sum_{n=-\infty}^{+\infty} H_n^{(1)'}(k\rho_0) J_n(k\rho_S)\mathrm{e}^{-\iota n\theta_S}\mathrm{e}^{\iota n\theta} \tag{2.98}$$

This equation is satisfied if and only if the equalities

$$a_n J_n'(k\rho_0) = H_n^{(1)'}(k\rho_0) J_n(k\rho_S)\mathrm{e}^{-\iota n\theta_S}$$

stand for any n. It is obvious that, if $J_n'(k\rho_0) \neq 0 \forall n$, that is if $k \neq k_{nm} \forall (n, m)$, then the coefficients a_n are all uniquely defined by

$$a_n = \frac{H_n^{(1)'}(k\rho_0) J_n(k\rho_S) \mathrm{e}^{-\iota n\theta_S}}{J_n'(k\rho_0)}$$

The solution of the problem is given by the series

$$p(M) = -\frac{\iota}{4} H_0^{(1)}(k \mid MS \mid) + \frac{\iota}{4} \sum_{n=-\infty}^{+\infty} \frac{H_n^{(1)'}(k\rho_0) J_n(k\rho_S) \mathrm{e}^{-\iota n\theta_S}}{J_n'(k\rho_0)} J_n(k\rho) \mathrm{e}^{\iota n\theta} \qquad (2.99)$$

This solution is defined for any value of k different from an eigenwavenumber. The series involved in the equality (2.99) represents a regular (analytical) function: it is convergent everywhere. From a numerical point of view, experience shows that the number of terms required for a given accuracy is almost independent of the point M at which it is evaluated and of the source position S. More precisely, if ρ_0 is about 2 to 3 times the wavelength, a 1 dB accuracy requires roughly 150 eigenmodes, while in the representation (2.99) the series can be reduced to about 10 terms.

2.5. Enclosures Bounded by Plane Surfaces: Introduction to the Method of Images

The method here proposed is identical to geometrical optics. It is well known that when a light source is placed in front of a plane mirror the total lighting is identical to what could be obtained by removing the mirror and placing a second light source in a location symmetrical to the original one with respect to the mirror plane. Though this solution is an approximation of the governing equations, the result is quite good. In many situations, optical phenomena are governed by the Helmholtz equation. This has led acousticians to adopt the methods which have proved to be efficient in optics. In particular, the principles of geometrical optics are used in acoustics as far as the wavelengths involved are short compared to the dimensions of the reflecting structures. The results so obtained cannot have the same accuracy as in optics because the ratio of wavelength to reflecting structure dimensions is always much larger. Nevertheless, geometrical acoustics is a powerful tool for sound level chart predictions: it enables low cost calculation of noise pollution in cities or factory halls where complex geometries as well as complex sources are involved (errors are less than about 3 to 4 dB); many efficient computation programs for concert-hall design are based on geometrical approximations of the wave propagation equation and of the energy attenuation by boundaries.

2.5.1. Reflection of a spherical wave by an infinite plane: The ‘plane wave’ and the ‘geometrical acoustics’ approximations

Let $\Omega = (z > 0)$ be the propagation domain of a harmonic ($\mathrm{e}^{-\iota\omega t}$) wave radiated by a point isotropic unit source located at $S(0, 0, s)$. The acoustic properties of the plane $\Sigma(z=0)$ are characterized by a normal specific impedance ζ. The acoustic pressure $p(M)$ at a point $M(x, y, z)$ is the solution of a Robin problem:

$$(\Delta + k^2)p(M) = \delta_S(M), \qquad M \in \Omega$$

$$\partial_z p(M) + \frac{\iota k}{\zeta} p(M) = 0, \qquad M \in \Sigma \tag{2.100}$$

Sommerfeld condition (all radiated energy goes towards infinity)

The solution of this problem can be decomposed into two terms:

$$p(M) = -\frac{e^{\iota kr(M,S)}}{4\pi r(M,S)} + \psi(M) \tag{2.101}$$

In this expression, the first term represents the acoustic pressure radiated by the source in free space; the second term is the sound pressure reflected by the boundary Σ of the half-space Ω.

Assume first that Σ is a perfectly hard surface ($1/|\zeta| = 0$). Then the function $\psi(M)$ is given by

$$\psi(M) = -\frac{e^{\iota kr(M,S')}}{4\pi r(M,S')}$$

where S', the point with coordinates $(0, 0, -s)$, is the image of S with respect the plane $z=0$. A simple result would be that, for any boundary conditions, the function $\psi(M)$ could be correctly approximated by a similar expression, that is by

$$\psi(M) \simeq -A\,\frac{e^{\iota kr(M,S')}}{4\pi r(M,S')}$$

where the amplitude coefficient A is a function of impedance. A simple argument can intuitively justify this approximation. Let λ be the wavelength. Assume that $s \gg \lambda$ and $z \gg \lambda$, that is both points S and M are 'far away' from the reflecting surface. Thus, when the wavefront reaches Σ, its radius of curvature is very large and the incident spherical wave can be considered as a plane wave with an angle of incidence θ (see Fig. 2.6). The reflected wave goes back in the specular direction, that is in the direction of the line symmetrical with the incident direction with respect to the normal to the plane $z=0$. The natural reflection coefficient to be adopted is that of a plane wave and it is given by

$$A = \frac{\zeta\cos\theta - 1}{\zeta\cos\theta + 1}$$

The approximation of the total acoustic pressure is written

$$p(M) \simeq -\frac{e^{\iota kr(M,S)}}{4\pi r(M,S)} - \frac{\zeta\cos\theta - 1}{\zeta\cos\theta + 1}\,\frac{e^{\iota kr(M,S')}}{4\pi r(M,S')} \tag{2.102}$$

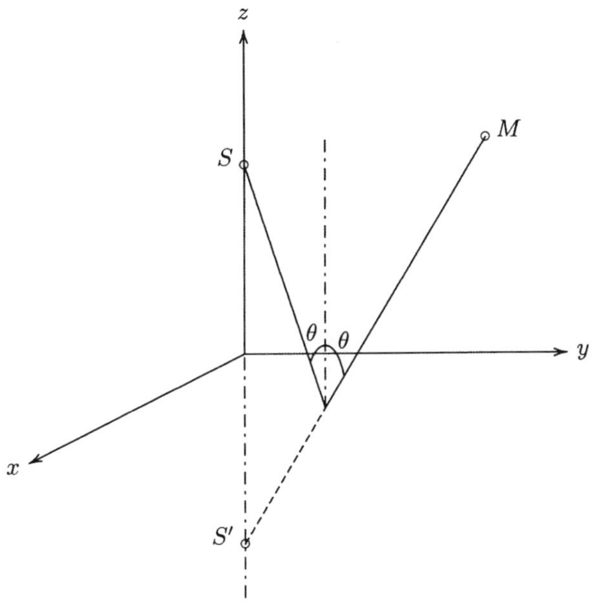

Fig 2.6. Geometry of the propagation domain.

and is often called the *plane wave approximation*. This result can be rigorously established by looking at an asymptotic series expansion of $p(M)$ in terms of the successive powers of $1/kr(M, S')$. It is then shown that the difference between the exact solution and approximation (2.102) decreases at least as fast as $1/[kr(M, S')]^2$.

If interference phenomena are not present (or are not relevant) – this is, for example, the case for traffic noise – the amplitude of the sound pressure field is sufficient information. The former formula can then be much simplified. An approximation of the form

$$|\, p(M)\,| \simeq \frac{1}{4\pi r(M, S)} + \frac{B}{4\pi r(M, S')} \tag{2.103}$$

is looked at. Energy considerations lead us to adopt for B the modulus of the complex reflection coefficient:

$$B = \left| \frac{\zeta \cos\theta - 1}{\zeta \cos\theta + 1} \right| \tag{2.104}$$

This last expression can again be simplified by neglecting the variations with respect to the angle of incidence:

$$B = \left| \frac{\zeta - 1}{\zeta + 1} \right|$$

Approximation (2.103) is very similar to those in use in optics.

2.5.2. Forced regime in a polyhedral enclosure: series representation of the response by the image method

For the sake of simplicity, let us consider a parallelepipedic domain Ω defined by

$$x \in \left]-\frac{a}{2}, +\frac{a}{2}\right[, \qquad y \in \left]-\frac{b}{2}, +\frac{b}{2}\right[, \qquad z \in \left]-\frac{c}{2}, +\frac{c}{2}\right[$$

The boundary of the domain is characterized by a normal specific impedance ζ, which is independent of the space variables. Let $S(x_0, y_0, z_0)$ be a point isotropic source with unit amplitude, which generates a signal containing all frequencies belonging to the interval $[f_0 - \Delta f_0, f_0 + \Delta f_0]$. The relative frequency bandwidth $2\Delta f_0/f_0$ is supposed to be small enough that the reflection coefficient $B = |\zeta - 1|/|\zeta + 1|$ of the boundary of the domain can be considered as independent of frequency.

The first order images are defined by

$$\begin{aligned} S_x^{1+} &= (a - x_0, y_0, z_0), & S_x^{1-} &= (-a - x_0, y_0, z_0) \\ S_y^{1+} &= (x_0, b - y_0, z_0), & S_y^{1-} &= (x_0, -b - y_0, z_0) \\ S_z^{1+} &= (x_0, y_0, c - z_0), & S_z^{1-} &= (x_0, y_0, -c - z_0) \end{aligned}$$

The corresponding approximation of the sound field is

$$|p(M)| \simeq \frac{1}{4\pi}\left\{\frac{1}{r(S, M)} + B\left[\frac{1}{r(S_x^{1+}, M)} + \frac{1}{r(S_x^{1-}, M)} + \frac{1}{r(S_y^{1+}, M)} + \frac{1}{r(S_y^{1+}, M)} \right.\right.$$
$$\left.\left. + \frac{1}{r(S_z^{1+}, M)} + \frac{1}{r(S_z^{1+}, M)}\right]\right\} \tag{2.105}$$

In this expression, the waves accounted for have been reflected only once on the boundary of the domain Ω. This approximation can be improved by adding waves which have been reflected twice. For that purpose, second order images must be used. As an example, the image of S_x^{1+} in the plane $y = b/2$ is located at $S_{xy}^{2++} = (a - x_0, b - y_0, z_0)$, and its contribution is $B^2/4\pi r(S_{xy}^{2++}, M)$. So, the contribution of all second order images is of the form $B^2 f_2(S, M)/4\pi$. Accounting for the images of successive orders, the acoustic pressure field is approximated by an expression of the form

$$|p(M)| \simeq \frac{1}{4\pi}\left\{\frac{1}{r(S, M)} + \sum_{n=1}^{\infty} B^n f_n(S, M)\right\} \tag{2.106}$$

It can be shown that if the boundaries absorb energy ($B < 1$), this series is convergent. It is shown, too, that as the frequency is increased, the error on $|p(M)|$ decreases: the image method thus provides a 'high frequency' or 'short wavelength' approximation. For instance, for a concert hall or a theatre, the prediction of the various acoustical quantities is almost perfect (errors less than a few dB) above 500 Hz.

In this method that we have briefly described, the acoustic properties of the boundaries of the domain can vary from one wall to another, each one being described by a given reflection coefficient. It is easy to consider any convex polyhedral boundary. Additional difficulties occur when the surface which bounds the propagation domain is no longer convex (this, for example, is the case in a concert hall with mezzanines). Not all the image sources are 'seen' from everywhere in the room: thus, the representation of the pressure field includes the contribution of sets of images which are different from one region to another. Finally, if the boundary is not a polyhedral surface, it is no longer possible to define image sources: an equivalent method is provided by the ray theory which is a simplification of the geometrical theory of diffraction due to J.B. Keller (see chapters 4 and 5).

2.6. General Case: Introduction to the Green's Representation of Acoustic Fields

The aim of this last section is to show how the acoustic field reflected by the boundary of an enclosure can be represented as the radiation of sources located on this boundary. It is established that the function which describes this source is the solution of a boundary integral equation: this solution exists and is unique if and only if the boundary value problem has one and only one solution. This means that the boundary integral equation has no solution if and only if the wavenumber of the excitation signal equals one of the eigenwavenumbers of the initial problem.

2.6.1. Green's representation of the acoustic pressure

Let Ω be a bounded domain of $\mathbb{R}^n$ ($n = 2$ or 3), with a boundary σ which is assumed to be 'sufficiently regular': in particular, this implies that a unit vector $\vec{n}$, normal to σ and pointing out to the exterior of Ω, can be defined almost everywhere; from a practical point of view, any piecewise twice differentiable surface fits the required conditions. Let $f(M)$ be the space density of a harmonic ($e^{-\iota\omega t}$) source and $p(M)$ the corresponding pressure field. This function is the solution of the following boundary value problem:

$$(\Delta + k^2)p(M) = f(M), \qquad M \in \Omega \tag{2.107}$$

$$\alpha\partial_n p(M) + \beta p(M) = 0, \qquad M \in \sigma \tag{2.107'}$$

A representation of $p(M)$ is sought as the sum of the free field pressure $p_0(M)$, that is the pressure radiated by the source $f(M)$ in an unbounded domain which is a particular solution of equation (2.107), and a function $\Phi(M)$, often called the 'diffracted field', which is a solution of the homogeneous Helmholtz equation. This last function must be such that the boundary condition (2.107′) is satisfied.

Let $G(M, M')$ be the pressure field radiated, in the unbounded space $\mathbb{R}^n$, by a point isotropic unit source $\delta_{M'}(M)$ located at point M'. $G(M, M')$ satisfies the equation

$$(\Delta_M + k^2)G(M, M') = \delta_{M'}(M), \qquad M \in \mathbb{R}^n$$

where the symbol Δ_M means that the derivatives are taken with respect to the variable M. It is proved that, with the time dependence which has been chosen, the solution of this equation which satisfies the principle of energy conservation (which the acoustics equations are based on) is

$$\begin{aligned} G(M, M') &= -\frac{\iota}{4} H_0^{(1)}[kr(M, M')], \qquad M \in \mathbb{R}^2 \\ &= -\frac{e^{\iota kr(M, M')}}{4\pi r(M, M')}, \qquad M \in \mathbb{R}^3 \end{aligned} \tag{2.108}$$

where $r(M, M')$ is the distance between the two points M and M'.

Let us first show briefly that, if $f(M)$ is an integrable function, $p_0(M)$ is given by

$$p_0(M) = \int_{\mathbb{R}^n} f(M')G(M, M')\, d\Omega(M') \tag{2.109}$$

It is intuitive that the source $f(M')$ can be decomposed into a set of elementary isotropic point sources with amplitudes $f(M')\, d\Omega(M')$ located at M'. Then, formula (2.109) looks natural. A more correct validation consists in applying the Helmholtz operator to this expression

$$\begin{aligned} (\Delta + k^2)p_0(M) &= (\Delta + k^2) \int_{\mathbb{R}^n} f(M')G(M, M')\, d\Omega(M') \\ &= \int_{\mathbb{R}^n} f(M')(\Delta_M + k^2)G(M, M')\, d\Omega(M') \end{aligned}$$

The second equality is obtained by inverting differentiation and integration and requires a mathematical justification. Using the equation satisfied by $G(M, M')$ one gets the following formal equalities:

$$\begin{aligned} (\Delta + k^2)p_0(M) &= \int_{\mathbb{R}^n} f(M')\delta_{M'}(M)\, d\Omega(M') \\ &= f(M) \qquad \text{[property of the Dirac measure]} \end{aligned}$$

The lack of rigour of this proof is that the integral of a Dirac measure is not defined; it can be used just as a symbolic expression. This formal proof can be justified rigorously with basic theorems of the theory of distributions as will be seen in chapter 3.

Let us now look for an integral representation of the diffracted field $\Phi(M)$. Let $\tilde{p}(M)$ be the function which is equal to $p(M)$ in Ω, and to zero in the exterior $\complement\overline{\Omega}$ of this domain. The proof is established for $n=3$, but remains valid for $n=2$. Let $I(M)$ be the integral defined by

$$I(M) = -\int_{\mathbb{R}^3} [G(M, M')(\Delta + k^2)\tilde{p}(M') - \tilde{p}(M')(\Delta_{M'} + k^2)G(M, M')]\, d\Omega(M') \tag{2.110}$$

Accounting for the equations satisfied by $p(M)$ and $G(M, M')$, this expression becomes (formally)

$$I(M) = -\int_{\mathbb{R}^3} [G(M, M')f(M') - \tilde{p}(M')\delta_M(M')]\, d\Omega(M') = \tilde{p}(M) - p_0(M) = \Phi(M)$$

Let us show now that $\Phi(M)$ can be expressed by surface integrals which involve the values of the functions $p(M)$ and $\partial_n p(M)$ on σ. To do so, expression (2.110) is integrated by parts. To get a simplified proof, the surface σ is assumed to be indefinitely differentiable and convex, which implies that each coordinate axis cuts it at two points only.

(a) Integration with respect to the variable x. Let $x_1(y, z)$ and $x_2(y, z)$ the intersection points between the line ($y = \text{constant}$, $z = \text{constant}$) and σ. One gets

$$\begin{aligned} I_1(M) &\equiv \int_{\mathbb{R}^3} \tilde{p}(M') \frac{\partial^2 G(M, M')}{\partial x'^2}\, d\Omega(M') \\ &= \int_{-\infty}^{+\infty} dy' \int_{-\infty}^{+\infty} dz' \int_{x_1'(y', z')}^{x_2'(y', z')} p(M') \frac{\partial^2 G(M, M')}{\partial x'^2}\, dx' \\ &= \int_{-\infty}^{+\infty} dy' \int_{-\infty}^{+\infty} dz' \left[p(M') \frac{\partial G(M, M')}{\partial x'} \right]_{x_1'(y', z')}^{x_2'(y', z')} \\ &\quad - \int_{\mathbb{R}^3} \frac{\partial \tilde{p}(M')}{\partial x'} \frac{\partial G(M, M')}{\partial x'}\, d\Omega(M') \end{aligned} \tag{2.111}$$

In this expression, the double integral is, in fact, an integral over σ; denoting by $\theta_1(M')$ the angle between the normal vector $\vec{n}(M')$ and the x-axis, the last expression becomes

$$\begin{aligned} I_1(M) &= \int_\sigma p(M') \frac{\partial G(M, M')}{\partial x'} \cos\theta_1(M')\, d\sigma(M') \\ &\quad - \int_{\mathbb{R}^3} \frac{\partial \tilde{p}(M')}{\partial x'} \frac{\partial G(M, M')}{\partial x'}\, d\Omega(M') \end{aligned}$$

We finally obtain the following equality:

$$\int_{\mathbb{R}^3} \left[\tilde{p}(M') \frac{\partial^2 G(M, M')}{\partial x'^2} - G(M, M') \frac{\partial^2 p(M')}{\partial x'^2} \right] d\Omega(M')$$

$$= \int_\sigma \left[p(M') \frac{\partial G(M, M')}{\partial x'} - G(M, M') \frac{\partial p(M')}{\partial x'} \right] \cos\theta_1(M')\, d\sigma(M') \quad (2.112)$$

(b) Integration by parts with respect to the three variables. Let $\theta_2(M')$ (resp. $\theta_3(M')$) be the angle between $\vec{n}$ and the y-coordinate axis (resp. the z-coordinate axis). Integrations by parts with respect to y' and z' finally lead to

$$I(M) = \int_\sigma \left\{ p(M') \left[\frac{\partial G(M, M')}{\partial x'} \cos\theta_1 + \frac{\partial G(M, M')}{\partial y'} \cos\theta_2 + \frac{\partial G(M, M')}{\partial z'} \cos\theta_3 \right] \right.$$

$$\left. - G(M, M') \left[\frac{\partial p(M')}{\partial x'} \cos\theta_1 + \frac{\partial p(M')}{\partial y'} \cos\theta_2 + \frac{\partial p(M')}{\partial z'} \cos\theta_3 \right] \right\} d\sigma(M')$$

(the dependence of the angles θ_i with the integration point M' has been omitted). The following notation is then introduced:

$$\partial_{n'} p(M') = \vec{n} \cdot \nabla p(M') = \frac{\partial p(M')}{\partial x'} \cos\theta_1 + \frac{\partial p(M')}{\partial y'} \cos\theta_2 + \frac{\partial p(M')}{\partial z'} \cos\theta_3$$

and then the integral $I(M)$ is written

$$I(M) = \int_\sigma [p(M') \partial_{n'} G(M, M') - G(M, M') \partial_{n'} p(M')]\, d\sigma(M') \quad (2.113)$$

(c) Green's representation of $p(M)$. This is obtained by gathering the equalities (2.110) and (2.113), which gives

$$p(M) = p_0(M) + \int_\sigma [p(M') \partial_{n'} G(M, M') - G(M, M') \partial_{n'} p(M')]\, d\sigma(M') \quad (2.114)$$

This expression represents the solution of the boundary value problem (2.107), when it exists. Three remarks must be made. The function given by (2.114) is defined in the whole space and is identically zero inside $\complement\overline{\Omega}$ the complement of $\Omega \cup \sigma$ (that is the space domain which is outside σ). It is completely determined as soon as the functions $p(M')$ and $\partial_{n'} p(M')$ are known on σ: it must be noticed that these two functions are not independent since they are related by the boundary condition (2.107′). Finally, the expression (2.114) remains valid when the boundary σ is a piecewise regular surface (or curve) as for example a polyhedron: for such a surface, the edges, being of zero measure, do not give any contribution to the surface integral (but their presence diminishes the regularity of the pressure field).

2.6.2. Simple layer and double layer potentials

The expression (2.114) of the acoustic pressure $p(M)$ contains two types of integrals:

$$\varphi_1(M) = \int_\sigma \partial_{n'} p(M') G(M, M')\, d\sigma(M') \tag{2.115}$$

$$\varphi_2(M) = -\int_\sigma p(M') \partial_{n'} G(M, M')\, d\sigma(M') \tag{2.116}$$

The function $\varphi_1(M)$ can be interpreted as the sound field radiated by point isotropic sources with amplitude $\partial_{n'} p(M')\, d\sigma(M')$ and supported by σ: for this reason, it is called *simple layer potential*; the function $\partial_{n'} p(M')$ is called the *density of the simple layer*.

The interpretation of the function $\varphi_2(M)$ is less straightforward: it is first necessary to define the radiation of a slightly more complicated point source, the *acoustic doublet*. Let $S^+(+\varepsilon, 0, 0)$ and $S^-(-\varepsilon, 0, 0)$ two infinitely closed points supporting isotropic point sources with respective amplitudes $1/2\varepsilon$ and $-1/2\varepsilon$. The resulting field at a point $M(x, y, z)$ of $\mathbb{R}^3$ is

$$\psi_\varepsilon(M) = -\frac{1}{2\varepsilon}\left(\frac{e^{\imath k r^+}}{4\pi r^+} - \frac{e^{\imath k r^-}}{4\pi r^-}\right)$$

with $r^\pm = \sqrt{(x \pm \varepsilon)^2 + y^2 + z^2}$. It is obvious that the function $\psi_\varepsilon(M)$ has a limit for $\varepsilon \longrightarrow 0$:

$$\psi(M) = \lim_{\varepsilon \longrightarrow 0} \psi_\varepsilon(M) = -\frac{\partial}{\partial x}\left(\frac{e^{\imath k r}}{4\pi r}\right), \qquad r = \sqrt{x^2 + y^2 + z^2}$$

The source which has this radiation is called the *acoustic doublet* (by analogy with electrostatics) and is represented by the x-derivative of a Dirac measure located at the origin. If such a source is located at point $S(x_0, y_0, z_0)$, the corresponding field is

$$\psi_S(M) = \frac{\partial G(M, S)}{\partial x} = -\frac{\partial G(M, S)}{\partial x_0}$$

It is now easy to give an interpretation of the function $\varphi_2(M)$: it is the radiation of a set of acoustic doublets, oriented along the unit vector $\vec{n}'$ normal to σ and with amplitude $p(M')\, d\sigma(M')$. For this reason, the function $\varphi_2(M)$ is called *double layer potential* and $p(M')$ is the density of the double layer.

The integrals which define $\varphi_1(M)$ and $\varphi_2(M)$ are Riemann integrals in so far as the point M does not belong to σ. If $M \in \sigma$, the functions $G(M, M')$ and $\partial_{n'} G(M, M')$ have a singularity at $M = M'$. The values on σ of the simple and double layer potentials, together with their normal derivatives, must be evaluated by a limit procedure. The following results can be established:

$$\lim_{M \in \Omega \to Q \in \sigma} \varphi_1(M) = \int_\sigma \partial_{n'} p(M') G(Q, M')\, d\sigma(M') \qquad \text{(a)}$$

where the integral is of Riemann type

$$\lim_{M \in \Omega \to Q \in \sigma} \varphi_2(M) = -\frac{p(Q)}{2} - \int_\sigma p(M') \partial_{n'} G(Q, M')\, d\sigma(M') \qquad \text{(b)}$$

$$\lim_{M \in \Omega \to Q \in \sigma} \partial_n \varphi_1(M) = -\frac{\partial_n p(Q)}{2} + \int_\sigma \partial_{n'} p(M') \partial_n G(Q, M')\, d\sigma(M') \qquad \text{(b')}$$

the integrals being Cauchy principal values;

$$\lim_{M \in \Omega \to Q \in \sigma} \partial_n \varphi_2(M) = -\text{Pf.} \int_\sigma p(M') \partial_n \partial_{n'} G(Q, M')\, d\sigma(M') \qquad \text{(c)}$$

is defined as a *Hadamard finite part* (Pf.): for this particular case, the standard definition given in classical textbooks (as, for example, L. Schwartz's textbook cited here) does not apply, and a limit process must be used for an analytical or a numerical evaluation.

2.6.3. Boundary integral equations associated with the Green's representation of the pressure field

The boundary condition (2.107′) shows that as soon as one of the two functions $p(M)$ or $\partial_{\vec{n}}(M)$ is known, the other one is known too; the Green's representation (2.114) of the acoustic pressure field is then completely determined. By introducing the boundary condition into the Green's formula, one of the two unknown functions is eliminated and an equation is needed to determine the remaining one. Assume, for example, $\alpha \neq 0$ and introduce the notation $\gamma = \alpha/\beta$. Expression (2.114) becomes

$$p(M) = p_0(M) + \int_\sigma p(M')[\partial_{n'} G(M, M') + \gamma G(M, M')] d\sigma(M'), \qquad M \in \Omega \qquad (2.114')$$

Taking the limit $M \in \Omega \to Q \in \sigma$ leads to:

$$\frac{p(Q)}{2} - \int_\sigma p(M')[\partial_{n'} G(Q, M') + \gamma G(Q, M')]\, d\sigma(M') = -p_0(Q), \qquad Q \in \sigma \qquad (2.117)$$

which equation is obtained by accounting for property (b) of the double layer potential. This equation determines the unique value on σ of the pressure $p(M)$ if the boundary value problem which it is deduced from has a unique solution. The following theorem can be established.

Theorem 2.2 (Existence and uniqueness of the solution of the B.I.E.)

(a) If k is an eigenwavenumber of the initial boundary value problem (2.107), with multiplicity order N (the homogeneous problem has N linearly independent solutions), the homogeneous equation (2.117) has N linearly independent solutions. The non-homogeneous equation ($p_0(Q) \neq 0$) has no solution.
(b) If k is not an eigenwavenumber of the initial boundary value problem, then equation (2.117) has a unique solution for any $p_0(Q)$.
(c) Conversely, there exists a countable sequence of wavenumbers k_n such that the homogeneous equation (2.117) has a finite number of linearly independent solutions. Each of these solutions defines a solution of the homogeneous boundary value problem by expression (2.114) with $p_0(M) \equiv 0$. If k is different from all the k_n, then the non-homogeneous equation (2.117) has a unique solution which, by (2.114), defines the corresponding unique solution of the boundary value problem.

If β is assumed to be different from zero, then it is possible to define $\delta = \alpha/\beta$ and one is left with the following Green's representation of $p(M)$:

$$p(M) = p_0(M) - \int_\sigma \partial_{n'} p(M')[\delta \partial_{n'} G(M, M') + G(M, M')]\, d\sigma(M'), \qquad M \in \Omega \tag{2.114'}$$

An integral equation to determine the function $\partial_{\vec{n}'} p(M')$ is obtained by taking the limit of this expression for $M \in \Omega \rightarrow Q \in \sigma$:

$$-\frac{\delta \partial_n p(Q)}{2} + \int_\sigma \partial_{n'} p(M')[\delta \partial_{n'} G(Q, M') + G(Q, M')]\, d\sigma(M') = p_0(Q), \qquad q \in \sigma \tag{2.117'}$$

The former theorem remains true for this last boundary integral equation.

We will see in the next chapter that it is possible to derive other boundary integral equations from the Green's formula. It will be shown too that different boundary integral representations of the diffracted field can be defined which lead to boundary integral equations obeying the same theorem of existence and uniqueness of the solution.

Bibliography

[1] BOWMAN, J.J., SENIOR, T.B.A. and USLENGHI, P.L.E., 1969. *Electromagnetic and acoustic scattering by simple shapes*. North Holland, Amsterdam.

[2] BRUNEAU, M., 1983. *Introduction aux théories de l'acoustique*. Université du Maine Editeur, Le Mans, France.

[3] JEFFREYS, H. and JEFFREYS, B., 1972. *Methods of mathematical physics*. University Press, Cambridge.

[4] COLTON, D. and KRESS, R., 1983. *Integral equations methods in scattering theory*. Wiley-Interscience, New York.

[5] COLTON, D. and KRESS, R., 1992. *Inverse acoustic and electromagnetic scattering theory*. Springer-Verlag, Berlin.

[6] MARSDEN, J.E. and HOFFMANN, M.J., 1993. *Elementary classical analysis*. Freeman, New York.

[7] MORSE, PH.M. and FESHBACH, H., 1953. *Methods of theoretical physics*. McGraw-Hill, New York.

[8] MORSE, PH.M. and INGARD, K.U., 1968. *Theoretical acoustics*. McGraw-Hill, New York.

[9] PIERCE, A.D., 1981. *Acoustics: an introduction to its physical principles and applications*. McGraw-Hill, New York.

[10] SCHWARTZ, L., 1961. *Méthodes mathématiques pour les sciences physiques*. Hermann, Paris.

CHAPTER 3

Diffraction of Acoustic Waves and Boundary Integral Equations

Paul J.T. Filippi

Introduction

This chapter is devoted to the most general representation of an acoustic field. The acoustic pressure radiated in free space by any source is expressed, in general, by a convolution product (source density described by a distribution), or by an integral extended to the space domain occupied by the source if the source density is a sufficiently regular function (at least a square integrable function). When the propagation domain is bounded or when obstacles are present, the influence of the boundaries can always be represented by the radiation of sources, the supports of which are the various surfaces which limit the propagation domain. These sources, often called *diffraction sources*, are chosen such that the total field satisfies the conditions that the physical properties of the boundaries material impose. Their determination requires the solution of a boundary integral equation (or a system of B.I.E. if the propagation domain is limited by several disjoint surfaces): it must be noticed that such boundary integral equations can be solved analytically when and only when there exists direct analytical methods (mainly, the method of separation of variables) which provide an exact solution.

The main interest of the approach here proposed is as follows.

- The boundary value problem to be solved is reduced to an equation along the boundary of the propagation domain; thus, the equation to be solved numerically extends along a surface (or a curve if a two-dimensional problem is considered), which, in general, is bounded while the propagation domain can extend up to infinity.
- The numerical approximations are made of functions defined on the boundary only; the acoustic pressure (as well as the particle velocity or the intensity vector) is given by an analytical expression which satisfies exactly the propagation equation and satisfies approximately the boundary condition.
- The analytical approximation of the acoustic field presents many advantages compared to other numerical approximations such as those obtained by finite

element methods: for example, it provides very easily an analytical approximation of the far field directivity pattern of a diffracting or a radiating object.

This chapter has four sections. First, the radiation of a few simple sources is described. Then, the Green's representation of the solution of any boundary value problem encountered in linear acoustics is presented: this is the most commonly used representation of a diffracted field by boundary integrals and, historically, the oldest. It is shown that the Green's representation is not the only possible boundary integral expression of a diffracted field. Finally, boundary integral equations are established which are governed by the same theorem of existence and uniqueness of the solution as the boundary value problem initially considered. This is illustrated by two examples: the propagation of a harmonic wave inside and outside a perfectly rigid cylinder (two-dimensional domains).

This chapter requires the knowledge of the basis of the theory of distributions. Among the French books which present this theory from a physical point of view, we must recommend *L'outil mathématique* by R. Petit. Most of the results are not proved here: the reader who is interested in the mathematics of the diffraction theory must refer to the books cited in the short bibliography at the end of this chapter or to any treatise on partial differential equations and integral equations.

3.1. Radiation of Simple Sources in Free Space

In Chapter 2, it was seen that a harmonic ($e^{-\iota\omega t}$) source radiates an acoustic pressure $p(M)$ which satisfies the Helmholtz equation:

$$(\Delta + k^2)p(M) = f(M) \tag{3.1}$$

where $f(M)$ is a function or, more generally, a distribution which describes the physical system which the acoustic wave comes from. The principle of energy conservation, which the wave equation is based upon, is expressed by the Sommerfeld condition:

$$\begin{aligned} \lim_{r\to\infty} p &= \mathcal{O}(r^{(1-n)/2}) \\ \lim_{r\to\infty} (\partial_r p - \iota k p) &= o(r^{(1-n)/2}) \end{aligned} \tag{3.2}$$

where r is the distance between the observation point M and the origin of the coordinates, and n is the dimension of the space ($n = 1$, 2 or 3). Let us recall the meaning of the notations used: $u = \mathcal{O}(\varepsilon)$ means that u tends to zero as fast as ε; $u = o(\varepsilon)$ means that u tends to zero faster than ε.

3.1.1. Elementary solution of the Helmholtz equation

A function ψ is called an *elementary solution of the Helmholtz equation* if it satisfies the non-homogeneous equation

$$(\Delta + k^2)\psi = \delta \tag{3.3}$$

where δ is the Dirac measure located at the origin of coordinates. A function ψ_S is called an *elementary kernel of the Helmholtz equation* if it satisfies

$$(\Delta + k^2)\psi_S = \delta_S \tag{3.3'}$$

where δ_S is the Dirac measure located at point S. The elementary kernel $G(S, M)$ of the Helmholtz equation which satisfies the Sommerfeld condition is called the *free space Green's function of the Helmholtz equation*. The following results can be established:

$$G(S, M) = \frac{e^{\iota kr(S, M)}}{2\iota k} \qquad \text{in } \mathbb{R} \tag{3.4}$$

$$G(S, M) = -\frac{\iota}{4} H_0^{(1)}(kr(S, M)) \qquad \text{in } \mathbb{R}^2 \tag{3.5}$$

$$G(S, M) = -\frac{e^{\iota kr(S, M)}}{4\pi r(S, M)} \qquad \text{in } \mathbb{R}^3 \tag{3.6}$$

In these expressions, $r(S, M)$ represents the distance between the two points S and M. $H_0^{(1)}(z)$ is the Hankel function of order zero and of the first kind: in what follows, unless confusion is possible, it will be simply denoted by $H_0(z)$. The function $G(S, M)$ describes the radiation, in free space, of a source whose dimensions are small compared to the wavelength and which emits the same energy flux density in any direction (isotropic source).

Let us show that expression (3.4) satisfies the one-dimensional form of equation (3.3′). Let s and x be the abscissae of points S and M respectively. One has

$$r(S, M) = |x - s|$$

The derivation rule in the distributions sense leads to

$$\begin{aligned} \frac{d}{dx}\frac{e^{\iota k|x-s|}}{2\iota k} &= \iota k \frac{e^{\iota k|x-s|}}{2\iota k}\,\mathrm{sgn}(x - s) \\ \frac{d^2}{dx^2}\frac{e^{\iota k|x-s|}}{2\iota k} &= -k^2 \frac{e^{\iota k|x-s|}}{2\iota k} + \delta_s \end{aligned} \tag{3.7}$$

The second equality is equation (3.3′).

To prove that expression (3.6) satisfies equation (3.3′) in $\mathbb{R}^3$, a less direct method must be used. There is no lack of generality in assuming that the point source S is located at the coordinates origin O: by a change of variables (a translation), this situation can always be obtained. The function G as given by (3.6) not being defined at point O, we first consider a domain Ω_ε which is the exterior of the sphere B_ε with radius ε and centred in O. In this domain, the function G is indefinitely differentiable. Let Φ be an indefinitely differentiable function with compact (bounded) support (test function); and denote by G_ε the function which is equal to G in Ω_ε, and to zero in the interior of B_ε. In the distributions sense, the Laplace operator is defined by

$$\langle \Delta G_\varepsilon, \Phi \rangle = \int_{\Omega_\varepsilon} G \, \Delta\Phi \, d\Omega$$

where $\langle \cdot, \cdot \rangle$ stands for the duality product. Thus, one has

$$\langle (\Delta + k^2) G_\varepsilon, \Phi \rangle = \int_{\Omega_\varepsilon} G(\Delta + k^2)\Phi \, d\Omega = I_\varepsilon$$

Let (r, θ, φ) be the spherical coordinates of the integration point, in a system centred at O. The integral I_ε is written

$$I_\varepsilon = -\int_0^{2\pi} d\varphi \int_0^{\pi} \sin\theta \, d\theta \int_\varepsilon^\infty \frac{e^{\iota kr}}{4\pi r} \left\{ \frac{1}{r^2} \frac{\partial}{\partial r}\left(r^2 \frac{\partial}{\partial r} \right) + \frac{1}{r^2 \sin\theta} \frac{\partial}{\partial \theta}\left(\sin\theta \frac{\partial}{\partial\theta} \right) + \frac{1}{r^2 \sin^2\theta} \frac{\partial^2}{\partial\varphi^2} + k^2 \right\} \Phi r^2 \, dr \tag{3.8}$$

A double integration by parts of the term

$$I_1 = -\int_\varepsilon^\infty \frac{e^{\iota kr}}{4\pi r} \frac{1}{r^2} \frac{\partial}{\partial r}\left(r^2 \frac{\partial}{\partial r} \right) \Phi r^2 \, dr$$

leads to

$$\begin{aligned} I_1 &= -\frac{e^{\iota kr}}{4\pi r} r^2 \frac{\partial\Phi}{\partial r}\Bigg|_\varepsilon^\infty + \int_\varepsilon^\infty r^2 \frac{\partial}{\partial r}\left(\frac{e^{\iota kr}}{4\pi r} \right) \frac{\partial\Phi}{\partial r} \, dr \\ &= -\frac{e^{\iota kr}}{4\pi r} r^2 \frac{\partial\Phi}{\partial r}\Bigg|_\varepsilon^\infty + r^2 \frac{\partial}{\partial r}\left(\frac{e^{\iota kr}}{4\pi r} \right)\Phi\Bigg|_\varepsilon^\infty - \int_\varepsilon^\infty \frac{1}{r^2} \frac{\partial}{\partial r}\left\{ r^2 \frac{\partial}{\partial r}\left(\frac{e^{\iota kr}}{4\pi r} \right) \right\} \Phi r^2 \, dr \end{aligned} \tag{3.9}$$

The function Φ being compactly supported, the upper bounds of the integrated terms are zero. Thus the integral I_ε is written

$$I_\varepsilon = -\int_0^{2\pi} d\varphi \int_0^{\pi} \varepsilon^2 \frac{e^{\iota k\varepsilon}}{4\pi\varepsilon} \left\{ \frac{\partial\Phi}{\partial\varepsilon}(\varepsilon, \theta, \phi) + \left(\iota k - \frac{1}{\varepsilon} \right) \Phi(\varepsilon, \theta, \varphi) \right\} \sin\theta \, d\theta + \int_{\Omega_\varepsilon} (\Delta + k^2)\left(-\frac{e^{\iota kr}}{4\pi r} \right) \Phi \, d\Omega \tag{3.10}$$

The integral over Ω_ε is zero because, in this domain, the function G satisfies a homogeneous Helmholtz equation. The function $\Phi(\varepsilon, \theta, \varphi)$ being indefinitely differentiable, it can be expanded into a Taylor series around the origin:

$$\Phi(\varepsilon, \theta, \varphi) = \Phi(0, 0, 0) + \varepsilon \frac{\partial \Phi}{\partial \varepsilon}(0, \theta, \varphi) + \mathcal{O}(\varepsilon^2)$$

This expansion is introduced into (3.10) and then the limit $\varepsilon \to 0$ is taken. The result is

$$\begin{aligned}
\lim_{\varepsilon \to 0} I_\varepsilon = &- \lim_{\varepsilon \to 0} \int_0^{2\pi} d\varphi \int_0^{\pi} \varepsilon^2 \frac{e^{ik\varepsilon}}{4\pi\varepsilon} \left\{ \frac{\partial \Phi}{\partial \varepsilon}(0, \theta, \phi) \right. \\
&\left. + \left(ik - \frac{1}{\varepsilon} \right) \left(\Phi(0, 0, 0) + \varepsilon \frac{\partial \Phi}{\partial \varepsilon}(0, \theta, \varphi) + \mathcal{O}(\varepsilon^2) \right) \right\} \sin\theta \, d\theta \\
= &\lim_{\varepsilon \to 0} \left\{ \int_0^{2\pi} d\varphi \int_0^{\pi} \frac{e^{ik\varepsilon}}{4\pi} \Phi(0, 0, 0) \sin\theta \, d\theta + \mathcal{O}(\varepsilon) \right\} = \Phi(0, 0, 0) \qquad (3.11)
\end{aligned}$$

So, the distribution $(\Delta + k^2)G$ satisfies

$$\langle (\Delta + k^2)G, \Phi \rangle = \lim_{\varepsilon \to 0} I_\varepsilon = \Phi(0, 0, 0) \qquad (3.12)$$

It is, thus, equal to the Dirac measure located at the coordinate's origin (definition of the distribution δ). A similar proof establishes that the Green's function given by (3.5) satisfies equation (3.3′) in $\mathbb{R}^2$.

3.1.2. Point sources

It has been mentioned that the elementary solution defined in the preceding subsection represents the sound field radiated by a small isotropic physical source. A source which can be represented by a Dirac measure is called a *point isotropic source*.

Experience shows that the small physical sources do not, in general, have isotropic radiation. It is thus necessary to pay attention to sources which radiate most of their energy in given directions. Let us consider the system composed of two harmonic isotropic sources, close to each other and having opposite phases. Assume that they are located at $S^-(x = -\varepsilon, y = 0, z = 0)$ and $S^+(x = +\varepsilon, y = 0, z = 0)$ and that their respective amplitudes are $-1/2\varepsilon$ and $+1/2\varepsilon$. In free space, the corresponding acoustic pressure is the solution of the equation

$$(\Delta + k^2)p_\varepsilon = \frac{1}{2\varepsilon}(\delta_{S^+} - \delta_{S^-})$$

and satisfies the Sommerfeld condition. It is given by

$$p_\varepsilon = -\frac{1}{2\varepsilon}\left(\frac{e^{ikr_+}}{4\pi r_+} - \frac{e^{ikr_-}}{4\pi r_-} \right) \qquad (3.13)$$

with $r_\pm^2 = (x \mp \varepsilon)^2 + y^2 + z^2$. If ε is small enough, each term in this expression is accurately approximated by the first two terms of its Taylor series:

$$\frac{e^{\iota k r_\pm}}{4\pi r_\pm} = \frac{e^{\iota k r}}{4\pi r}\left[1 \mp \varepsilon\left(\iota k - \frac{1}{r}\right)\frac{x}{r}\right] + \mathcal{O}(\varepsilon^2)$$

with $r^2 = x^2 + y^2 + z^2$. This leads to the approximation of p_ε:

$$p_\varepsilon = \left(\iota k - \frac{1}{r}\right)\frac{e^{\iota k r}}{4\pi r}\frac{x}{r} + \mathcal{O}(\varepsilon) \tag{3.14}$$

It appears clearly that p_ε is close to zero in the plane $x = 0$ while it is maximum along the x-axis. For this reason, the limit, for $\varepsilon \to 0$, of p_ε is called the dipole radiation, which is given by

$$\lim_{\varepsilon \to 0} p_\varepsilon = \left(\iota k - \frac{1}{r}\right)\frac{e^{\iota k r}}{4\pi r}\frac{x}{r} = \frac{\partial}{\partial x}\left(-\frac{e^{\iota k r}}{4\pi r}\right) \tag{3.15}$$

The corresponding source is called the dipole oriented along the x-axis; it is represented by $-\partial\delta/\partial x$, the opposite of the derivative, with respect to x, of the Dirac measure. Indeed, one has

$$(\Delta + k^2)\left[\frac{\partial}{\partial x}\left(-\frac{e^{\iota k r}}{4\pi r}\right)\right] = \frac{\partial}{\partial x}\left[(\Delta + k^2)\left(-\frac{e^{\iota k r}}{4\pi r}\right)\right] = \frac{\partial\delta}{\partial x}$$

Obviously, a dipole can have any orientation and can be located anywhere. Let S be the location of a dipole and $\vec{u}$ the unit vector which defines its orientation. The acoustic pressure radiated is given by

$$p(M) = -\vec{u} \cdot \vec{\nabla}_M \frac{e^{\iota k r(S, M)}}{4\pi r(S, M)}$$

where ∇_M means that the gradient of the function is taken with respect to the coordinates of the point M.

The modulus of $p(M)$ is given by

$$|\, p(M)\,| = \left(k^2 - \frac{1}{r^2}\right)^{1/2}\frac{|\,x\,|}{4\pi r^2}$$

For fixed r, it is equal to zero in the x-plane and it has a maximum along the x-axis: the energy flux density is maximum in both directions $x < 0$ and $x > 0$.

The acoustic sources encountered in physics have, very often, much more complicated directivity patterns. The number of directivity lobes (directions in which the energy flux density reaches a maximum) can be rather high. For sources with small dimensions, it is necessary to introduce the notion of a *multipolar point source* which is a linear combination of derivatives of the Dirac measure:

$$\sum_{m,n,q} A_{mnq} \frac{\partial^{mnq} \delta}{\partial x^m \, \partial y^n \, \partial z^q}$$

Each term generates the acoustic field

$$P_{m,n,q} = -A_{mnq} \frac{\partial^{mnq}}{\partial x^m \, \partial y^n \, \partial z^q} \frac{e^{\iota kr}}{4\pi r} \tag{3.16}$$

with $r^2 = x^2 + y^2 + z^2$. When the source is located at a point S, r is the distance between this point and the observation point.

3.1.3. Cylindrical and spherical harmonics

There exists a particular class of multipolar sources which have extremely useful mathematical properties (they will be given at the end of this subsection). Among their various advantages, it must be mentioned that they can easily describe the radiation of any physical source when the observation point is far enough from it.

In $\mathbb{R}^2$, let (r, φ) be the cylindrical coordinates of a point M. The functions $e^{\iota n\varphi}$, $(n = -\infty, \ldots, -1, 0, +1, \ldots, +\infty)$ are called *cylindrical harmonics*. The following set of linearly independent functions, which satisfy a homogeneous Helmholtz equation in the complement of the origin, can be associated with the cylindrical harmonics:

$$\psi_n^{(1)}(M) = H_n^{(1)}(kr)e^{\iota n\varphi}, \qquad \psi_n^{(2)}(M) = H_n^{(2)}(kr)e^{\iota n\varphi} \tag{3.17}$$

where the function $H_n^{(1)}(u) = J_n(u) + \iota Y_n(u)$ (resp. $H_n^{(2)}(u) = J_n(u) - \iota Y_n(u)$) is the Hankel function of order n and of the first (resp. second) kind. Using the expression of the Laplace operator in cylindrical coordinates and the definition of the Hankel functions, it is easily shown that these functions are solutions of the homogeneous Helmholtz equation in any domain which does not contain the origin. Furthermore, $\psi_n^{(1)}(M)$ satisfies the Sommerfeld condition (3.2), while $\psi_n^{(2)}(M)$ satisfies the complex conjugate of (3.2). These elementary proofs are left to the reader. The source distribution associated to either of these two pressure fields is a linear combination of the first n derivatives of the Dirac measure located at $r = 0$.

Similarly in $\mathbb{R}^3$, let (r, θ, φ) be the spherical coordinates of a point M. The functions

$$U_n^m(\theta, \varphi) = P_n^{|m|}(\cos\theta)e^{\iota m\varphi}$$

$$n = -\infty, \ldots, -1, 0, +1, \ldots, +\infty, \qquad m = -\infty, \ldots, -1, 0, +1, \ldots, +\infty$$

where $P_n^{|m|}(\cos\theta)$ is the regular Legendre function of degree n and order $|m|$, are called *spherical harmonics*. To each spherical harmonic, two linearly independent functions are associated:

$$\psi_{nm}^{(1)} = h_n^{(1)}(kr)P_n^{|m|}(\cos\theta)e^{\iota m\varphi}, \qquad \psi_{nm}^{(2)} = h_n^{(2)}(kr)P_n^{|m|}(\cos\theta)e^{\iota m\varphi} \tag{3.18}$$

where $h_n^{(1)}(u) = j_n(u) + \iota y_n(u)$ (resp. $h_n^{(2)}(u) = j_n(u) - \iota y_n(u)$) is the spherical Hankel function of order n and of the first (resp. second) kind. The functions (3.18) satisfy a homogeneous Helmholtz equation in the complement of the origin: the first one satisfies the Sommerfeld condition at infinity (3.2), while the second one satisfies the complex conjugate condition. In the whole space, they satisfy a non-homogeneous equation, the source term of which is a linear combination of the successive derivatives of the Dirac measure up to order n.

If there is no ambiguity, we will refer to $\psi_n^{(1,2)}$ (resp. $\psi_{nm}^{(1,2)}$) as cylindrical (resp. spherical) harmonics. These functions have the following fundamental property:

Theorem 3.1 (Solutions of the homogeneous Helmholtz equation) Let $\phi(M)$ be a function which satisfies the homogeneous Helmholtz equation

$$(\Delta + k^2)\phi(M) = 0$$

inside the domain limited by the two circles (resp. spheres) with radii r_0 and $r_1 > r_0$ and centred at the coordinates origin. Then, for any r such that $r_0 < r < r_1$, $\phi(M)$ can be expanded into a convergent series of cylindrical (resp. spherical) harmonics:

$$\phi(M) = \sum_{n=-\infty}^{\infty} a_n H_n^{(1)}(kr)\mathrm{e}^{\iota n\varphi} + \sum_{n=-\infty}^{\infty} b_n H_n^{(2)}(kr)\mathrm{e}^{\iota n\varphi}, \qquad M \in \mathbb{R}^2 \tag{3.19}$$

$$\phi(M) = \sum_{n=0}^{\infty} h_n^{(1)}(kr) \sum_{m=-n}^{n} a_m^n P_n^{|m|}(\cos\theta)\mathrm{e}^{\iota m\varphi}$$

$$+ \sum_{n=0}^{\infty} h_n^{(2)}(kr) \sum_{m=-n}^{n} b_m^n P_n^{|m|}(\cos\theta)\mathrm{e}^{\iota m\varphi}, \qquad M \in \mathbb{R}^3 \tag{3.19$'$}$$

For fixed r, the expansion (3.19) is a Fourier series with respect to the variable φ; the expansion (3.19$'$) behaves like a Fourier series with respect to the two variables φ and θ. Thus, these two expansions are $\mathbf{L}^2$-convergent for fixed r: this means that the speed of convergence depends on the value of the radial variable.

Throughout this book, the time dependence of a harmonic acoustic field is assumed to be $\mathrm{e}^{-\iota\omega t}$. Thus the functions $H_n^{(1)}(kr)\mathrm{e}^{\iota n\varphi}$ and $h_n^{(1)}(kr)P_n^{|m|}(\cos\theta)\mathrm{e}^{\iota m\varphi}$ represent elementary waves which carry energy towards infinity (outgoing waves). The functions $H_n^{(2)}(kr)\mathrm{e}^{\iota n\varphi}$ and $h_n^{(2)}(kr)P_n^{|m|}(\cos\theta)\mathrm{e}^{\iota m\varphi}$ represent elementary waves which carry energy from infinity (incoming waves). These two types of waves are necessary to represent the pressure field inside a bounded domain whose boundaries reflect a part of the energy. If all the sources are inside the domain $r < r_0$, the total field can be split into two components: a series of outgoing elementary waves with amplitudes $(a_n - b_n)$ which is the field that the sources produce in free space; and a series of regular terms (Bessel functions $J_n(kr)$ or $j_n(kr)$) with amplitudes $2b_n$ which represents the sound field reflected by the boundaries of the propagation domain.

If the propagation domain is unbounded, and does not contain any source or obstacle outside the domain $r < r_0$, the former expansions contain Hankel functions of the first kind only and are valid for $r_0 < r < \infty$.

Notation. In what follows, the symbols $H_n(u)$ and $h_n(u)$ stand for the Hankel functions of the first kind; only in case of a possible confusion are the superscripts used.

3.1.4. Surface sources

Sounds and noise are often generated by vibrating structures: the exterior boundary of a machine, the table of a stringed musical instrument, etc. Thus, it appears necessary to give a mathematical description of such sources. An intuitive method is to try to extend to a surface the notion of monopole and dipole point sources. Denote by σ a radiating surface which is assumed to be composed of N elementary surfaces σ_j with areas $\tilde{\sigma}_j$ and centre of gravity P_j. A point isotropic harmonic source with amplitude $\mu_j \tilde{\sigma}_j$ is located at each point P_j. The corresponding acoustic field is given by

$$\phi(M) = -\sum_{j=1}^{N} \mu_j \frac{e^{\iota k r(M, P_j)}}{4\pi r(M, P_j)} \tilde{\sigma}_j$$

A natural limit process appears clearly: an isotropic point source is associated to the infinitesimal surface element $d\sigma(P)$, its amplitude being $\mu(P)\, d\sigma(P)$. The function $\mu(P)$ is called the source surface density. The corresponding acoustic radiation is given by the integral

$$\psi_1(M) = -\int_{\sigma} \mu(P) \frac{e^{\iota k r(M, P)}}{4\pi r(M, P)} d\sigma(P), \qquad M \notin \sigma \qquad (3.20)$$

The function $\psi_1(M)$ is called the *simple layer potential* (see chapter 2, section 6.2); the function $\mu(P)$ is called the *simple layer density*. The definition of expression (3.20) is meaningful if both the surface σ and the function $\mu(P)$ are sufficiently smooth. In practice, this does not induce any physical restriction: the required smoothness is deduced from the necessity to attach a finite radiated energy to such surface sources.

Assume that σ is a piecewise smooth surface (piecewise smooth implies that σ can have edges). Thus, a unit normal vector $\vec{n}(P)$ can be defined almost everywhere (in practice, everywhere but along the edges). Using the same set of surface elements σ_j, a dipole point source with amplitude $\nu_j \tilde{\sigma}_j$ is located at each point P_j. This system radiates the following pressure:

$$\phi(M) = \sum_{j=1}^{N} \nu_j \vec{n}(P_j) \cdot \nabla_{P_j} \frac{e^{\iota k r(M, P_j)}}{4\pi r(M, P_j)} \tilde{\sigma}_j$$

(∇_{P_j} means that the derivatives are taken with respect to the coordinates of P_j.) Using the same limit process, we obtain the definition of the *double layer potential* (see chapter 2, section 6.2):

$$\psi_2(M) = \int_{\sigma} \nu(P)\partial_{n(P)} \frac{e^{\iota k r(M, P)}}{4\pi r(M, P)} d\sigma(P), \qquad M \notin \sigma \qquad (3.21)$$

in which expression the normal derivative of the kernel is defined by

$$\partial_{n(P)} \frac{e^{ikr(M,P)}}{4\pi r(M,P)} = \vec{n}(P) \cdot \nabla_P \frac{e^{ikr(M,P)}}{4\pi r(M,P)}$$

Remark. If two-dimensional problems are looked at, the notion of surface sources is replaced by that of curve sources. Simple and double layer potentials are defined in the same way by using the two-dimensional elementary kernel. The properties of the three-dimensional layer potentials that are presented in the next subsection remain unchanged in $\mathbb{R}^2$.

3.1.5. Properties of the simple and double layer potentials

Let us recall a classical definition. A surface σ splits the neighbourhood Ω of any regular point P (that is a point where a normal unit vector is defined) into two domains: Ω^+ which contains the unit normal vector and Ω^- which does not contain it. The side of σ in contact with Ω^+ is called the positive side, the other one being called the negative side of the surface.

The source which generates a simple layer potential will be denoted by $\mu \otimes \delta_\sigma$, where δ_σ is the Dirac measure attached to the surface σ and $\otimes$ stands for the tensor product. The source which generates the double layer potential will be denoted by $\nu \otimes \delta_\sigma'$. These notations are defined by:

$$\langle \mu \otimes \delta_\sigma, f \rangle \equiv \int_\sigma \mu(P) f(P) \, d\sigma(P)$$

$$\langle \nu \otimes \delta_\sigma', f \rangle \equiv - \int_\sigma \nu(P) \partial_{n(P)} f(P) \, d\sigma(P)$$

where f is any indefinitely differentiable function with compact support. The functions $\psi_1(M)$ and $\psi_2(M)$ satisfy non-homogeneous Helmholtz equations:

$$(\Delta + k^2)\psi_1 = \mu \otimes \delta_\sigma \tag{3.22}$$

$$(\Delta + k^2)\psi_2 = \nu \otimes \delta_\sigma' \tag{3.23}$$

They are defined in the whole space. It can be shown that they are indefinitely differentiable in any domain which does not intersect the surface σ. As will be shown, they have a discontinuous behaviour across σ.

Let ϕ be a function which is twice differentiable outside a given surface σ. The Laplacian, in the distributions sense, of this function is shown to be

$$\Delta\phi = \{\Delta\phi\} + [\mathrm{Tr}^+ \phi - \mathrm{Tr}^- \phi] \otimes \delta_\sigma' + [\mathrm{Tr}^+ \partial_n\phi - \mathrm{Tr}^- \partial_n\phi] \otimes \delta_\sigma$$

In this expression, $\{\Delta\phi\}$ is the Laplacian of ϕ in the classical sense which is defined outside of σ only; the symbol $\mathrm{Tr}^+ \phi$ (resp. $\mathrm{Tr}^- \phi$) is the limit of $\phi(M)$ when the point M reaches σ by its positive (resp. negative) side. Similarly, one defines the symbols $\mathrm{Tr}^+ \, \partial_n\phi$ and $\mathrm{Tr}^- \, \partial_n\phi$ by

$$\mathrm{Tr}^\pm \, \partial_n\phi(P) = \lim_{M \in \Omega^\pm \to P \in \sigma} \vec{n}(P) \cdot \nabla\phi(M)$$

Outside σ, the functions ψ_1 and ψ_2 satisfy a homogeneous Helmholtz equation, that is

$$\{\Delta\psi_{1,2}\} + k^2\psi_{1,2} = 0$$

Thus, they can have a singularity along the layer support only. Using the definition of the Laplacian in the distributions sense, we have

$$(\Delta + k^2)\psi_1 = [\mathrm{Tr}^+ \, \psi_1 - \mathrm{Tr}^- \, \psi_1] \otimes \delta_\sigma' + [\mathrm{Tr}^+ \, \partial_n\psi_1 - \mathrm{Tr}^- \, \partial_n\psi_1] \otimes \delta_\sigma \qquad (3.24)$$

$$(\Delta + k^2)\psi_2 = [\mathrm{Tr}^+ \, \psi_2 - \mathrm{Tr}^- \, \psi_2] \otimes \delta_\sigma' + [\mathrm{Tr}^+ \, \partial_n\psi_2 - \mathrm{Tr}^- \, \partial_n\psi_2] \otimes \delta_\sigma \qquad (3.25)$$

By comparing these equations with (3.22) and (3.23), we can state the following result.

Theorem 3.2 (Discontinuities of the simple and double layer potentials)

1. The simple layer potential is a continuous function. Its normal derivative is discontinuous across the layer support σ and is equal to the layer density μ:

$$\mathrm{Tr}^+ \, \psi_1 - \mathrm{Tr}^- \, \psi_1 = 0, \qquad \mathrm{Tr}^+ \, \partial_n\psi_1 - \mathrm{Tr}^- \, \partial_n\psi_1 = \mu$$

2. The double layer potential has a continuous gradient. It has a discontinuity across the layer support σ which is equal to the layer density ν:

$$\mathrm{Tr}^+ \, \psi_2 - \mathrm{Tr}^- \, \psi_2 = \nu, \qquad \mathrm{Tr}^+ \, \partial_n\psi_2 - \mathrm{Tr}^- \, \partial_n\psi_2 = 0$$

It is now necessary to give the expressions of these various limits. This is not a trivial task: due to the singularity of the Green's kernels given by (3.5) and (3.6), the values on σ of the layer potentials and of their normal derivatives must be defined by limit procedures.

On the layer support σ, the singularity of the Green's kernel is integrable in the Riemann sense. Thus the value of a simple layer potential is given by:

$$\mathrm{Tr}^+ \, \psi_1(P) = \mathrm{Tr}^- \, \psi_1(P) = \mathrm{Tr}\psi_1(P) = \int_\sigma \mu(P')G(P, P') \, d\sigma(P') \qquad (3.26)$$

$$\text{with} \begin{cases} G(P, P') = -\dfrac{\imath}{4} H_0[kr(P, P')] & \text{in } \mathbb{R}^2 \\[2ex] G(P, P') = -\dfrac{e^{\imath kr(P, P')}}{4\pi r(P, P')} & \text{in } \mathbb{R}^3 \end{cases}$$

The expressions of the double layer potential and of the normal gradient of the simple layer potential involve the normal derivative of the kernel which can present a stronger singularity. Thus, the calculation of the values on σ of these functions requires a little care and attention must be paid to the fact that the limits on the positive and negative sides of the layer support are different.

We first consider the normal gradient of the simple layer potential. Let $\mathfrak{D}$ be the straight line which is normal to σ at a point P, and M and M' be two points on $\mathfrak{D}$ symmetrical to each other with respect to P. Let $\phi(M)$ be the function defined by

$$\phi(M) = \int_\sigma \mu(P')\vec{n}(P) \cdot [\nabla_M G(M, P') + \nabla_{M'} G(M', P')]\, d\sigma(P')$$

When M crosses over the surface σ from the negative side to the positive one, the point M' crosses it over from the positive side to the negative one. The first term in $\phi(M)$ has a jump equal to $+\mu$, while the second term has a jump equal to $-\mu$. So the function $\phi(M)$ is continuous, that is

$$\mathrm{Tr}^+\ \phi(P) = \mathrm{Tr}^-\ \phi(P) = \mathrm{Tr}\ \phi(P)$$

But the value of $\phi(M)$ on σ is the sum $\mathrm{Tr}^+\ \partial_n\psi(M) + \mathrm{Tr}^-\ \partial_n\psi(M)$. It is shown that $\mathrm{Tr}\ \phi(M)$ is expressed by a Cauchy principal value:

$$\mathrm{Tr}\ \phi(P) = 2\int_\sigma \mu(P')\partial_{n(P)}G(P, P')\, d\sigma(P')$$

This leads to the following result:

$$\begin{aligned} \mathrm{Tr}^+\ \partial_n\psi_1(P) &= +\frac{\mu(P)}{2} + \int_\sigma \mu(P')\partial_{n(P)}G(P, P')\, d\sigma(P') \\ \mathrm{Tr}^-\ \partial_n\psi_1(P) &= -\frac{\mu(P)}{2} + \int_\sigma \mu(P')\partial_{n(P)}G(P, P')\, d\sigma(P') \end{aligned} \tag{3.27}$$

In a similar way, the values on σ of a double layer potential are shown to be given by

$$\begin{aligned} \mathrm{Tr}^+\ \psi_2(P) &= +\frac{\nu(P)}{2} - \int_\sigma \nu(P')\partial_{n(P')}G(P, P')\, d\sigma(P') \\ \mathrm{Tr}^-\ \psi_2(P) &= -\frac{\nu(P)}{2} - \int_\sigma \nu(P')\partial_{n(P')}G(P, P')\, d\sigma(P') \end{aligned} \tag{3.28}$$

The last expression to be established is that of the value on σ of the normal derivative of a double layer potential. The normal gradient of a double layer potential involves a double derivative of the Green's kernel: it is easily shown that this double derivative is not integrable on σ. Nevertheless, it can be established that the limit for $M \in \Omega \to P \in \sigma$ (Ω is the propagation domain) exists, that is

$$\mathrm{Tr}\ \partial_n\psi_2(P) = -\lim_{M \in \Omega \to P \in \sigma} \vec{n}(P) \cdot \int_\sigma \nabla_M[\partial_{n(P')}G(M, P')]\, d\sigma(P') \tag{3.29}$$

If σ is a closed curve (in $\mathbb{R}^2$) or a closed surface (in $\mathbb{R}^3$), this function can be expressed by convergent integrals. For simplicity we will show it in $\mathbb{R}^2$. Let $r(M, P')$ be the distance between a point M outside σ and a point P on σ. The Hankel function of the first kind $H_0(kr)$ has a logarithmic singularity given by

$$H_0(kr) = \frac{2\iota}{\pi} \ln r + \text{regular function}$$

The double layer potential can thus be written

$$\begin{aligned}\psi_2(M) = \nu(P) \int_\sigma \partial_{n(P')} \frac{\ln r(M, P')}{2\pi} \, d\sigma(P') \\ + \int_\sigma \partial_{n(P')} \frac{\ln r(M, p')}{2\pi} [\nu(P') - \nu(P)] \, d\sigma(P') \\ + \int_\sigma \nu(P')\partial_{n(P')} \left[G(M, P') - \frac{\ln r(M, P')}{2\pi} \right] d\sigma(P')\end{aligned} \tag{3.30}$$

In this expression, P is the point which M will tend to. The kernel which appears in the third integral has such a regularity that the gradient of this term is defined everywhere by a convergent integral, in particular for $M = P$. In the second integral, the function $\nu(P') - \nu(P)$ is zero when P coincides with P': it is shown that, if $\nu(P)$ is sufficiently regular (at least continuously differentiable) the value on σ of the normal derivative of this integral is expressed by a Cauchy principal value. Let us now examine the first integral. One has

$$\int_\sigma \partial_{n(P')} \frac{\ln r(M, P')}{2\pi} \, d\sigma(P') = \int_\sigma \frac{\cos(\vec{r}, \vec{n})}{r(M, P')} \, d\sigma(P') \tag{3.31}$$

where $(\vec{r}, \vec{n})$ is the angle between the vector $\vec{r} = \overrightarrow{P'M}$ and the unit vector $\vec{n}$ normal to σ at P'. The quantity $\cos(\vec{r}, \vec{n}) \, d\sigma(P')/r$ is the elementary angle under which the elementary arc $d\sigma(P')$ is seen from the point M. As a consequence, the integral (3.31) is the angle under which the closed contour σ is seen from M: it is equal to 2π if M lies in the domain inside this contour, and to 0 if M lies inside the outer domain. Its gradient is thus identically zero. The final result is

$$\begin{aligned}\text{Tr } \partial_n \psi_2(P) = \int_\sigma \partial_{n(P)} \partial_{n(P')} \tilde{G}(P, P')[\nu(P') - \nu(P)] \, d\sigma(P') \\ + \int_\sigma \nu(P')\partial_{n(P)}\partial_{n(P')}[G(P, P') - \tilde{G}(P, P')] \, d\sigma(P')\end{aligned} \tag{3.32}$$

$$\text{with} \quad \tilde{G}(P, P') = \frac{\ln r(P, P')}{2\pi}$$

In $\mathbb{R}^3$, the same result is valid for a closed surface σ, the functions G and $\tilde{G}$ being given by

$$G(P, P') = -\frac{e^{\iota kr(M,M')}}{4\pi r(M, M')}, \qquad \tilde{G}(M, M') = -\frac{1}{4\pi r(M, M')}$$

Expression (3.32) is interesting for numerical purposes: it provides an analytical expression for the finite part of the integral. Roughly speaking, the other methods consider a geometrical approximation of a boundary element and a local approximation of the unknown function; then use is made of the necessary limit procedure to get an approximation – and not an exact value – of the finite part of the integral; furthermore, this always requires some care.

3.2. Green's Representation of the Solution of Linear Acoustics Boundary Value Problems

In chapter 2, various analytical methods were presented which provide either exact or approximate representations of the solutions of boundary value problems which describe the sound field due to a harmonic ($e^{-\iota\omega t}$) excitation. The disadvantage of these methods is that they can be used for simple geometrical configurations only. This chapter deals with the most general representations of a harmonic sound field: the sound pressure is sought as the sum of the incident pressure (the pressure field that the sources generate in free space) and a pressure field due to fictitious sources, the support of which is the boundary of the propagation domain (usually called the reflected or diffracted or scattered field). These sources are chosen so that the total sound pressure satisfies the boundary conditions which are imposed.

3.2.1. Statement of the problem

Let Ω be a domain of $\mathbb{R}^n$ ($n = 1$, 2 or 3), with a regular boundary σ. It is assumed that a unit normal vector $\vec{n}$, pointing out to the exterior of Ω, can be defined almost everywhere on σ. If Ω is bounded, we have an interior problem; if Ω extends up to infinity, we have an exterior problem. Let f be the distribution which represents the energy harmonic sources. The acoustic pressure p is the solution of the non-homogeneous equation:

$$(\Delta + k^2)p(M) = f(M), \qquad M \in \Omega \tag{3.33}$$

where k is, as usual, the wavenumber. The boundary σ is assumed to satisfy a local boundary condition:

$$\alpha \text{Tr } \partial_n p(M) + \beta \text{Tr } p(M) = 0, \qquad M \in \sigma \tag{3.34}$$

Very often, the symbol 'Tr' can be omitted. But, as has been seen already, the sound field can be represented by surface integrals which are discontinuous or have a discontinuous normal gradient or involve, on σ, a finite part of the integral: in these cases, the symbol 'Tr' is absolutely necessary to indicate that the value on σ of the function is obtained as the limit of a Riemann integral. The possible existence of a solution of this boundary value problem is beyond the scope of this book; among the best textbooks dealing with boundary value problems of classical physics and in which such a proof is given, we can mention *Methods of Mathematical Physics* by R. Courant and D. Hilbert [7].

Some remarks must be made on the mathematical requirements on which mathematical physics is based. The basic hypothesis which is always made in physics is that any system has a finite energy density and its behaviour is based on the Hamilton principle (energy conservation). In acoustics, this implies that the sound pressure and its gradient are described by functions which are (at least locally) square integrable. Furthermore, an energy flux density across any elementary surface of the propagation domain boundary must be defined. The consequences are: (a) the value on the boundary of the function which represents the sound pressure inside the propagation domain can be different from the value of the sound pressure; (b) the value of the pressure field (or of its gradient) on the boundary of the propagation domain is obtained as a limit of a function defined inside the propagation domain; (c) the limits on the boundary of the functions representing the pressure field and its gradient need to be (at least locally) square integrable functions. These remarks justify the necessity to use the symbol 'Tr' as a reminder of the mathematical requirements which must be respected to get a description of the physical phenomena.

If $\alpha = 1$ and $\beta = 0$, we are left with the Neumann problem (cancellation of the normal component of the particle velocity); $\alpha = 0$ and $\beta = 1$ describe the Dirichlet problem (cancellation of the sound pressure). Finally, let us recall that, in most situations, an absorbing boundary can be characterized by a specific normal impedance ζ which is defined by

$$\zeta = -\iota k \frac{\mathrm{Tr}\, p}{\mathrm{Tr}\, \partial_n p} \qquad \left(\frac{\beta}{\alpha} = \frac{\iota k}{\zeta}\right)$$

This corresponds to $\alpha = 1$ and $\beta \neq 0$. Such a boundary condition is generally called the Robin condition.

In what follows, it is assumed that α and β can be piecewise continuous functions: this allows the boundary σ to be made of different materials (for example, one part is rather hard, another one being highly absorbent). To ensure the uniqueness of the solution for any real frequency when the domain Ω is unbounded, the principle of energy conservation must be respected: this is done by adding a Sommerfeld condition at infinity (which has already been given) or by using either the principle of limit absorption or the principle of limit amplitude.

Limit absorption principle. Let ε be a positive parameter and p_ε be the unique bounded solution corresponding to a wavenumber $k_\varepsilon = k(1 + \iota\varepsilon)$. For a real wavenumber k, the solution which satisfies the energy conservation principle is the limit, for $\varepsilon \longrightarrow 0$, of p_ε.

Limit amplitude principle. Let us consider the unique solution $P_\omega(M; t)$ of the wave equation which corresponds to a harmonic source with angular frequency ω starting at a time $t = 0$. The solution of the Helmholtz equation which satisfies the energy conservation principle is the limit for $t \to \infty$ of $P_\omega(M; t)$.

3.2.2. Green's representation of the acoustic pressure

Let $\tilde{p}$ be the function which is equal to p in Ω and to zero in its complement $\complement\Omega$. Out of the sources support, the possible discontinuities of $\tilde{p}$ and of its gradient are located on σ. The equation satisfied (in the distributions sense) by $\tilde{p}$ is

$$(\Delta + k^2)\tilde{p} = f + (\mathrm{Tr}^+\ \tilde{p} - \mathrm{Tr}^-\ \tilde{p}) \otimes \delta'_\sigma + (\mathrm{Tr}^+\ \partial_n\tilde{p} - \mathrm{Tr}^-\ \partial_n\tilde{p}) \otimes \delta_\sigma$$

This equation is valid in the whole space $\mathbb{R}^n$. But $\tilde{p}$ is identically zero in $\complement\Omega$; we are thus left with

$$(\Delta + k^2)\tilde{p} = f - \mathrm{Tr}\ p \otimes \delta'_\sigma - \mathrm{Tr}\ \partial_n p \otimes \delta_\sigma \tag{3.35}$$

where $\mathrm{Tr}^-\ \tilde{p}$ (resp. $\mathrm{Tr}^-\ \partial_n\tilde{p}$) has been replaced by $\mathrm{Tr}\ p$ (resp. $\mathrm{Tr}\ \partial_n p$). Let G be the elementary solution of the Helmholtz equation which satisfies the required Sommerfeld condition. Then, the solution of equation (3.35) is expressed by a space convolution product. In Ω, we can write

$$p = G * (f - \mathrm{Tr}\ p \otimes \delta'_\sigma - \mathrm{Tr}\ \partial_n p \otimes \delta_\sigma)$$

where $*$ is the convolution product symbol. Elementary solutions and Green's functions have the following property:

$$G * \phi(M) = \langle G(M, P), \phi(P)\rangle = \int_{\mathbb{R}^n} G(M, P)\phi(P)\ d\Omega(P)$$

(the last form is valid as far as the function $G(M, P)\phi(P)$ is integrable). Let recall the expressions of the Green's function used here:

$$G(M, P) = \frac{e^{\iota k r(M, P)}}{2\iota k} \qquad \text{in } \mathbb{R}$$

$$= -\frac{\iota}{4} H_0(kr(M, P)) \qquad \text{in } \mathbb{R}^2$$

$$= -\frac{e^{\iota k r(M, P)}}{4\pi r(M, P)} \qquad \text{in } \mathbb{R}^3$$

where $r(M, P)$ is the distance between the points M and P. Let $p_0(M)$ be the incident field, that is

$$p_0(M) = G * f(M)$$

The total acoustic pressure can be represented by the following expression which is called the Green's representation:

$$p(M) = p_0(M) - \int_\sigma [\text{Tr}\ \partial_{n'} p(P')G(M, P') - \text{Tr}\ p(P')\partial_{n'} G(M, P')]\ d\sigma(P') \qquad (3.36)$$

($\partial_{n'}$ means that the derivative is taken with respect to the variable P'). This expression is valid in $\mathbb{R}^2$ and $\mathbb{R}^3$. In one dimension, the boundary reduces to one or two points; if, for example, $\Omega = x \in]a, +\infty[$, Green's formula gives

$$p(x) = p_0(x) + \partial_x p(a)G(x-a) - p(a)\partial_a G(x-a) \qquad (3.36^*)$$

In the Green's representation, the diffracted field is the sum of a simple layer potential with density $-\text{Tr}\ \partial_{n'} p(P')$ and a double layer potential with density $-\text{Tr}\ p(P')$. These two functions are not independent of each other: they are related through the boundary condition.

Expression (3.36) simplifies for the following two boundary conditions:

1. Dirichlet problem ($\text{Tr}\ p = 0$):

$$p(M) = p_0(M) - \int_\sigma \text{Tr}\ \partial_{n'} p(P')G(M, P')\ d\sigma(P') \qquad (3.36')$$

2. Neumann problem ($\text{Tr}\ \partial_n p = 0$):

$$p(M) = p_0(M) + \int_\sigma \text{Tr}\ p(P')\partial_{n'} G(M, P')\ d\sigma(P') \qquad (3.36'')$$

The function p is known once the layer densities are known.

3.2.3. Transmission problems and non-local boundary condition

Energy is absorbed by a boundary because the incident wave generates a motion of the boundary material. Due to this motion, the boundary gives back a part of its vibratory energy to the fluid. But it must be noted that the motion of the fluid/boundary material interface depends on the motion of the whole volume of material which constitutes the frontier of the fluid domain. If, inside the boundary, the vibrations can propagate without too much damping, it is not possible to use a local boundary condition: a non-local boundary condition is required.

Let us consider a simple example. The propagation domain Ω is bounded and contains a fluid with density ρ and sound speed c. Its complement $\complement\Omega$ is occupied by an isotropic homogeneous porous medium. For a harmonic excitation with angular frequency ω, the acoustic pressure in the fluid satisfies a Helmholtz equation:

$$(\Delta + k^2)p(M) = f(M), \qquad M \in \Omega, \qquad \text{with } k^2 = \frac{\omega^2}{c^2} \qquad (3.37)$$

A porous medium is composed of a solid matrix with pores which are small compared to the wavelength and filled up with a fluid. A detailed description of such a complex system is quite impossible and totally useless. To describe the propagation of a wave inside a porous medium, various approximations are made to replace this non-homogeneous system by an equivalent homogeneous medium. The simplest models describe a porous medium as a homogeneous fluid with a complex density ρ' and a complex sound speed c', which are both frequency dependent. Thus the acoustic pressure inside $\complement\Omega$ is the solution of a Helmholtz equation:

$$(\Delta + k'^2)p'(M) = 0, \qquad M \in \Omega, \qquad \text{with } k'^2 = \frac{\omega^2}{c'^2} \tag{3.38}$$

Furthermore, it is shown that the energy transfer between the external fluid and the porous medium is governed by the continuity, along the common boundary of the two media, of the acoustic pressures and of the normal accelerations:

$$\text{Tr } p(M) = \text{Tr } p'(M), \qquad \frac{\text{Tr } \partial_n p(M)}{\rho} = \frac{\text{Tr } \partial_n p'(M)}{\rho'}, \qquad M \in \sigma \tag{3.39}$$

We are going to show that the system (3.37, 3.38, 3.39) can be replaced by a boundary value problem for p only, with a non-local boundary condition. Let $G'(M, P)$ be the Green's kernel of equation (3.38) which is bounded at infinity: it represents the radiation of a point isotropic source located at P inside the porous medium occupying the whole space. Using the result of the former subsection, the Green's representation of p' is

$$p'(M) = \int_\sigma [\text{Tr } \partial_{n'} p'(P')G'(M, P') - \text{Tr } p'(P')\partial_{n'} G(M, P')] \, d\sigma(P')$$

Accounting for the continuity conditions, this last expression becomes

$$p'(M) = \int_\sigma \left[\frac{\rho'}{\rho} \text{Tr } \partial_{n'} p(P')G'(M, P') - \text{Tr } p(P')\partial_{n'} G(M, P') \right] d\sigma(P')$$

Let us now take the value on σ of p' and of its normal derivative and use the continuity conditions; one gets two expressions which the pressure p must satisfy on the boundary of the propagation domain:

$$\frac{\text{Tr } p(P)}{2} - \int_\sigma \left[\frac{\rho'}{\rho} \text{Tr } \partial_{n'} p(P')G'(P, P') - \text{Tr } p(P')\partial_{n'} G(P, P') \right] d\sigma(P') = 0, \qquad P \in \sigma \tag{3.40}$$

$$\frac{\rho'}{\rho}\frac{\text{Tr }\partial_n p(P)}{2} - \text{Tr }\partial_n\left\{\int_\sigma\left[\frac{\rho'}{\rho}\text{ Tr }\partial_{n'}p(P')G'(P,P')\right.\right.$$
$$\left.\left. - \text{Tr }p(P')\partial_{n'}G(P,P')\right] d\sigma(P')\right\} = 0, \qquad P\in\sigma \quad (3.40')$$

It can be proved that equation (3.37) with either of the two boundary conditions (3.40) or (3.40′) has a unique solution for any real angular frequency ω (the resonance frequencies of the physical system have a non-zero imaginary part due to the energy loss into the porous medium).

3.3. Representation of a Diffracted Field by a Layer Potential

The Green's representation is a particular form of integral representation of the acoustic field diffracted by a boundary σ: indeed, it consists in building a function which is defined in the whole space, which is equal to the diffracted field inside the propagation domain and which is zero outside. But it is not necessary at all that an integral representation of the diffracted field fulfil this last condition: any representation which is defined in the whole space and is equal to the diffracted field inside the propagation domain is quite convenient.

3.3.1. Representation of the field diffracted by a closed surface in $\mathbb{R}^3$ or a closed curve in $\mathbb{R}^2$

Let consider a domain Ω, with boundary σ which is either a bounded domain or the exterior of a 'thick' bounded domain: by 'thick' we exclude the acoustic thin screens which are modelled by surfaces in $\mathbb{R}^3$ or curves in $\mathbb{R}^2$ (this particular case is examined in the next subsection). It is *a priori* possible to consider three types of integral representations of the sound field reflected by σ:

(a) Simple layer potential:

$$p(M) = p_0(M) + \int_\sigma \mu(P)G(M,P)\,d\sigma(P), \qquad M\in\Omega \qquad (3.41)$$

(b) Double layer potential:

$$p(M) = p_0(M) - \int_\sigma \mu(P)\partial_{n(P)}G(M,P)\,d\sigma(P), \qquad M\in\Omega \qquad (3.42)$$

(c) Hybrid layer potential:

$$p(M) = p_0(M) + \int_\sigma \mu(P)[-\partial_{n(P)}G(M,P) + \gamma G(M,P)]\,d\sigma(P),$$
$$M\in\Omega \qquad (3.43)$$

In this last expression, γ is *a priori* an arbitrary constant. In general, the functions defined by (3.41, 3.42, 3.43) are not identically zero outside the propagation domain. The only condition which is required is that the layer density μ is such that the total field satisfies the boundary condition imposed on σ.

It must be recalled that these representations have discontinuities across σ: the first one is continuous with a discontinuous normal gradient; the second one is discontinuous with a continuous normal gradient; the third one is discontinuous together with its normal gradient.

3.3.2. Diffraction by an infinitely thin screen

There are many real life cases of diffracting obstacles whose thickness is small compared to the other dimensions and to the wavelength (acoustic screens along roads or train tracks, diffracting structures in concert halls, etc.) This leads to modelling of such structures as infinitely thin obstacles: the diffracted effect is represented by a layer potential which must have a discontinuity or a discontinuous normal gradient (or both). Let us give a formal justification of such a model. For simplicity, let us consider a two-dimensional obstacle defined as follows. Let σ_0 be a curve segment parametrized by a curvilinear abscissa $-a < t < +a$; and let $\vec{n}(t)$ be the unit vector normal to σ_0 at point t. This unit vector enables us to define a positive side of σ_0 and a negative side. Let $h(t)$ be a positive function which is zero at $t = -a$ and $t = +a$ and ε a positive parameter. They define two curves σ^+ and σ^- by (see Fig. 3.1):

$$\sigma^+ = \text{set of points } P^+(t) \text{ such that } \overrightarrow{P^+P} = \varepsilon h(t)\vec{n}(t)$$

$$\sigma^- = \text{set of points } P^-(t) \text{ such that } \overrightarrow{P^-P} = -\varepsilon h(t)\vec{n}(t)$$

It is assumed that $\varepsilon h(t)$ is small compared to the total length of the curve arc σ_0 and sufficiently regular so that an exterior unit vector $\vec{\nu}$ normal to $\sigma = \sigma^+ \cup \sigma^-$ can be defined everywhere. Let Ω_σ be the exterior of the bounded domain limited by σ and

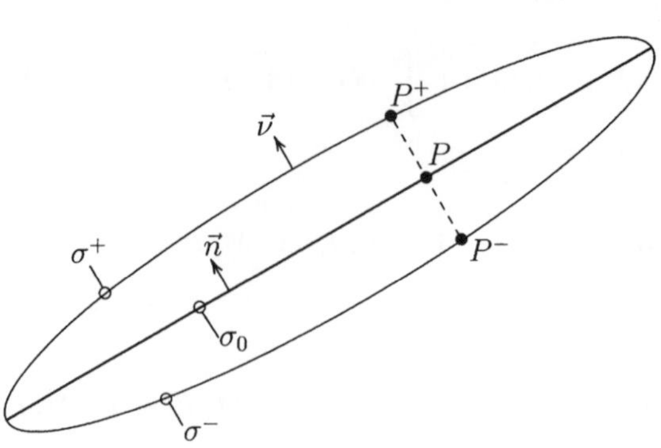

Fig. 3.1. The infinitely thin screen as the limit of a screen with small thickness.

consider the Neumann problem

$$\begin{aligned} &(\Delta + k^2)p_\sigma(M) = f(M), \qquad M \in \Omega_\sigma \\ &\mathrm{Tr}\ \partial_\nu p_\sigma(M) = 0, \qquad M \in \sigma \\ &\text{Sommerfeld condition} \end{aligned} \tag{3.44}$$

The solution p_σ is unique. Let us see if it has a limit when the parameter ε tends to zero. If such a limit $p(M)$ exists, it satisfies the following boundary value problem:

$$\begin{aligned} &(\Delta + k^2)p(M) = f(M), \qquad M \in \Omega = \complement\sigma_0 \\ &\mathrm{Tr}\ \partial_n p = 0, \qquad M \in \sigma_0 \\ &\text{Sommerfeld condition} \end{aligned} \tag{3.45}$$

We just present here a formal proof, since a rigorous proof requires too much functional analysis. It can be immediately remarked that p must have a continuous gradient: this implies that it can be represented by a double layer potential only.

The Green's representation of p_σ is

$$\begin{aligned} p_\sigma(M) &= p_0(M) - \int_\sigma \mathrm{Tr}\ p_\sigma(P)\partial_{\nu(P)}G(M, P)\ d\sigma(P) \\ &= p_0(M) - \int_{\sigma^-} \mathrm{Tr}\ p_\sigma(P^-)\partial_{\nu(P^-)}G(M, P^-)\ d\sigma(P^-) \\ &\quad - \int_{\sigma^+} \mathrm{Tr}\ p_\sigma(P^+)\partial_{\nu(P^+)}G(M, P^+)\ d\sigma(P^+) \end{aligned} \tag{3.46}$$

For $\varepsilon \to 0$, the vector $\vec{\nu}(P^+)$ tends to $\vec{n}(P)$ while $\vec{\nu}(P^-)$ tends to $-\vec{n}(P)$. Using a Taylor series of the kernel $G(M, P^\pm)$ for $\varepsilon h/a \ll 1$, it is easily seen that the sum of the integrals over σ^- and σ^+ in (3.46) is close to

$$-\int_{\sigma_0} [\mathrm{Tr}\ p_\sigma(P^+) - \mathrm{Tr}\ p_\sigma(P^-)]\partial_{n(P)}G(M, P)\ d\sigma_0(P)$$

It is proved that the functions $\mathrm{Tr}\ p_\sigma(P^+)$ and $\mathrm{Tr}\ p_\sigma(P^-)$ have different limits $\mathrm{Tr}^+ p(P)$ and $\mathrm{Tr}^- p(P)$ respectively, and that the solution $p(M)$ (3.45) exists and is given by

$$p(M) = p_0(M) - \int_{\sigma_0} [\mathrm{Tr}^+\ p(P) - \mathrm{Tr}^-\ p(P)]\partial_{n(P)}G(M, P)\ d\sigma_0(P) \tag{3.47}$$

This shows that a thin screen can be geometrically approximated by a curve (diffracting obstacle with zero thickness) and that the corresponding diffracted field is described by a double layer potential, the layer support being this curve. Another important result is the behaviour of the solution of equation (3.45). It is shown that, close to the edges $t = \pm a$ of the screen, the diffracted pressure is zero but that its tangent gradient is singular. More precisely, one has

$$p(M) - p_0(M) = r \ln r + \text{regular function}$$

where r is the distance between the point of abscissa t and any of the two edges. Thus, to define rigorously the boundary value problem (3.45) an edge condition must be added. If the solution is sought as a double layer potential as in (3.47), it is sufficient to state that the layer density has the behaviour indicated above.

These results remain in $\mathbb{R}^3$: it is shown that the pressure field diffracted by an infinitely thin and perfectly rigid screen can be expressed by a double layer potential, the density of which cancels along the screen edge. For the Dirichlet problem, the integral representation of the diffracted field is a simple layer potential. The layer density is singular along the screen edges. A general theory of the singular behaviour of the solution induced by discontinuities of the boundary of the domain can be found in the book by P. Grisvard; the results concerning the theory of diffraction are presented in many classical textbooks and articles.

3.4. Boundary Integral Equations

We have given various possible representations of the sound field diffracted by the boundary of a propagation domain. They all involve a boundary source which must be determined. The aim of this section is to establish the equations which determine the layer density of the different boundary sources and to state theorems of existence and uniqueness of the solutions. The results are illustrated by a simple two-dimensional example for which an analytical solution is known: the boundary is a circle and the propagation domain is either the interior of the circle or its exterior.

3.4.1. Integral equations deduced from the Green's representation of the pressure field

Let σ be a closed surface (or curve). It splits the space into an interior domain Ω_i, which is bounded, and an unbounded exterior domain Ω_e. A unit vector $\vec{n}$, normal to σ and pointing out to Ω_e, is defined everywhere but along the edges (if σ has any): thus, $\vec{n}$ is an exterior normal for Ω_i and an interior normal for Ω_e. We consider both an interior problem and an exterior problem.

- Interior problem:

$$\begin{aligned} (\Delta + k^2)p_i(M) &= f_i(M), \qquad M \in \Omega_i \\ \alpha \text{Tr } \partial_n p_i(M) + \beta \text{Tr } p_i(M) &= 0, \qquad M \in \sigma \end{aligned} \tag{3.48}$$

- Exterior problem:

$$\begin{aligned} (\Delta + k^2)p_e(M) &= f_e(M), \qquad M \in \Omega_e \\ \alpha \text{Tr } \partial_n p_e(M) + \beta \text{Tr } p_e(M) &= 0, \qquad M \in \sigma \\ &\text{Sommerfeld condition} \end{aligned} \tag{3.49}$$

The Green's representations of p_i and p_e are

$$p_i(M) = \psi_i(M) - \int_\sigma [\text{Tr } \partial_{n(P)} p_i(P) G(M,P) - \text{Tr } p_i(P) \partial_{n(P)} G(M,P)] \, d\sigma(P), \qquad M \in \Omega_i \tag{3.50}$$

$$p_e(M) = \psi_e(M) + \int_\sigma [\text{Tr } \partial_{n(P)} p_e(P) G(M,P) - \text{Tr } p_e(P) \partial_{n(P)} G(M,P)] \, d\sigma(P), \qquad M \in \Omega_e \tag{3.51}$$

In these expressions, ψ_i (resp. ψ_e) represents the free field produced by the source density f_i (resp. f_e).

Assume now $\alpha \neq 0$ everywhere on σ. The boundary conditions lead us to express the normal derivative of each pressure field in terms of the pressure itself; thus, the Green's representations are written

$$p_i(M) = \psi_i(M) + \int_\sigma \text{Tr } p_i(P) \left[\frac{\beta}{\alpha} G(M,P) + \partial_{n(P)} G(M,P) \right] d\sigma(P), \qquad M \in \Omega_i \tag{3.52}$$

$$p_e(M) = \psi_e(M) - \int_\sigma \text{Tr } p_e(P) \left[\frac{\beta}{\alpha} G(M,P) + \partial_{n(P)} G(M,P) \right] d\sigma(P), \qquad M \in \Omega_e \tag{3.53}$$

Let us take the values, on σ of p_i and p_e; accounting for the discontinuity of the double layer potentials involved, one gets:

$$\frac{\text{Tr } p_i(M)}{2} - \int_\sigma \text{Tr } p_i(P) \left[\frac{\beta}{\alpha} G(M,P) + \partial_{n(P)} G(M,P) \right] d\sigma(P) = \psi_i(M), \qquad M \in \sigma \tag{3.54}$$

$$\frac{\text{Tr } p_e(M)}{2} + \int_\sigma \text{Tr } p_e(P) \left[\frac{\beta}{\alpha} G(M,P) + \partial_{n(P)} G(M,P) \right] d\sigma(P) = \psi_e(M), \qquad M \in \sigma \tag{3.55}$$

These are the boundary integral equations which Tr p_i and Tr p_e satisfy. The value, on σ, of the normal derivatives of the pressure fields can equally be calculated, leading to a second set of boundary integral equations:

$$-\frac{\beta}{\alpha} \frac{\text{Tr } p_i(M)}{2} - \int_\sigma \text{Tr } p_i(P) \frac{\beta}{\alpha} \partial_{n(M)} G(M,P) \, d\sigma(P) - \text{Tr } \partial_{n(M)} \int_\sigma \text{Tr } p_i(P) \frac{\beta}{\alpha} \partial_{n(P)} G(M,P) \, d\sigma(P) = \partial_{n(M)} \psi_i(M), \qquad M \in \sigma \tag{3.56}$$

$$-\frac{\beta}{\alpha}\frac{\text{Tr } p_e(M)}{2}+\int_\sigma \text{Tr } p_e(P)\frac{\beta}{\alpha}\,\partial_{n(M)}G(M,P)\,d\sigma(P)$$

$$+\text{Tr } \partial_{n(M)}\int_\sigma \text{Tr } p_e(P)\frac{\beta}{\alpha}\,\partial_{n(P)}G(M,P)\,d\sigma(P)=\partial_{n(M)}\psi_e(M), \qquad M\in\sigma \quad (3.57)$$

Equations (3.54) to (3.57) describe both the Robin problem ($\alpha=1$, $\beta\neq 0$) and the Neumann problem ($\alpha=1$, $\beta=0$). For the Dirichlet problem ($\alpha=0$, $\beta=1$), the Green's representations are

$$p_i(M)=\psi_i(M)-\int_\sigma \text{Tr } \partial_{n(P)}p_i(P)G(M,P)\,d\sigma(P), \qquad M\in\Omega_i \qquad (3.50')$$

$$p_e(M)=\psi_e(M)+\int_\sigma \text{Tr } \partial_{n(P)}p_e(P)G(M,P)\,d\sigma(P), \qquad M\in\Omega_e \qquad (3.51')$$

This leads to the following boundary integral equations:

$$\int_\sigma \text{Tr } \partial_{n(P)}p_i(P)G(M,P)\,d\sigma(P)=\psi_i(M), \qquad M\in\sigma \qquad (3.58)$$

$$\int_\sigma \text{Tr } \partial_{n(P)}p_e(P)G(M,P)\,d\sigma(P)=-\psi_e(M), \qquad M\in\sigma \qquad (3.59)$$

$$\frac{\text{Tr } \partial_n p_i(M)}{2}+\int_\sigma \text{Tr } \partial_{n(P)}p_i(P)\text{Tr } \partial_{n(M)}G(M,P)\,d\sigma(P)=\partial_n\psi_i(M), \qquad M\in\sigma \qquad (3.60)$$

$$\frac{\text{Tr } \partial_n p_e(M)}{2}-\int_\sigma \text{Tr } \partial_{n(P)}p_e(P)\text{Tr } \partial_{n(M)}G(M,P)\,d\sigma(P)=\partial_n\psi_e(M), \qquad M\in\sigma \qquad (3.61)$$

3.4.2. Existence and uniqueness of the solutions

For the interior problems, it can be shown that for equations (3.54), (3.56), (3.58) and (3.60) the conditions of existence and uniqueness of the solution are the same as for the boundary value problem (3.48). We can state the following theorem:

Theorem 3.3 (Green's representation for the interior problem and B.I.E.) The boundary integral equations deduced from the Green's representation of the solution of the interior problem satisfy the following properties:

(a) There exists a countable sequence of wavenumbers k_i, called 'eigenwavenumbers', (real if α/β is real) for which the homogeneous B.I.E. have a finite number of non-identically zero, linearly independent solutions; to each of these solutions, called 'eigensolutions', there corresponds a solution of the homogeneous boundary value problem (3.48), called again 'eigensolutions' of the boundary value problem. The sequence k_i is identical to the sequence of eigenwavenumbers of the boundary value problem; the eigensolutions of the B.I.E. generate all the eigensolutions of the boundary value problem, and vice versa.

(b) If k is equal to one of the eigenwavenumbers, the non-homogeneous boundary integral equations have no solution. If this is not the case, they have a unique solution for any second member.

(c) The solution of the boundary value problem (with a zero or a non-zero source term) is given by expression (3.50).

For the exterior problem, things are a little different. We know that the boundary value problem (3.49) has one and only one solution whatever the source term is. It would be useful if the boundary integral equations deduced from the Green's representation of the sound pressure – equations (3.55), (3.57), (3.59) and (3.61) – had the same property. Unfortunately, this is not true. Let us consider, for example, equation (3.59) which corresponds to the exterior Dirichlet problem: it involves the same boundary integral operator as equation (3.58) which corresponds to the interior Dirichlet problem. Thus, it has the same eigenwavenumbers. A classical result of the theory of operators states that when k is equal to any of these eigenwavenumbers, the equation has a (non-unique) solution for any second member which is orthogonal to the eigensolutions of the adjoint equation: this is always the case for any $\psi_e(M)$ which is the value on σ of an incident field (that is the value on σ of a function which satisfies a homogeneous Helmholtz equation in its neighbourhood). Any solution Tr $\partial_n p_e(P)$ of equation (3.59) generates a unique pressure field $p_e(M)$ given by (3.51′) and which is the solution of the exterior Dirichlet boundary value problem. Furthermore, two solutions which differ by a solution of the homogeneous B.I.E. generate the same exterior pressure field. The following theorem can be stated:

Theorem 3.4 (Green's representation for the exterior problem and B.I.E.) The boundary integral equations deduced from the Green's representation of the solution of the exterior problem have at least one solution for any second member which is the value on the boundary of either an incident pressure or its normal derivative, depending on the equation considered. More precisely, they satisfy the following properties:

(a) The integral operators defined by (3.55), (3.57), (3.59) and (3.61) have a countable sequence of eigenwavenumbers which are real if the ratio α/β is real; to each of these eigenwavenumbers there corresponds a finite number of linearly independent eigensolutions.

(b) If k is equal to any of these eigenwavenumbers, the boundary integral equation has a solution which is determined up to an arbitrary linear combination of the corresponding eigensolutions.

(c) The unique solution of the boundary value problem is always given by (3.51).

3.4.3. Boundary integral equations deduced from the representation of the diffracted field by a hybrid layer potential

For exterior problems, it has been seen that the Green's representation of the diffracted field leads to boundary integral equations which have real eigenwavenumbers for a real ratio α/β. This induces instabilities in numerical procedures: it is, indeed, never easy to get an approximate solution of an equation which has an infinite number of solutions. Among all the methods which have been proposed to overcome this difficulty, the simplest one is to adopt an integral representation of the diffracted field which leads to a boundary integral equation with non-real eigenwavenumbers: thus, the wavenumber of a physical excitation being real, such B.I.E. can be solved for any physical data.

Let us consider the exterior Dirichlet problem and look for a representation of the diffracted field as a hybrid layer potential:

$$p_e(M) = \psi_e(M) - \int_\sigma \mu(P)[\partial_{n(P)}G(M, P) + \iota G(M, P)]\, d\sigma(P), \qquad M \in \Omega_e \quad (3.62)$$

The boundary condition leads to

$$-\frac{\mu(M)}{2} + \int_\sigma \mu(P)[\partial_{n(P)}G(M, P) + \iota G(M, P)]\, d\sigma(P) = \psi_e(M), \qquad M \in \sigma \quad (3.63)$$

This equation involves the same operator as equation (3.54) for $\beta/\alpha = \iota$. So it has the eigenwavenumbers of the interior problem with the boundary condition

$$\partial_n p_i(M) + \iota p(M) = 0$$

These wavenumbers all have a non-zero imaginary part. Thus, for any real frequency, equation (3.63) has a unique solution. This result is valid for any boundary condition:

Theorem 3.5 (Hybrid potential for the exterior problem and B.I.E.) For any real wavenumber k, the solution of the exterior boundary value problem (3.49) can always be expressed with a hybrid layer potential by formula (3.62). The boundary condition leads to a boundary integral equation which solution μ exists and is unique whatever the source term is.

Remark. The solution of any interior problem can also be expressed with a hybrid layer potential. The B.I.E. thus obtained has the following convenient properties: the sequence of eigenwavenumbers is identical to that of the boundary value problem; the solution is unique if the excitation wavenumber differs from any of the eigenwavenumbers. Nevertheless, this representation is, in general, less interesting than the Green's formula: indeed, for the Neumann and the Robin boundary conditions, the B.I.E. involves the normal derivative of a double layer potential, which is a highly singular integral.

3.5. Two-dimensional Neumann Problem for a Circular Boundary

Let us consider a circle σ with radius R_0 and centred at the coordinates origin; it splits the plane into an interior domain Ω_i, which is bounded, and an exterior unbounded domain Ω_e. Using a cylindrical coordinate system with origin the centre of σ, a point M is defined by (R, θ).

We consider both the interior and the exterior Neumann problems. It is assumed that point isotropic sources are located at $S_i(\rho_i < R_0, \varphi_i)$ (interior problem) and at $S_e(\rho_e > R_0, \varphi_e)$ (exterior problem).

3.5.1. Interior problem and Green's representation

We look for the solution of the following equations:

$$\begin{aligned} (\Delta + k^2) p_i(M) &= \delta_{S_i}, \qquad M \in \Omega_i \\ \text{Tr}\ \partial_n p_i(M) &= 0, \qquad M \in \sigma \end{aligned} \tag{3.64}$$

The Green's representation of the solution is written

$$p_i(M) = -\frac{\iota}{4} H_0(k \mid S_i M \mid) - \frac{\iota}{4} \int_\sigma \text{Tr}\ p_i(P) \partial_{n(P)} H_0(k \mid MP \mid)\, d\sigma(P), \qquad M \in \Omega_i \tag{3.65}$$

where $\mid S_i M \mid$ (resp. $\mid MP \mid$) is the distance between the two points S_i and M (resp. M and P).

Let us first recall the expansion formulae of the Hankel function $H_0(k \mid MM' \mid)$ where the coordinates of the two points M and M' are (R, θ) and (R', θ'), respectively:

$$\begin{aligned} H_0(k \mid MM' \mid) &= \sum_{n=-\infty}^{+\infty} H_n(kR) J_n(kR') e^{\iota n(\theta - \theta')}, \qquad \text{if } R > R' \\ &= \sum_{n=-\infty}^{+\infty} H_n(kR') J_n(kR) e^{\iota n(\theta - \theta')}, \qquad \text{if } R < R' \end{aligned} \tag{3.66}$$

Let (R_0, φ) be the coordinates of a point P on σ. The kernel which appears in (3.65) is written

$$\partial_{n(P)} H_0[k \mid MP \mid] = k \sum_{n=-\infty}^{+\infty} H_n'(kR_0) J_n(kR) e^{\iota n(\theta - \varphi)} \tag{3.67}$$

where $H_n'(u)$ is the derivative of $H_n(u)$ with respect to u. The integral thus becomes

$$\int_\sigma \mathrm{Tr}\, p_i(P)\partial_{n(P)}H_0(k\,|\,MP\,|)\, d\sigma(P)$$

$$= k \sum_{n=-\infty}^{+\infty} H_n'(kR_0)J_n(kR)\mathrm{e}^{\iota n\theta} \int_0^{2\pi} \mathrm{Tr}\, p_i(P)\mathrm{e}^{-\iota n\varphi} R_0\, d\varphi$$

$$= 2\pi R_0 k \sum_{n=-\infty}^{+\infty} a_n H_n'(kR_0)J_n(kR)\mathrm{e}^{\iota n\theta} \tag{3.68}$$

where the a_n are the Fourier coefficients of $\mathrm{Tr}\, p_i(P)$ defined by

$$\mathrm{Tr}\, p_i(P) = \sum_{n=-\infty}^{+\infty} a_n \mathrm{e}^{\iota n\varphi}, \qquad a_n = \frac{1}{2\pi}\int_0^{2\pi} \mathrm{Tr}\, p_i(R_0, \varphi)\mathrm{e}^{-\iota n\varphi}\, d\varphi$$

The expansion of the incident field for $R > \rho_i$ is given by

$$-\frac{\iota}{4} H_0(k\,|\,S_iM\,|) = -\frac{\iota}{4} \sum_{n=-\infty}^{+\infty} J_n(k\rho_i)H_n(kR)\mathrm{e}^{\iota n(\theta-\varphi_i)}$$

Finally, for $\rho_i < R < R_0$, expression (3.65) becomes:

$$p_i(M) = -\frac{\iota}{4} \sum_{n=-\infty}^{+\infty} \{J_n(k\rho_i)H_n(kR)\mathrm{e}^{\iota n(\theta-\varphi_i)} + 2\pi k R_0 a_n H_n'(kR_0)J_n(kR)\mathrm{e}^{\iota n\theta}\} \tag{3.69}$$

Now, let us set the normal derivative of $p_i(M)$, that is its derivative with respect to R, to zero on σ. One gets

$$\sum_{n=-\infty}^{+\infty} \{J_n(k\rho_i)H_n'(kR_0)\mathrm{e}^{-\iota n\varphi_i} + 2\pi k R_0 a_n H_n'(kR_0)J_n'(kR_0)\}\mathrm{e}^{\iota n\theta} = 0, \qquad \forall\theta \tag{3.70}$$

This equation is satisfied if and only if each term is zero, that is if

$$J_n(k\rho_i)H_n'(kR_0)\mathrm{e}^{-\iota n\varphi_i} + 2\pi k R_0 a_n H_n'(kR_0)J_n'(kR_0) = 0, \qquad \forall n \tag{3.71}$$

Assume that k is real (this is always the case in physics); then the functions $H_n'(kR_0)$ never cancel (their roots have non-zero imaginary parts). Thus equation (3.71) is equivalent to

$$2\pi k R_0 a_n J_n'(kR_0) = -J_n(k\rho_i)\mathrm{e}^{-\iota n\phi_i}, \qquad \forall n \tag{3.72}$$

(a) Eigenwavenumbers and eigenfunctions. Each function $J_n'(kR_0)$ has a countable sequence of real zeros $z_{np}(p = 1, 2, \ldots, \infty)$ which define a countable sequence of eigenwavenumbers $k_{np} = z_{np}/R_0$. For each of these eigenwavenumbers, one of the equations (3.72) has no bounded solution a_n. Thus, the non-homogeneous problem

has no solution, while the homogeneous one has the following non-trivial solution:

$$\psi_{np} = J_n(k_{np}R)\mathrm{e}^{\iota n\theta}$$

(b) Solution of the non-homogeneous problem. If k is not equal to an eigenwavenumber, the coefficients a_n are all defined and given by

$$a_n = -\frac{J_n(k\rho_i)\mathrm{e}^{-\iota n\varphi_i}}{2\pi k R_0 J_n'(kR_0)}$$

The solution of the boundary integral equation takes the form

$$\mathrm{Tr}\ p_i(P) = -\frac{1}{2\pi k R_0}\sum_{n=-\infty}^{+\infty}\frac{J_n(k\rho_i)}{J_n'(kR_0)}\mathrm{e}^{\iota n(\varphi - \varphi_i)} \tag{3.73}$$

and the Green's representation of the solution leads to

$$p_i(M) = -\frac{\iota}{4}H_0(k \mid S_iM \mid) + \frac{\iota}{4}\sum_{n=-\infty}^{+\infty}\frac{J_n(k\rho_i)}{J_n'(kR_0)}H_n'(kR_0)J_n(kR)\mathrm{e}^{\iota n(\theta - \varphi_i)} \tag{3.74}$$

3.5.2. Exterior problem and Green's representation

We are looking for the pressure field $p_e(M)$ which satisfies the system

$$\begin{aligned} (\Delta + k^2)p_e(M) &= \delta_{S_e}, \qquad M \in \Omega_e \\ \mathrm{Tr}\ p_e(M) &= 0, \qquad M \in \sigma \end{aligned} \tag{3.75}$$

and the convenient Sommerfeld condition. The following representation is adopted:

$$p_e(M) = -\frac{\iota}{4}H_0(k \mid S_eM \mid) + \frac{\iota}{4}\int_\sigma \mathrm{Tr}\ p_e(M)\partial_{n(P)}H_0(k \mid MP \mid)\, d\sigma(P), \qquad M \in \Omega_e \tag{3.76}$$

Following the same method as in the former subsection, the pressure field is represented, for $R < \rho_e$, by the following series:

$$\begin{aligned} p_e(M) &= -\frac{\iota}{4}\sum_{n=-\infty}^{+\infty}\{H_n(k\rho_e)J_n(kR)\mathrm{e}^{-\iota n\varphi_e} - 2\pi k R_0 a_n J_n'(kR_0)H_n(kR)\}\mathrm{e}^{\iota n\theta} \\ &\text{with } a_n = \frac{1}{2\pi}\int_0^{2\pi}\mathrm{Tr}\ \partial_n p_e(P)\mathrm{e}^{-\iota n\varphi}\, d\varphi \end{aligned} \tag{3.77}$$

The boundary condition leads to the following equalities:

$$\{H_n(k\rho_e)\mathrm{e}^{-\iota n\varphi_e} - 2\pi k R_0 a_n H_n'(kR_0)\}J_n'(kR_0) = 0, \qquad \forall n \tag{3.78}$$

It is obvious that, if k is not equal to an eigenwavenumber of the interior Neumann problem (that is if $J_n'(kR_0) \neq 0, \forall n$), expression (3.78) determines all the coefficients by

$$a_n = \frac{H_n(k\rho_e)\mathrm{e}^{-\iota n\varphi_e}}{2\pi k R_0 H_n'(kR_0)} \tag{3.79}$$

Then, the Green's representation of $p_e(M)$ takes the form

$$p_e(M) = -\frac{\iota}{4} H_0(k \mid S_e M \mid) + \frac{\iota}{4} \sum_{n=-\infty}^{+\infty} \frac{J_n'(kR_0)}{H_n'(kR_0)} H_n(k\rho_e) H_n(kR)\mathrm{e}^{\iota n(\theta - \varphi_e)} \tag{3.80}$$

It must be remarked that, in accordance with the existence and uniqueness theorem which has been given, this expression is defined for any real k.

Assume now that k is equal to an eigenwavenumber k_{qr} of the interior Neumann problem. Then, for $n = q$, expression (3.78) is satisfied whatever the value of a_q is; thus, this coefficient is arbitrary. The solution of the boundary integral equation is the sum of

$$\sum_{\substack{n \neq q \\ n=-\infty}}^{+\infty} \frac{H_n(k\rho_e)}{2\pi k R_0 H_n'(kR_0)} \mathrm{e}^{\iota n(\varphi - \varphi_e)}$$

and an arbitrary function proportional to $\mathrm{e}^{\iota q\varphi}$: it is thus not uniquely determined. Nevertheless, in the series (3.80), the coefficient of the qth term is zero. Thus, the expansion (3.80) of the solution is uniquely determined. This shows that the Green's representation of the solution of the exterior problem leads to a unique solution even though the corresponding B.I.E. has eigenwavenumbers.

3.5.3. Exterior problem and hybrid layer potential

Let us now look for a representation of the diffracted field as a hybrid layer potential:

$$p_e(M) = -\frac{\iota}{4} H_0(k \mid S_e M \mid)$$
$$- \frac{\iota}{4} \int_\sigma \mu(P)[\partial_{n(P)} H_0(k \mid MP \mid) + \iota H_0(k \mid MP \mid)]\, d\sigma(P), \qquad M \in \Omega_e \tag{3.81}$$

As done in the former subsections, the two kernels and the incident field are expanded into series:

$$\partial_{n(P)} H_0(k \mid MP \mid) = k \sum_{n=-\infty}^{+\infty} J_n'(kR_0) H_n(kR)\mathrm{e}^{\iota n(\theta - \varphi)}$$

$$H_0(k \mid MP \mid) = \sum_{n=-\infty}^{+\infty} J_n(kR_0)H_n(kR)e^{\iota n(\theta - \varphi)}$$

$$H_0(k \mid S_eM \mid) = \sum_{n=-\infty}^{+\infty} J_n(kR)H_n(k\rho_e)e^{\iota n(\theta - \varphi_e)}, \qquad \text{for } R < \rho_e$$

The unknown function $\mu(P)$ is expanded into a Fourier series of the angular variable:

$$\mu(P) = \sum_{n=-\infty}^{+\infty} b_n e^{\iota n\varphi}, \qquad b_n = \frac{1}{2\pi}\int_0^{2\pi} \mu(P)e^{-\iota n\varphi}\, d\varphi$$

Expression (3.81) of the pressure field reduces thus to

$$p_e(M) = -\frac{\iota}{4}\sum_{n=-\infty}^{+\infty} J_n(kR)H_n(k\rho_e)e^{\iota n(\theta - \varphi_e)}$$
$$-\frac{\iota}{4}\, 2\pi R_0 \sum_{n=-\infty}^{+\infty} b_n[kJ_n'(kR_0) + \iota J_n(kR_0)]H_n(kR)e^{\iota n\theta}, \qquad \text{for } R < \rho_e \tag{3.82}$$

The boundary condition leads to the following equations:

$$J_n'(kR_0)H_n(k\rho_e)e^{-\iota n\varphi_e} + 2\pi R_0 b_n[kJ_n'(kR_0) + \iota J_n(kR_0)]H_n'(kR_0) = 0 \tag{3.83}$$

For real k, the functions $J_n'(kR_0)$ and $J_n(kR)$ are real; thus the linear combination $kJ_n'(kR_0) + \iota J_n(kR_0)$ cannot cancel. This implies that the coefficients b_n are always uniquely determined by

$$b_n = -\frac{J_n'(kR_0)H_n(k\rho_e)e^{-\iota n\varphi_e}}{2\pi R_0[kJ_n'(kR_0) + \iota J_n(kR_0)]H_n'(kR_0)}$$

When this value is introduced into the expression of the diffracted field given in (3.82), one gets, of course, the same expression as in (3.80). But none of the coefficients b_n has an undetermined form.

Finally, let us remark that the denominator of b_n is zero for the non-real values k_{nr} of the wavenumber defined by

$$k_{nr}J_n'(k_{nr}R_0) + \iota J_n(k_{nr}R_0) = 0$$

The function

$$p_{nr}(R, \theta) = J_n(k_{nr}R)e^{\iota n\theta}$$

satisfies, in Ω_i, a homogeneous Helmholtz equation and, on σ, the following Robin condition:

$$\text{Tr } \partial_R p_{nr}(R, \theta) + \iota p_{nr}(R, \theta) = 0, \qquad \text{for } R = R_0$$

This is the result which has been given previously.

Bibliography

[1] ABRAMOWITZ, M. and STEGUN, L.A., 1970. *Handbook of mathematical functions*. National Bureau of Standards, Washington D.C.

[2] BOCCARA, N., 1984. *Analyse fonctionnelle*. Ellipses, Editions Marketing, Paris.

[3] BOWMAN, J.J., SENIOR, T.B.A. and USLENGHI, P.L.E., 1969. *Electromagnetic and acoustic scattering by simple shapes*. North Holland, Amsterdam.

[4] BRUNEAU, M., 1983. *Introduction aux théories de l'acoustique*. Université du Maine Editeur, Le Mans, France.

[5] COLTON, D. and KRESS, R., 1983. *Integral equations methods in scattering theory*. Wiley-Interscience Publication, New York.

[6] COLTON, D. and KRESS, R., 1992. *Inverse acoustic and electromagnetic scattering theory*. Springer-Verlag, Berlin.

[7] COURANT, R. and HILBERT, D., 1962. *Methods of mathematical physics*. Interscience Publishers–John Wiley & Sons, New York.

[8] DELVES, L.M. and WALSH, J., 1974. *Numerical solution of integral equations*. University Press, Oxford.

[9] FILIPPI, P., 1983. *Theoretical acoustics and numerical techniques*, CISM Courses and Lecture Notes no. 277. Springer-Verlag, Wien, New York.

[10] JEFFREYS, H. and JEFFREYS, B., 1972. *Methods of mathematical physics*. University Press, Cambridge.

[11] LANDAU, L. and LIFSCHITZ, E., 1971. *Mécanique des fluides*. Editions Mir, Moscow.

[12] MARSDEN, J.E. and HOFFMANN, M.J., 1993. *Elementary classical analysis*. Freeman, New York.

[13] MORSE, PH.M. and FESHBACH, H., 1953. *Methods of theoretical physics*. McGraw-Hill, New York.

[14] MORSE, PH.M. and INGARD, K.U., 1968. *Theoretical acoustics*. McGraw-Hill, New York.

[15] PETIT, R., 1983. *L'outil mathématique*. Masson, Paris.

[16] PIERCE, A.D., 1981. *Acoustics: an introduction to its physical principles and applications*. McGraw-Hill, New York.

[17] SCHWARTZ, L., 1966. *Théorie des distributions*. Hermann, Paris.

CHAPTER 4

Outdoor Sound Propagation

Dominique Habault

Introduction

Studies of outdoor sound propagation are essential nowadays since they are linked to problems of environmental noise pollution which appear everywhere (transportation noise, plant and factory noise, etc.).

The prediction of sound levels emitted around a factory which is to be set up is a classical example of engineering acoustics. Such a problem is in general quite complicated. It must take into account various phenomena which can be divided into three groups: ground effect, diffraction by obstacles, propagation in an inhomogeneous medium (i.e. characterized by a varying sound speed). Each of these three aspects corresponds to a section of this chapter.

Obviously, studying each aspect separately leads to a great simplification of the complete problem and to a more detailed analysis. Indeed, there are two ways to solve it:

- taking all the aspects into account; this leads to heavy, highly time-consuming computer programs, which are not suitable for quick studies of the respective effect of the (acoustic, geometrical) parameters of the problem;
- dividing the problem into several sub-problems for which simple solutions or at least general characteristics can be obtained.

Explicit solutions or simple computation methods are now well known for each of the three aspects but they still correspond to elementary problems. The computer programs developed for the prediction of outdoor propagation are still quite heavy (both for computing time and storage).

It is then essential to get *a priori* estimations of the relative influence of each phenomenon, in order to neglect those with minor effect. For instance, for the prediction of sound levels emitted close to the ground, the ground effect dominates and the propagation medium can be modelled as a homogeneous medium (constant sound speed) if the wind speed and temperature gradient are weak. Obviously, the choice of the phenomena to be neglected depends on the accuracy required for the prediction of the sound levels. Generally speaking, the accuracy does not need to be higher than the accuracy obtained from experiment.

For a given problem, a highly accurate method is generally not required, because of all the neglected phenomena, the accuracy of the experimental results or the kind of results needed. For example, to evaluate the efficiency of a barrier, the interesting curve is mainly the envelope of the maximum sound levels obtained behind the barrier (opposite side to the source): an accurate description of the interference pattern is not necessary.

In this chapter, only harmonic signals are studied. For non-harmonic signals, it is possible to solve the wave equation (d'Alembert equation) directly or to solve the Helmholtz equation and use a Fourier transform.

4.1. Ground effect in a homogeneous atmosphere

4.1.1. Introduction

The study of the ground effect has straightforward applications, such as noise propagation along a traffic axis (road or train).

The propagation medium is air; in the simplest cases, it is modelled as a homogeneous fluid (characterized by a constant density and a constant sound speed). In the case of a wind or temperature gradient, the sound speed is a function of the space variables. This leads to a Helmholtz equation with varying coefficients (see Section 4.3). Furthermore, if turbulence phenomena cannot be neglected, a deterministic model is inadequate and random processes must be included.

In this section, only the homogeneous model is considered. In a quiet atmosphere and if there are no obstacles on the ground, the sound propagation problem can be modelled and solved rather simply.

The aim of a complete study of the ground effect is to determine the acoustical characteristics of the ground and then to evaluate the sound field for these characteristics. The propagation above a homogeneous plane ground is presented in Section 4.1.2. The propagation above an inhomogeneous plane ground (that is a plane described by discontinuous boundary conditions) is presented in Section 4.1.3.

4.1.2. Propagation above a homogeneous plane ground

The most classical ground model (the simplest) is the locally reacting surface. The ground is characterized by a complex parameter which depends on frequency: the normal specific impedance. The sound pressure emitted above the ground is then the solution of the following system:

$$\begin{cases} \text{Helmholtz equation above the plane} \\ \text{impedance condition on the plane} \\ \text{Sommerfeld conditions} \end{cases}$$

The solution can be obtained by using a space Fourier transform. Depending on the technique chosen to evaluate the inverse Fourier transform, several kinds of representations of the sound pressure (exact or approximate) are found. The approximations are often available for large values of kR, the ratio of the distance between source and observation point and the wavelength.

When an exact representation of the sound pressure is available (and numerically efficient) new techniques can be developed to evaluate the ground impedance. In the case of a homogeneous plane ground, the study then leads to a complete description of the ground effect: expression of the sound field for any position of the source and the observation point, identification of the acoustic parameters of the ground.

Analytic expressions of the sound pressure emitted above the ground

In the case of a plane surface, the classical method is based on a space Fourier transform. It is indeed easy to obtain the Fourier transform of the sound pressure. The most difficult part is to evaluate its inverse transform to obtain a representation of the sound pressure.

(a) Expression of the Fourier transform of the sound pressure. In the three-dimensional space (O, x, y, z), the plane $(z = 0)$ represents the ground surface. The propagation medium $(z > 0)$ is characterized by a constant density ρ_1 and a constant sound speed c_1. Let S be an omnidirectional point source located at $(0, 0, z_0 > 0)$. The emitted signal is harmonic $((\exp(-\iota\omega t))$. The sound pressure $p(x, y, z)$ is the solution of the following system:

$$\begin{cases} (\Delta + k^2)p(x, y, z) = \delta(x)\delta(y)\delta(z - z_0) \text{ for } z > 0 \\ \dfrac{\partial p(x, y, z)}{\partial \vec{n}} + \dfrac{\iota k}{\zeta} p(x, y, z) = 0 \text{ on } z = 0 \\ \text{Sommerfeld conditions} \end{cases} \tag{4.1}$$

where $\vec{n}$ is the unit vector, normal to $(z = 0)$, interior to the propagation domain. ζ is the reduced specific normal impedance, that is the ratio between the normal impedance of the ground and the product $\rho_1 c_1$, impedance of a plane wave in the fluid.

Because $\overrightarrow{Oz}$ is a symmetry axis, the sound pressure only depends on z and the radial coordinate $\rho = \sqrt{x^2 + y^2}$. Then $\hat{p}(\xi, z)$, the Fourier transform of p with respect to ρ, is defined by

$$\hat{p}(\xi, z) = 2\pi \int_0^{+\infty} p(\rho, z) J_0(\xi\rho)\rho \, d\rho \tag{4.2}$$

and conversely

$$p(\rho, z) = \frac{1}{4\pi} \int_{-\infty}^{+\infty} \hat{p}(\xi, z) H_0^{(1)}(\xi\rho)\xi \, d\xi \tag{4.3}$$

J_0 is the Bessel function of zero order. $H_0^{(1)}$ is the Hankel function of first kind and zero order, which satisfies $H_0^{(1)}(-z) = H_0^{(1)}(ze^{\iota\pi})$. $\hat{p}(\xi, z)$, also called Hankel transform of p, is the solution of the Fourier transform of system (4.1). The Helmholtz equation becomes a one-dimensional differential equation. $\hat{p}(\xi, z)$ can then be written as the sum of a particular solution of the inhomogeneous equation and the general solution of the homogeneous equation:

$$\hat{p}(\xi, z) = \frac{e^{\iota K|z-z_0|}}{2\iota K} + \hat{A}(\xi)\frac{e^{\iota K|z+z_0|}}{2\iota K} + \hat{B}(\xi)\frac{e^{\iota K|z+z_0|}}{2\iota K} \qquad \text{for } z \geqslant 0 \tag{4.4}$$

with $K^2 = (k^2 - \xi^2)$ and $\Im(K) > 0$. Because of Sommerfeld conditions, $\hat{B}(\xi) = 0$. $\hat{A}(\xi)$ is deduced from the boundary condition on $(z = 0)$. Here [2]:

$$\hat{A}(\xi) = \frac{K - k/\zeta}{K + k/\zeta} \tag{4.5}$$

$\hat{p}(\xi, z)$ is then written as the sum of a plane wave emitted by the source S and a plane wave emitted by a source $S' = (0, 0, -z_0)$, the image of S relative to the plane $(z = 0)$. $\hat{A}$ is the plane wave reflection coefficient.

(b) Exact representations of the sound pressure. Several kinds of representations can be obtained, depending on the method used to evaluate the inverse Fourier transform of $\hat{p}$. They are equivalent but have different advantages.

$\hat{p}(\xi, z)$ can be written as a sum of Fourier transforms of known functions; the expression for $p(M)$ then includes a sum of layer potentials [2]:

$$\begin{aligned} p(M) = &-\frac{e^{\iota kR(S,M)}}{4\pi R(S,M)} - \frac{e^{\iota kR(S',M)}}{4\pi R(S',M)} \\ &+ \frac{\iota k^2}{2\zeta^2}\int_{z'=-z_0} H_0^{(1)}(\alpha\rho(P))\frac{e^{\iota kr(M,P)}}{4\pi r(M,P)}\,d\sigma(P) \\ &- \frac{k}{2\zeta}\frac{\partial}{\partial z}\int_{z'=-z_0} H_0^{(1)}(\alpha\rho(P))\frac{e^{\iota kr(M,P)}}{4\pi r(M,P)}\,d\sigma(P) \end{aligned} \tag{4.6}$$

with $\alpha^2 = k^2(1 - 1/\zeta^2)$ and $\Im(\alpha) > 0$. $M = (x, y, z)$ is a point of the half-space $(z \geqslant 0)$, P is a point of the plane $(z' = -z_0)$, with coordinates $(\rho(P), z')$.

The pressure $p(M)$ is written as the sum of an incident wave emitted by S, a reflected wave 'emitted' by S', a simple layer potential and the derivative of a simple layer potential. This representation is quite convenient for obtaining analytic approximations. From a numerical point of view, it is not suitable because it contains double integrals on an infinite domain.

By using integration techniques in the complex ξ plane, it is also possible to express p as in [3]:

$$p(M) = -\frac{e^{\iota kR(S,M)}}{4\pi R(S,M)} - \frac{e^{\iota kR(S',M)}}{4\pi R(S',M)} + \frac{k}{4\zeta}[1+\operatorname{sgn}(\hat{\zeta})]\,Y(\theta-\theta_0)H_0^{(1)}(\alpha\rho(M))e^{-\iota k(z+z_0)/\zeta} + \frac{k}{2\zeta}\frac{e^{\iota kR(S',M)}}{\pi}\int_0^{+\infty}\frac{e^{-kR(S',M)t}}{\varepsilon\sqrt{W(t)}}\,dt \tag{4.7}$$

where $\zeta = \dot{\zeta} + \iota\hat{\zeta}$; $\operatorname{sgn}(\hat{\zeta})$ is the sign of $\hat{\zeta}$. θ is the angle of incidence, measured from the normal $\vec{n}$. θ_0 is defined by

$$-\frac{\dot{\zeta}}{|\zeta^2|}\cos\theta_0 + \frac{\Re(\alpha)}{k}\sin\theta_0 = 1$$

Y is the Heaviside function:

$$Y(\theta) = \begin{vmatrix} 1 & \text{if } y > 0 \\ 0 & \text{if not} \end{vmatrix}$$

$$W(t) = \left(\frac{1}{\zeta} + \cos\theta\right)^2 + 2\iota t\left(1 + \frac{\cos\theta}{\zeta}\right) - t^2$$

$$t_0 = \frac{\hat{\zeta}}{|\zeta|^2}\left(\frac{\dot{\zeta}}{|\zeta^2|} + \cos\theta\right)\left(1 + \frac{\dot{\zeta}}{|\zeta^2|}\cos\theta\right)^{-1}$$

$$\varepsilon = \begin{cases} -1 & \text{if } \Re(W(t_0)) < 0,\ t > t_0,\ \hat{\zeta} > 0 \\ +1 & \text{if not} \end{cases}$$

ε is introduced in order to ensure the continuous determination chosen for $\sqrt{W(t)}$.

In (4.7), the pressure is written as the sum of an incident wave, a reflected wave, a surface wave term and a Laplace type integral.

The surface wave term has the form $aH_0^{(1)}(\alpha\rho(M))e^{\iota\beta(z+z_0)}$ where α and β are complex numbers with a positive imaginary part. It represents a wave with an amplitude which is an exponentially decreasing function of both the horizontal distance between source and observation point and the heights z and z_0. If the radial coordinate $\rho(M)$ is fixed, the amplitude is maximum on the ground surface.

The expression (4.7) is particularly convenient from a numerical point of view. For the surface wave term only the computation of the Hankel function $H_0^{(1)}$ is needed. For example, if $|u| > 1$, $H_0^{(1)}(u)$ can be computed by using Padé approximations:

$$H_0^{(1)}(u) \simeq \frac{e^{\iota(u-\pi/4)}}{\sqrt{u}}\frac{P(u)}{Q(u)}$$

where $P(u)$ and $Q(u)$ are polynomials of order n ($n=3$ leads to a correct approximation for this type of problem). In [4], Y. Luke gives the values of the coefficients of the polynomials, along with the accuracy obtained for some values of the complex variable u.

Furthermore, for values of kR 'not too small', the Laplace type integral can be computed by using a 4-point integration technique based on Laguerre polynomials [5]. For small values of kR, it can be computed through a Gauss integration technique with a varying step.

The expression (4.7) can then be computed very quickly.

(c) Approximations of the sound pressure. For practical applications, the most interesting geometries correspond to large distances ($kR \gg 1$) from the source, whether at grazing incidence ($\theta \simeq 90°$) or not. It is then useful to obtain very simple approximations.

As seen previously, the expression (4.7) can be computed very quickly for $kR \gg 1$. However, analytic approximations can also be deduced from the exact expressions; they give the analytic behaviour of the sound pressure at large distances.

A classical method to obtain these approximations is the steepest descent method. It is applied to the Fourier transform $\hat{p}(\xi, z)$. One of the oldest articles devoted to sound propagation above an impedance plane is the one by Ingard [6]. p is approximated by:

$$p(M) \simeq -\frac{e^{\iota kR(S,M)}}{4\pi R(S,M)} - [R_0 + (1-R_0)F]\frac{e^{\iota kR(S',M)}}{4\pi R(S',M)}$$

where R_0 is the plane wave reflection coefficient $(\zeta\cos\theta - 1)/(\zeta\cos\theta + 1)$ and

$$1 - F = kR(S', M)\left(\frac{1}{\zeta} + \cos\theta\right) \times \int_0^{+\infty} \frac{e^{-kR(S',M)t}}{\left[\left(1 + \frac{\cos\theta}{\zeta} + \iota t\right)^2 - (1 - \cos^2\theta)\left(1 - \frac{1}{\zeta^2}\right)\right]^{1/2}}\, dt$$

Simpler approximations have also been obtained (see Brekhovskikh [7], for example) for various types of boundary conditions:

$$p(M) \simeq -\frac{e^{\iota kR(S,M)}}{4\pi R(S,M)} - \frac{e^{\iota kR(S',M)}}{4\pi R(S',M)}\left[V(\theta) - \frac{\iota}{2kR(S',M)}(V''(\theta) + V'(\theta)\cot\theta)\right] \tag{4.8}$$

where $V(\theta)$ is the plane wave reflection coefficient, and V' and V'' its derivatives with respect to θ. The same approximations can also be deduced from the exact expression (4.6) by using an asymptotic approximation of the Green's kernel (see chapter 5, section 5.1). For a locally reacting surface, p is obtained as

$$p(M) = -\frac{e^{\iota kR(S,M)}}{4\pi R(S,M)} - \frac{\zeta\cos\theta - 1}{\zeta\cos\theta + 1}\frac{e^{\iota kR(S',M)}}{4\pi R(S',M)} - \frac{2\iota}{k}\frac{\zeta(\zeta+\cos\theta)}{(\zeta\cos\theta+1)^3}\frac{e^{\iota kR(S',M)}}{4\pi R^2(S',M)} + \mathbb{O}(R^{-3}(S',M)) \tag{4.9}$$

This approximation is valid for $kR \gg 1$ but the exact limit of kR for which it is valid is not known. It depends on the ground properties. At grazing incidence (source and observation point on the ground), expression (4.9) becomes

$$p(M) = -\frac{\iota\zeta^2}{k}\frac{e^{\iota kR(S',M)}}{2\pi R^2(S',M)} + \mathbb{O}(R^{-3}(S',M)) \tag{4.10}$$

The terms in $1/R$ disappear.

The efficiency of these approximations is shown in some examples presented in the next section.

Description of the acoustic field

In this section, the sound levels presented in the figures are obtained from the formula

$$N \text{ dB} = 20 \log \frac{|p_a|}{|p_R|} \tag{4.11}$$

$|p_a|$ is the modulus of the pressure measured above the absorbing plane. $|p_R|$ is the modulus of the pressure measured above a perfectly reflecting plane (i.e. described by a Neumann condition) or in infinite space, for the same source and the same position of the observation point. With this choice, it is possible to eliminate the influence of the characteristics of the source. Obviously, N does not depend on the amplitude of the source. Let us emphasize that $(-N)$ corresponds to the definition of 'excess attenuation' sound levels. In the following, p_R is the sound pressure obtained above a perfectly reflecting plane:

$$p_R(M) = -\frac{e^{\iota kR(S,M)}}{4\pi R(S,M)} - \frac{e^{\iota kR(S',M)}}{4\pi R(S',M)}$$

For the particular case of grazing incidence, the curves which present sound levels versus the distance $R(S,M)$ for a given frequency have the general behaviour shown in Fig. 4.1. In region 1, close to the source, the sound levels are zero, as if the ground were perfectly reflecting. In region 2, they begin to decay. In region 3, the asymptotic region, they decay by 6 dB per doubling distance. In this region, expression (4.10) gives a correct description of the sound field. The lengths of these three regions depend on the frequency and the ground properties.

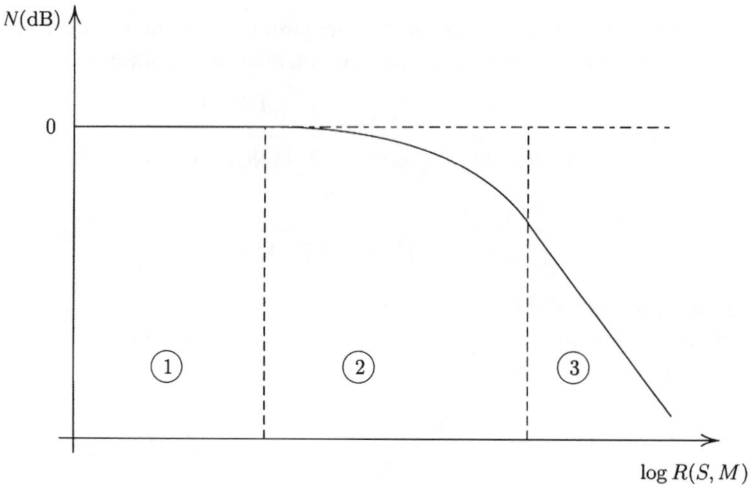

Fig. 4.1. Example of curve of sound levels obtained at grazing incidence, versus distance $R(S, M)$.

Figures 4.2 and 4.3 present two examples of curves obtained above grass. In Fig. 4.2 (580 Hz), the three regions clearly appear, the asymptotic region begins further than 32 m (about 55λ). In Fig. 4.3, at higher frequency (1670 Hz), the ground is much more absorbing, there is no region 1 and the asymptotic region begins at 2 m (about 10λ).

Determination of the normal impedance of the ground

The impedance of an absorbing material (rock wool, fibreglass, etc.) can generally be measured by using a Kundt's tube (or stationary wave tube). It is not so easy for

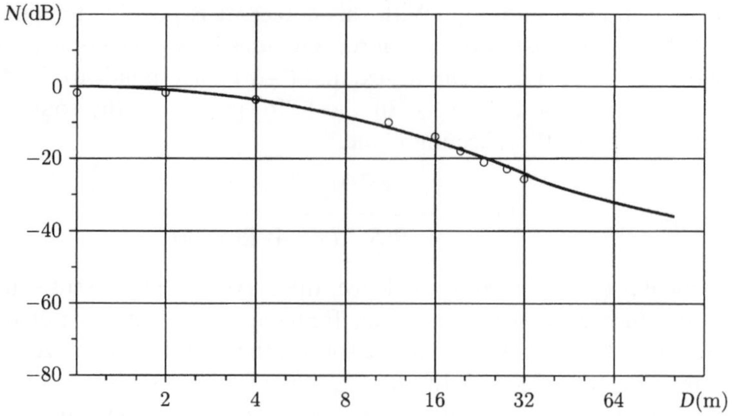

Fig. 4.2. $F = 580$ Hz, $\zeta = 2.3 + \iota 3.3$. Sound levels versus distance between source and receiver. Source and receiver on the ground. (O) measured; (—) computed.

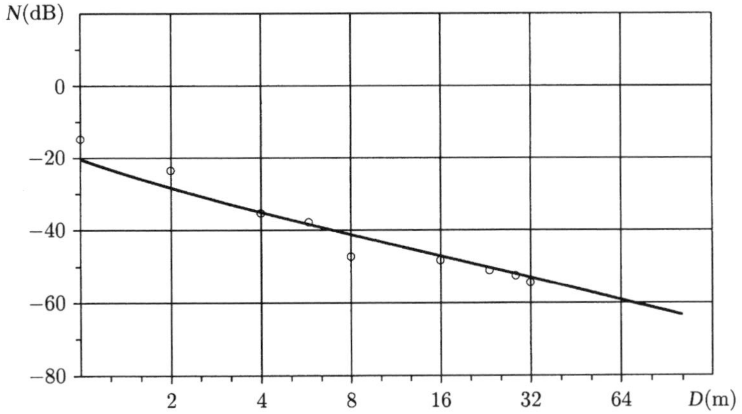

Fig. 4.3. $F = 1670$ Hz, $\zeta = 1.0 + \iota$. Sound levels versus distance between source and receiver. Source and receiver on the ground. (O) measured; (—) computed.

the ground. Many studies have been dedicated to this problem. Several methods have been suggested and tested: Kundt's tube, sound pressure measurements in free-field (for plane wave or spherical wave), and so on. The impedance values are deduced from measurements of the plane wave reflection coefficient.

The Kundt's tube method consists in placing a sample of material at the end of the tube and measuring the minima and maxima inside the tube created by the harmonic motion of a membrane. The diameter of the tube must be smaller than a quarter of a wavelength so that the plane wave assumption can be applied. This method has been applied *in situ* to evaluate the impedance of the ground. P. Dickinson and P.E. Doak [8] present measurements carried out *in situ* and describe the drawbacks of the method: the tube, when dug into the ground, changes the properties of the ground; the measurements are made only on a very small sample of ground; the exact position of the surface of the ground cannot be determined if the ground is grassy or not flat, and this leads to an error in the evaluation of the phase of the impedance.

Free-field methods have also been tested. One of them [9, 10] consists in emitting a transient wave above the ground. The incident wave and the wave reflected by the ground are recorded with a microphone located between the source and the ground. Both signals (direct and reflected) are separated by time-filtering and the reflection coefficient is deduced from the ratio of the spectra. The authors point out that this is again a local method which leads to errors when the ground is not too absorbing.

Global methods, taking into account a larger ground surface, have also been proposed [11]. Those based on a least-squares minimization consists in evaluating the value ζ of the impedance which makes the function $F(\zeta)$ minimum:

$$F(\zeta) = \sum_{i=1}^{N} (Y_i - 20 \log |\, p(X_i, \zeta)\,|)^2 \qquad (4.12)$$

ζ is the specific normal impedance. (X_i), $i = 1, \ldots, N$ represent the measurement points. Y_i is the sound level measured at X_i. $g(X_i, \zeta) = 20 \log |p(X_i, \zeta)|$ is the sound level computed at X_i. $F(\zeta)$ represents the difference between the measured and computed levels at N points above the ground.

The first step is then to measure, for a given frequency, N sound levels at N points above the ground. These sound levels and the coordinates of the measurement points are the input data of a minimization algorithm which provides the impedance value ζ. The simplest way, if possible, is to choose the points X_i on the ground surface; then, no more than six or eight points are needed. From this value ζ, the sound pressure can be computed for any position of the source and the observation point.

Such a method has several advantages: a larger sample of ground is taken into account; it does not change the properties of the ground; it does not require any measurements of the phase of the pressure.

This method can be modified [12] to determine the impedance on a frequency band from only one experiment. A large frequency band signal is emitted and the sound levels are measured at one or several points. An impedance model versus frequency (4.13) is then needed. The parameters of the model are computed in order to minimize the difference between measured and computed levels at one or several points and for N' frequencies. The numerical and experimental procedure is then faster but can lead to less accuracy (depending on the limitations of the impedance model, for example).

Figures 4.4 and 4.5 show an example of results, at frequency 1298 Hz. By taking into account six measurement points located on the ground (source located on the ground as well), the minimization algorithm provides the value $\zeta = 1.4 + \iota 1.8$. The measured sound levels are close to the theoretical curve, which is computed with formula (4.7). The same value ζ also provides an accurate description of the sound field, when source and measurement points are 22 cm above the surface. These

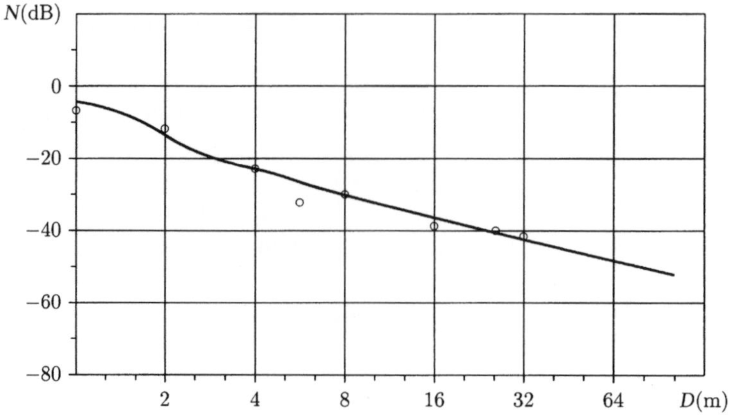

Fig. 4.4. Sound levels versus horizontal distance between source and receiver. $F = 1298$ Hz. Source and receiver on the ground. (O) measured; (—) computed with $\zeta = 1.4 + \iota 1.8$.

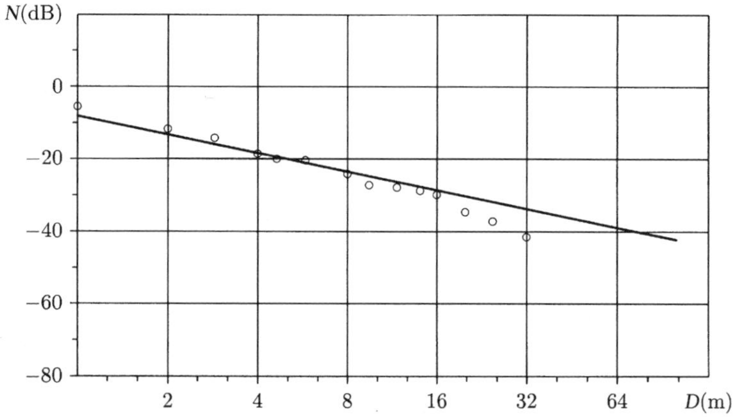

Fig. 4.5. Sound levels versus horizontal distance between source and receiver. $F = 1298$ Hz. Height of source and receiver $h = 0.22$ m. (O) measured; (—) computed with $\zeta = 1.4 + \iota 1.8$.

results show that, for this ground (grassy field), and this frequency, the local reaction model is correct.

Ground models

The 'local reaction' model, which is characterized by a complex parameter ζ, a function of frequency, is very often satisfactory for many kinds of ground (grassy, hard, sandy ground, and others) on large frequency bands, that is it provides a prediction of the sound field with an error no greater than the measurement errors.

This local reaction model is often used along with the empirical model, proposed by Delany and Bazley [13], which expresses the impedance as a function of frequency. The reduced normal impedance is given by

$$\zeta = 1 + 9.08\left(\frac{\sigma}{f}\right)^{0.75} + \iota 11.9\left(\frac{\sigma}{f}\right)^{0.73} \tag{4.13}$$

where f is the frequency and σ the flow-resistivity. This model has been tested for classical absorbing materials and several frequency bands. Its efficiency for real grounds can only be deduced from comparisons between measurements and theoretical results.

For 'particular' ground (deep layer of grass, several layers of snow, etc.), the model can be inaccurate; for example, there are situations for which it cannot describe an interference pattern above a layered ground. Other models have been studied, mostly a layer of porous material, with finite depth or not, with constant porosity or porosity as a function of depth. They can provide a better prediction of the sound field. However, because they are based on more parameters (2 or 4), their use is not so straightforward.

4.1.3. Propagation above an inhomogeneous plane ground

In this chapter, an inhomogeneous ground is a plane ground characterized by discontinuous boundary conditions, for example a plane made of two surfaces described by different impedances. The sound field emitted by a point source cannot be obtained as simply as in the case of a homogeneous surface. The aim of the prediction methods is to find analytic approximations (as rigorous as possible) or to use numerical methods (often based on boundary integral equations).

Boundary integral equation methods

The boundary integral equation methods are described in detail in chapter 3 and their solution techniques in chapter 6.

The sound pressure at any point above the ground can be written as an integral which involves the pressure and its normal derivative on the surface. The unknown functions on the surface are obtained by using the boundary conditions and solving an integral equation.

(a) An example of an integral equation. Let us assume that the plane $(z = 0)$ is made of two half-planes, each one described by a constant reduced specific normal impedance. The half-plane $\mathcal{P}_1$, $(x, y < 0)$ is described by an impedance ζ_1, the half-plane $\mathcal{P}_2$, $(x, y > 0)$ is described by an impedance ζ_2. Let G_1, be the Green's function such that

$$\left\{\begin{array}{ll} (\Delta + k^2)G_1(P, M) = \delta_P(M) & \text{for } z > 0 \\ \dfrac{\partial G_1(P, M)}{\partial \vec{n}} + \dfrac{\iota k}{\zeta_1} G_1(P, M) = 0 & \text{on } z = 0 \\ \text{Sommerfeld conditions} & \end{array}\right. \tag{4.14}$$

where $M = (x, y, z)$ and $\vec{n}$ is the inward unit vector normal to $(z = 0)$. The Green's representation of $p(M)$, in the case of a point source, gives

$$p(M) = G_1(S, M) + \iota k \left(\frac{1}{\zeta_1} - \frac{1}{\zeta_2} \right) \int_{\mathcal{P}_2} p(P')G_1(P', M)\, d\sigma(P') \tag{4.15}$$

at any point M of the half-space $(z > 0)$.

$p(M)$ is then related to the value of $p(P')$ on $\mathcal{P}_2$. When M tends towards a point P_0 of $\mathcal{P}_2$, (4.15) becomes

$$p(P_0) = G_1(S, P_0) + \iota k \left(\frac{1}{\zeta_1} - \frac{1}{\zeta_2} \right) \int_{\mathcal{P}_2} p(P')G_1(P', P_0)\, d\sigma(P') \tag{4.16}$$

This is an integral equation. Exact representations of G_1 can be found in Section 4.1.2. The unknown is the value of p on $\mathcal{P}_2$. Once it is obtained by solving the integral equation, the Green's representation (4.15) can be used to evaluate the sound pressure at any point of $(z > 0)$.

Let G_2 be the Green's function which satisfies (4.14) with ζ_1 replaced by ζ_2. By using G_2 instead of G_1, the Green's representation and the integral equation will include an integral on $\mathcal{P}_1$ instead of $\mathcal{P}_2$. For one problem, several integral equations can be obtained. They are equivalent. Depending on the boundary conditions, the Green's function can be chosen in order to simplify the numerical solution of the integral equation (see example in chapter 6, section 6.1).

(b) An example of results. Let us consider the case of a plane ground made of a perfectly reflecting strip, of constant width, which separates two half-planes described by the same normal impedance ζ [14] (see Fig. 4.6). The sound source is cylindrical, with axis parallel to the symmetry axis of the strip. The signal is harmonic.

This example is a very simple model of sound propagation due to a traffic road separating two grassy fields. The strip represents the traffic road. It is characterized by a homogeneous Neumann condition. The sound pressure can be obtained by a boundary integral equation method. The theoretical and numerical aspects of this problem are presented in detail in chapter 6.

The experiment was carried out in an anechoic room. The ground surface was made of two absorbing surfaces (polyether foam) separated by a perfectly reflecting strip of 38 cm width. Nine omnidirectional point sources (pressure drivers) were located on the symmetry axis of the strip, regularly spaced. The measurement microphone was moved on the surface, on the axis orthogonal to the symmetry axis of the strip and passing through the central source (see Fig. 4.7).

Two series of experiments were conducted: one with the nine sources turned on, the other one with only the central source turned on. Figure 4.8 presents a comparison between theoretical and experimental results at 5840 Hz. The sound levels are computed as in the previous section. When the nine sources are turned on,

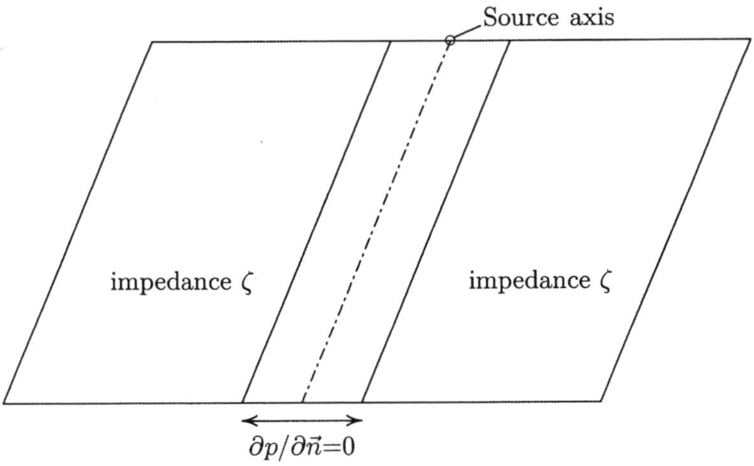

Fig. 4.6. Inhomogeneous plane ground.

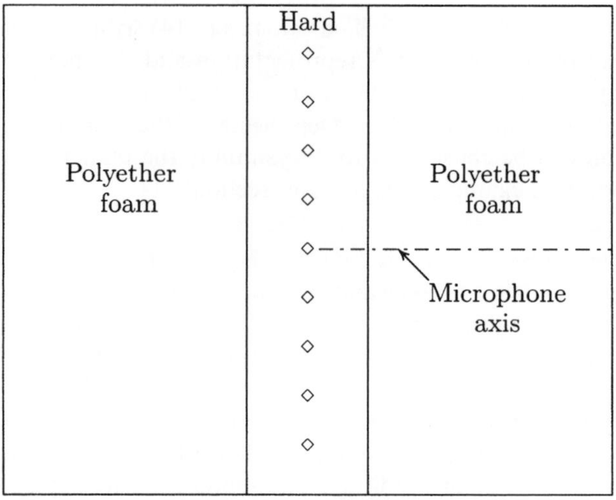

Fig. 4.7. Experiment conducted in an anechoic room.

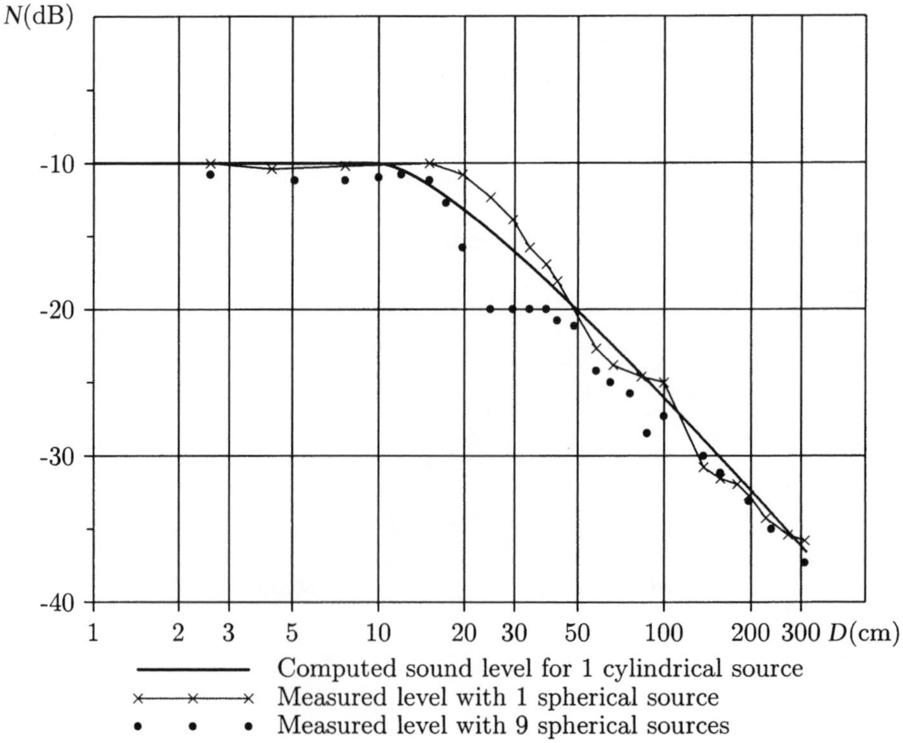

Fig. 4.8. Sound levels versus distance between source and receiver. Sources and microphones on the ground. $F = 5840$ Hz.

an interference pattern is obtained: the curve presented in Fig. 4.8 is the difference of the envelopes of the maxima measured above the inhomogeneous plane and a perfectly reflecting plane. The experimental curves obtained with one and nine sources are quite similar and are close to the theoretical curve: except for a few points, the difference between measured and computed levels is less than 1.5 dB. Above the strip, the sound levels are equal to zero (this is not surprising), then they decay above the absorbing surface by 7 dB/doubling distance. The integral equation techniques provide a good description of the sound field above an inhomogeneous plane.

Approximation techniques

(a) Approximation in the integral representation. To avoid solving the integral equation, one can imagine replacing the unknown function in the Green's representation by an approximate expression (first terms of a Neumann's series, for example) as described in chapter 5, section 5.3. Let us consider the sound propagation above the $(z = 0)$ plane made of a perfectly reflecting half-plane $\mathcal{P}_1(x, y < 0)$, with a homogeneous Neumann condition, and an absorbing half-plane $\mathcal{P}_2(x, y > 0)$ described by a normal impedance ζ. The sound source is an omnidirectional point source S located on $\mathcal{P}_1$. Let G_R be the Green's function which satisfies a homogeneous Neumann condition on the whole plane $(z = 0)$. The Green's representation of the sound pressure p is

$$p(M) = G_R(S, M) - \frac{\iota k}{\zeta} \int_{\mathcal{P}_2} p(P')G_R(P', M)\, d\sigma(P')$$

Since the integration domain is the half-plane characterized by an impedance, one can imagine replacing $p(P')$ by $p_a(P')$, the sound pressure that would exist on $\mathcal{P}_2$ if the whole plane were characterized by the same impedance ζ. The validity of the approximation then obtained for $p(M)$ is difficult to estimate. It is a kind of 'geometrical optics' approximation. A similar example is presented in [15] Durnin and Bertoni who have tested several types of approximations for several integral equations.

(b) Wiener–Hopf method. This method is presented in detail in chapter 5, section 5.7. It can be seen that it is well adapted to propagation above planes with discontinuous boundary conditions. However, even for the simple case of two half-planes, it is generally not possible to obtain an expression of the pressure suitable for numerical computations. The decomposition problems, described in chapter 5, which are simple in the case of a plane wave, become much more difficult in the case of a spherical source. Nevertheless, approximations of the sound field can be obtained close to the discontinuities or at large distances from the source.

In the case of two half-planes characterized by two different impedance values, Heins and Feschbach [16] present a far-field approximation, for the plane wave case. The Wiener–Hopf method leads to an approximation of the Fourier transform of the pressure. The approximation of the pressure is then deduced through a stationary phase method. When the Wiener–Hopf method provides an expression of the Fourier transform of the pressure, it is possible to obtain an approximation of the pressure, by using the results of chapter 5 on asymptotic expansions.

4.2. Diffraction by an Obstacle in Homogeneous Atmosphere

4.2.1. Introduction

A well-known application of outdoor diffraction is the evaluation of the efficiency of acoustical barriers which protect the inhabitants from traffic noise.

Even in the case of a homogeneous atmosphere, an exact expression of the sound field diffracted by an obstacle can only be obtained for very simple cases: particular geometry of the obstacle (cylinder, ellipse), local boundary conditions, etc. It is then possible to use a method of separation. The sound pressure is obtained as a series, whose convergence is not always as fast as suitable for numerical computations.

For more complex problems, there are mainly two kinds of methods:

- methods which are here termed 'numerical' such as integral equations or finite element methods; they can provide an accurate evaluation of the sound field, for any kind of geometry and any frequency band;
- the geometrical theory of diffraction (G.T.D.) which provides analytic approximations of the sound field, at high frequency.

For some particular problems, other kinds of approximations have been proposed. Some of them are presented for the case of the acoustical thin barriers in Section 4.2.4.

The efficiency and the advantages of these methods generally depend on the frequency band, the dimensions of the obstacle compared with the wavelength and the distance between the obstacle and the observation point.

In Section 4.2.2, an example of application of the separation method is presented. Section 4.2.3 is devoted to the diffraction by a convex cylinder, using the G.T.D., and Section 4.2.4 is devoted to the particular case of screens and barriers.

4.2.2. Diffraction of a plane wave by a circular cylinder

This two-dimensional example is chosen to present an application of the separation method. Many other examples can be found in the books by Morse and Feschbach [17] and Bowman *et al.* [18].

Let us consider a circular cylinder of radius $r = a$. Its boundary is assumed to be described by a homogeneous Dirichlet condition. An incident wave emitted in a plane perpendicular to the cylinder axis can be represented by

$$p_{inc}(M) = e^{\iota kx} = \sum_{m=0}^{\infty} \varepsilon_m \iota^m J_m(kR) \cos(m\phi)$$

$M = (r, \phi)$ is a point outside the cylinder. $\varepsilon_m = \pm 1$. J_m is the Bessel function of order m. The diffracted field is expressed as

$$p_{dif}(M) = \sum_{m=0}^{\infty} A_m H_m(kR) \cos(m\phi)$$

The coefficients A_m are deduced from the boundary condition $p_{inc} + p_{dif} = 0$. Because the cosines are orthogonal functions

$$A_m = \frac{J_m(ka)}{H_m(ka)} \varepsilon_m \iota^m$$

4.2.3. Geometrical theory of diffraction: Diffraction by a convex cylinder

The geometrical theory of diffraction (G.T.D.) was established by J. Keller (see [19] for example). It can be seen as an extension of the laws of geometrical optics to take into account diffraction phenomena. Indeed, the principles of geometrical optics (which are briefly summarized in chapter 5, section 5.5.2) do not provide a correct description of the sound field in the shadow zone behind an obstacle. For example, in Fig. 4.9, the geometrical optics laws imply that the sound pressure is equal to zero in Ω', which is obviously wrong.

Like geometrical optics, G.T.D. assumes that the wave propagation can be modelled by rays but it includes new types of rays: diffracted rays, curved rays, etc. The basic equations of G.T.D. are presented in chapter 5 (section 5.5.3) for

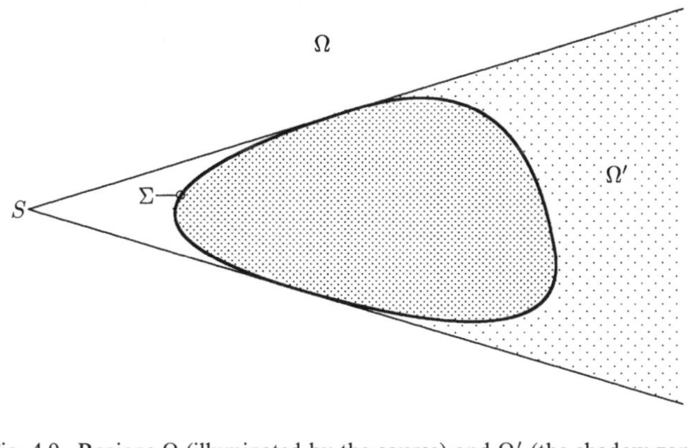

Fig. 4.9. Regions Ω (illuminated by the source) and Ω′ (the shadow zone).

the general case of an inhomogeneous propagation medium (the homogeneous case is a particular case). By solving these equations, one obtains an asymptotic expansion of the sound pressure (in $1/k^n$). The approximations are then valid at high frequency.

G.T.D. is presented here for the particular case of diffraction by a convex cylinder in a homogeneous medium. The sound source is assumed to be cylindrical, with the axis parallel to the axis of the cylinder. The problem is then reduced to a two-dimensional problem. The boundary of the cylinder is characterized by a homogeneous Dirichlet condition. The sound pressure p satisfies the following equations:

$$\begin{cases} (\Delta + k^2)p(M) = \delta_S(M) & \text{in } \mathcal{V} \\ p(M) = 0 & \text{on } \Sigma \\ \text{Sommerfeld conditions} & \end{cases} \tag{4.17}$$

where S is the point source, Σ is the boundary of the obstacle and $\mathcal{V}$ is the region of the plane outside the cylinder ($\mathcal{V} = \Omega \cup \Omega'$ in Fig. 4.9). The shadow zone Ω' is the region located between the two tangents to Σ, passing through S.

Outside the shadow zone Ω', the sound pressure is written as $p = p_{inc} + p_{ref} + p_{dif}$. p_{inc}, p_{ref} and p_{dif} are respectively the incident, reflected and diffracted pressure. In the shadow zone, $p = p_{dif}$.

Each of these terms represents the sum of the sound pressures corresponding to incident, reflected and diffracted rays, respectively.

To evaluate the pressure at a point M, it is then necessary to determine the paths of all the rays emitted by S and passing through M and the sound pressure (amplitude and phase) corresponding to each ray. In the homogeneous case (constant sound speed), rays are straight lines and the phase of the pressure is the product of the wavenumber and the distance (see chapter 5).

Incident field

Let us consider a thin beam composed of incident rays emitted by S (Fig. 4.10). This thin beam passes through M and, because it is incident, does not strike the obstacle. Let ρ be the distance SM. Let A_0 and A be the amplitudes on this beam at distances 1 and ρ respectively. The law of energy conservation on the beam

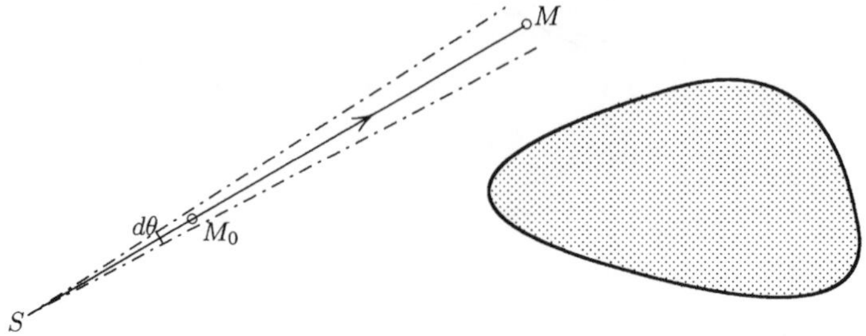

Fig. 4.10. Incident ray.

leads to

$$A^2 \rho \, d\theta = A_0^2 \, d\theta$$

where $d\theta$ is the angle of the beam.

Let us assume for simplicity that the phase of the pressure is zero at S; the incident pressure is then given by

$$p_{inc} = A_0 \frac{e^{\iota k \rho}}{\sqrt{\rho}} \tag{4.18}$$

It must be noted that p_{inc} is not a solution of the Helmholtz equation. It is the first term of the asymptotic expansion of the solution $(-\iota H_0^{(1)}(k\rho)/4)$, for $k\rho \gg 1$.

Reflected field

In Fig. 4.11, the ray PM represents a reflected ray. φ is the incidence angle of the ray SP made with the normal to the boundary Σ. Then, according to the laws of geometrical optics, $(\vec{n}, \overrightarrow{PM}) = -\varphi$.

In order to calculate the amplitude at point M, let us consider a thin beam of parallel rays ($S_1P_1 - S_2P_2$ in Fig. 4.12). When reflected on the obstacle, they lead to a divergent beam ($P_1M_1 - P_2M_2$). P_1M_1 and P_2M_2 generally cross 'inside' the object at a point I, because of the curvature of Σ. Let b be the radius of curvature of Σ at P_1; a, ρ', ρ'' are the distances IP_1, S_1P_1 and P_1M respectively; $d\theta$ is the angle $\widehat{P_1IP_2}$. $d\theta$ is assumed to be small enough that points P_1 and P_2 are close to a middle point P and points M_1 and M_2 are close to a middle point M.

The principle of energy conservation leads to

$$A(M)^2(a + \rho'')d\theta = A(P)^2 a \, d\theta \, \mathfrak{R}^2$$

$A(M)$ is the amplitude at point M. $\mathfrak{R}$ is the reflection coefficient, which depends on the boundary condition on Σ. For example, the homogeneous Dirichlet (resp. Neumann) condition corresponds to $\mathfrak{R} = -1$ (resp. $\mathfrak{R} = 1$).

Also, from the previous paragraph,

$$A(P)^2 = \frac{A_0^2}{\rho'}$$

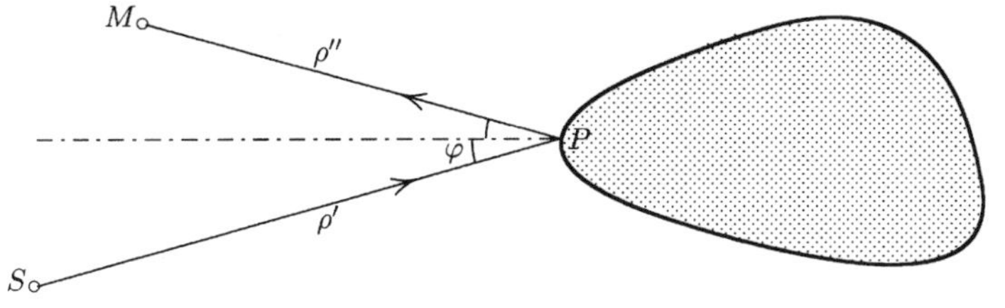

Fig. 4.11. Reflected ray.

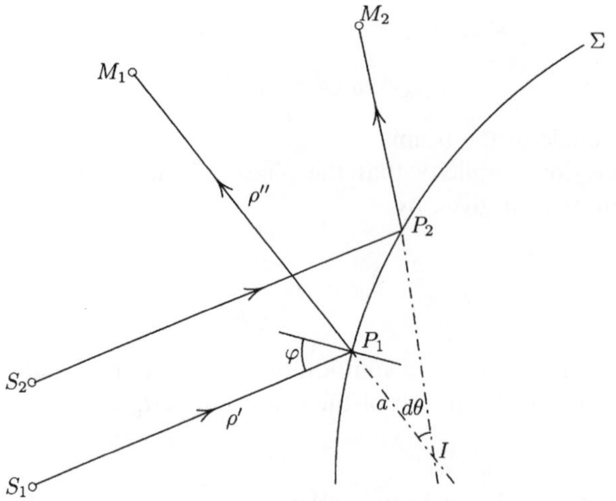

Fig. 4.12. Reflected rays.

Then

$$A(M) = \frac{A_0}{\sqrt{\rho'}} \mathcal{R} \sqrt{\frac{1}{1 + \rho''/a}} \tag{4.19}$$

In formula (4.19), ρ' and ρ'' are known and a is obtained from

$$\frac{1}{a} = \frac{2}{b \cos \varphi} + \frac{1}{\rho'}$$

The phase on the ray SPM has increased from 0 to $k(\rho' + \rho'')$ since $(\rho' + \rho'')$ is the total length of the ray. Then

$$p_{ref} = A_0 \mathcal{R} \sqrt{\frac{1}{\rho'(1 + \rho''/a)}}\, e^{\iota k(\rho' + \rho'')} \tag{4.20}$$

It must be noted that, for the particular case of an infinite plane boundary described by a homogeneous Dirichlet or Neumann condition, this formula provides the first term of the asymptotic expansion of the exact reflected pressure.

Diffracted field

Figure 4.13 presents two diffracted rays emitted by S and arriving at M. The two rays SQ_i arrive tangentially to the obstacle, pass along the boundary and part tangentially. The total diffracted field is obtained as the sum of the fields corresponding to these two rays and the rays which follow the same path and include $1, 2, \ldots, N, \ldots$ turns around the obstacle.

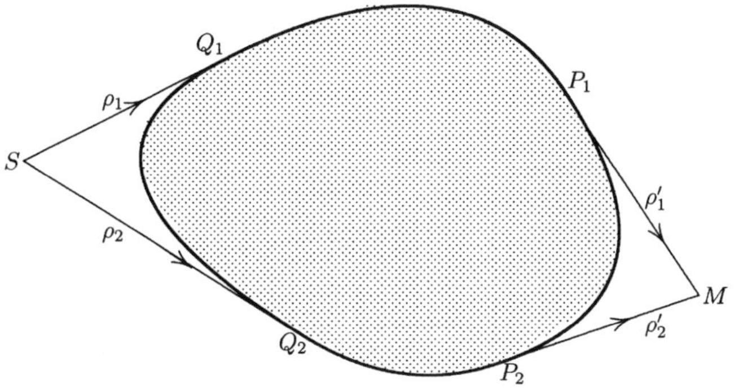

Fig. 4.13. Diffracted rays.

On each ray, the amplitude is again calculated by using the law of energy conservation.

Let us consider the ray SQ_1P_1M. This ray arrives at Q_1 with an amplitude equal to $A_0/\sqrt{\rho_1}$ where ρ_1 is the distance SQ_1 and A_0 is the amplitude at distance unity from the source. At point Q_1, this amplitude is multiplied by a coefficient $\mathfrak{D}(Q_1)$ called diffraction coefficient (by analogy with the reflection coefficient). The function $\mathfrak{D}$ will be determined later. For symmetry reasons, the same coefficient is also introduced at point P_1 and denoted $\mathfrak{D}(P_1)$. Then if A is the amplitude at point P_1, the amplitude at point M will be:

$$\mathfrak{D}(P_1)A/\sqrt{\rho'_1}$$

where ρ'_1 is the distance P_1M.

The behaviour of the amplitude along Q_1P_1 is still to be found. Let t be the abscissa along the obstacle, such that $t = 0$ at Q_1 and $t = t_1$ at P_1. It is assumed that the energy along this path decreases in accordance with

$$A^2(t + dt) = A^2(t) - 2\alpha(t)A^2(t)\ dt$$

where α is a proportionality factor still to be found. This means that between t and $t + dt$, a part of the energy goes away (on other rays) and is assumed to be proportional to the amount of energy at t. Then it is easy to find

$$A(t) = A(0)\exp\left(-\int_0^t \alpha(\tau)\ d\tau\right)$$

Finally, it is shown that the diffracted pressure corresponding to the ray SQ_1P_1M is given by:

$$p^1_{dif}(M) = \frac{A_0}{\sqrt{\rho_1}}\frac{1}{\sqrt{\rho'_1}}\,e^{ik(\rho_1 + t_1 + \rho'_1)}\mathfrak{D}(P_1)\mathfrak{D}(Q_1)\exp\left(-\int_0^{t_1} \alpha(\tau)\ d\tau\right)$$

By summing all the diffracted rays, the total diffracted pressure is obtained as

$$p_{dif}(M) = \frac{A_0}{\sqrt{\rho_1 \rho_1'}} \, e^{\iota k(\rho_1 + t_1 + \rho_1')} \mathfrak{D}(P_1)\mathfrak{D}(Q_1)$$
$$\times \exp\left[-\int_0^{t_1} \alpha(\tau)\, d\tau\right]\left[1 - \exp\left(-\int_0^{T} \alpha(\tau)\, d\tau + \iota kT\right)\right]^{-1}$$
$$+ \frac{A_0}{\sqrt{\rho_2 \rho_2'}} \, e^{\iota k(\rho_2 + t_2 + \rho_2')} \mathfrak{D}(P_2)\mathfrak{D}(Q_2)$$
$$\times \exp\left[-\int_0^{t_2} \alpha(\tau)\, d\tau\right]\left[1 - \exp\left(-\int_0^{T} \alpha(\tau)\, d\tau + \iota kT\right)\right]^{-1} \quad (4.21)$$

where index 2 corresponds to the ray SQ_2P_2M. T is the length of the boundary of the obstacle. The next step is to determine $\mathfrak{D}$ and α. They are obtained by comparing, for the canonical case of a disc, the expansion (4.21) and the asymptotic expansion for large k of the exact solution obtained by the separation method.

For α, there exists an infinite number of solutions $\alpha_n = -\iota k^{1/3} b^{-2/3} \tau_n$ with

$$\tau_n = \tfrac{1}{2}[3\pi(n + 3/4)]^{2/3} e^{\iota\pi/3} \qquad \text{for } n \geqslant 0$$

Keller [19] uses the term ‘cylinder mode’ for each diffracted field corresponding to one value α_n. For each α_n, the expression $\mathfrak{D}$, $\mathfrak{D}_n$, is obtained. It is proportional to an Airy integral.

The total diffracted field is then the sum of all these modes for all the rays passing along Q_1P_1 and all the rays passing along Q_2P_2. Finally

$$p_{dif}(M) \simeq \frac{A_0 e^{\iota\pi/12}}{\sqrt{2k\rho_1\rho_1'}} \, k^{1/3} e^{\iota k(\rho_1 + t_1 + \rho_1')} (b(P_1)b(Q_1))^{1/6}$$
$$\times \sum_{n=0}^{\infty} C_n \exp\left(\iota \tau_n k^{1/3} \int_0^{t_1} b^{-2/3}(\tau)\, d\tau\right)$$
$$\times \left[1 - \exp\left(\iota kT + \iota\tau_n k^{1/3} \int_0^{T} b^{-2/3}(\tau)\, d\tau\right)\right]^{-1} \quad (4.22)$$
$$+ \text{ analogous terms for the rays passing through } P_2 \text{ and } Q_2$$

$b(Q)$ is the radius of curvature at point Q. The coefficients C_n are given in [19]. The first coefficients are

$$\tau_0 = 1.855\,7571\, e^{\iota\pi/3} \qquad \text{and} \qquad \tau_1 = 3.244\,6076\, e^{\iota\pi/3}$$
$$C_0 = 0.910\,7193 \qquad \text{and} \qquad C_1 = 0.694\,2728$$

It must be noted that this expression for p_{dif} becomes infinite when ρ_1 or ρ_1' tends to zero, that is for S or M on the obstacle, but later results have been obtained to avoid these drawbacks ([20], U.T.D., etc). G.T.D. can similarly be used to describe the diffraction (in 2 or 3 dimensions) of a wave by a convex or concave obstacle, a half-plane, a wedge, etc. (see [21], for example).

4.2.4. Diffraction by screens

Many studies have been devoted to this problem, since screens and barriers are useful tools against noise. The aim of these studies is to evaluate the efficiency of the screen, which is defined as the difference of sound levels obtained with and without the screen.

A priori, the laws of geometrical optics imply that a region will be well protected against noise if it is located in the shadow zone of the screen. In practice, this shadow zone phenomenon exists but it is not so sharp.

Diffraction by a thin screen

Most studies have been devoted to the infinitely thin screen since it is the simplest case. The term 'infinitely thin' means that the thickness of the screen is small compared with the acoustic wavelength.

(a) Comparison between a boundary integral equation method and analytic approximations. The example chosen here is diffraction by an infinitely thin screen on a perfectly reflecting plane (see [22]). For simplicity, only the two-dimensional model is considered.

The sound pressure satisfies the following system:

$$\begin{cases} (\Delta + k^2)p(M) = \delta_S(M) & \text{in } ((y > 0) - \Sigma_0) \\ \dfrac{\partial p(M)}{\partial \vec{n}} = 0 & \text{on } \Sigma_0 \text{ and on } (y = 0) \\ \text{Sommerfeld conditions} \end{cases} \tag{4.23}$$

where $S = (x_0, y_0)$, Σ_0 represents the screen and $\vec{n}$ is the unit vector normal to Σ_0 or to the axis $(y = 0)$ (see Fig. 4.14).

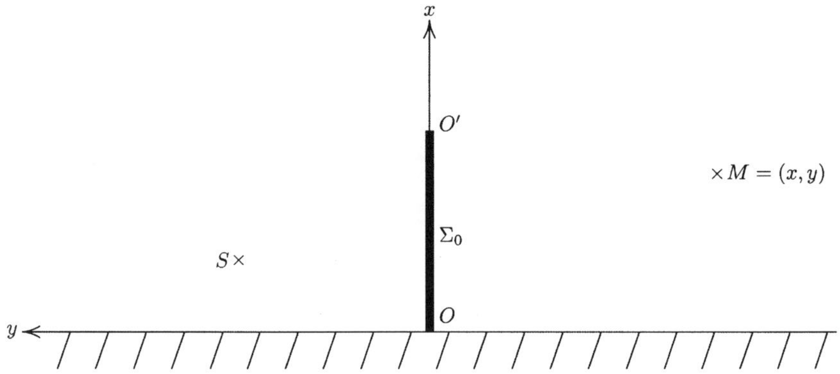

Fig. 4.14. Geometry of the problem.

Since the screen is characterized by a Neumann condition, $p(M)$ is equal to the pressure emitted by two sources S and $S' = (x_0, -y_0)$, in the presence of a screen Σ, centred at 0 and with double height (see Fig. 4.15).

The total sound pressure can be written $p = p_1 + p_2$, where p_1 is the incident pressure emitted by S and S' when there is no screen:

$$\begin{aligned} p_1 &= -\frac{\iota}{4}\,[H_0^{(1)}(kd(S, M)) + H_0^{(1)}(kd(S', M))] \\ &= p_{OS}(M) + p_{OS'}(M) \end{aligned} \tag{4.24}$$

The diffracted field is obtained by solving an integral equation. p_2 is written as a double layer potential (see chapter 3 and [23, 24]):

$$p_2(M) = \int_{\Sigma_0 \cup \Sigma_0'} \mu(P)\,\frac{\partial G(M, P)}{\partial \vec{n}(P)}\,d\sigma(P) \tag{4.25}$$

μ is the density of the potential; it is determined by writing the Neumann condition on $\Sigma_0 \cup \Sigma_0'$. This leads to the following integral equation:

$$\frac{\partial}{\partial \vec{n}(P)}\,p_1(P) + P.F. \int_{\Sigma_0 \cup \Sigma_0'} \mu(P')\,\frac{\partial^2 G(P, P')}{\partial \vec{n}(P')\partial \vec{n}(P)}\,d\sigma(P') = 0 \tag{4.26}$$

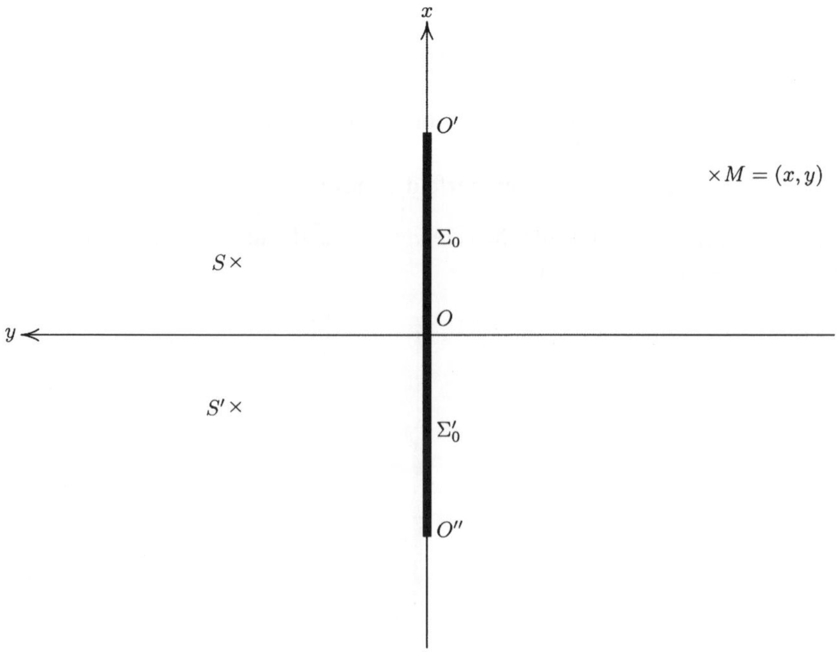

Fig. 4.15. Geometry of the modified problem.

P and P' are points of $\Sigma_0 \cup \Sigma_0'$. The term *P.F.* indicates that the second derivative of $G = -\iota(H_0^{(1)}(kr))/4$ is not integrable and that the integral is only defined as a finite part. The difficulties with and results of this type of equation are presented in [22, 23, 24].

The integral equation is solved by a collocation method (see chapter 6). μ is obtained everywhere on $\Sigma_0 \cup \Sigma_0'$ and the sound field can be computed at any point of the half-plane by using (4.25).

An approximate expression of $p(M)$ has also been obtained using the exact formulae obtained for the diffraction of a wave by a semi-infinite plane [18]. This kind of approximation has been proposed by Maekawa [25] for example. The total field $p(M)$ is approximated by the asymptotic behaviour (at large distance) of the two fields $\phi_1(M)$ and $\phi_2(M)$ where ϕ_i $(i = 1, 2)$ is the field emitted by both sources S and S' in the presence of the semi-infinite screen Σ_i (see Fig. 4.16).

Each ϕ_i can be written $\phi_i = \phi(S, \Sigma_i) + \phi(S', \Sigma_i)$ where each term represents the field emitted by S or S' in the presence of Σ_i.

Using the results obtained for the diffraction of a cylindrical wave by a semi-infinite screen, for distances $\mid OS + OM \mid \gg \lambda$, it is found that the field emitted by S in the presence of Σ_1 can be written as [26]

$$p_S^1(M) = \pm \frac{e^{-\iota\pi/4} e^{\iota k d(S, M)}}{\sqrt{k(d(S, M) + \Delta)}} \times \left[\left(\frac{1}{2} - \int_0^{\sqrt{2/\pi k\delta}} \cos \frac{\pi v^2}{2} dv \right) + \iota \left(\frac{1}{2} - \int_0^{\sqrt{2/\pi k\delta}} \sin \frac{\pi v^2}{2} dv \right) \right] + (1 - C_S(M)) \frac{e^{\iota k d(S, M)}}{\sqrt{kd(S, M)}} \tag{4.27}$$

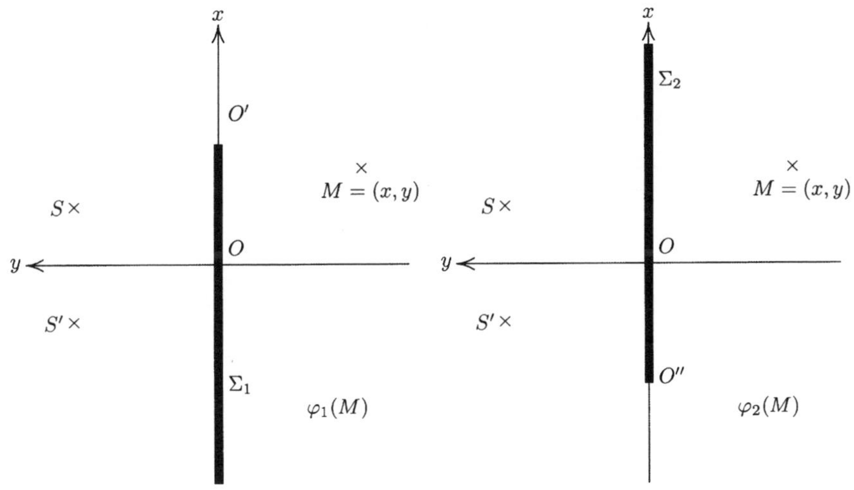

Fig. 4.16. Screens Σ_1 and Σ_2.

with $\Delta = d(S, O') + d(O', M)$; $\delta = \Delta - d(S, M)$

$$C_S(M) = \begin{cases} 1 & \text{if M is in the geometrical shadow of } \Sigma_1 \\ 0 & \text{if not} \end{cases}$$

The + (resp. −) sign corresponds to $\cos\alpha < 0$ (resp. $\cos\alpha > 0$), with $\alpha = (\theta - \beta)/2$ defined as in Fig. 4.17.

By adding the four terms corresponding to both sources S and S' and both screens Σ_1 and Σ_2, $p(M)$ is obtained as

$$p(M)\sqrt{\frac{\pi}{2}}\,e^{\iota\pi/4} \simeq (1 - C_S(M))\,\frac{e^{\iota k d(S,M)}}{\sqrt{kd(S,M)}} + (1 - C_{S'}(M))\,\frac{e^{\iota k d(S',M)}}{\sqrt{kd(S',M)}}$$
$$+ \sum_{j=1}^{4} \frac{-\cos\alpha_j}{|\cos\alpha_j|} \times \sqrt{\frac{D_j}{D_j + \Delta_j}}\,\frac{e^{\iota k D_j}}{\sqrt{kD_j}}$$
$$\times \left[\left(\frac{1}{2} - \int_0^{\sqrt{2/\pi k\delta}} \cos\frac{\pi v^2}{2}\,dv\right) + \iota\left(\frac{1}{2} - \int_0^{\sqrt{2/\pi k\delta}} \sin\frac{\pi v^2}{2}\,dv\right)\right] \tag{4.28}$$

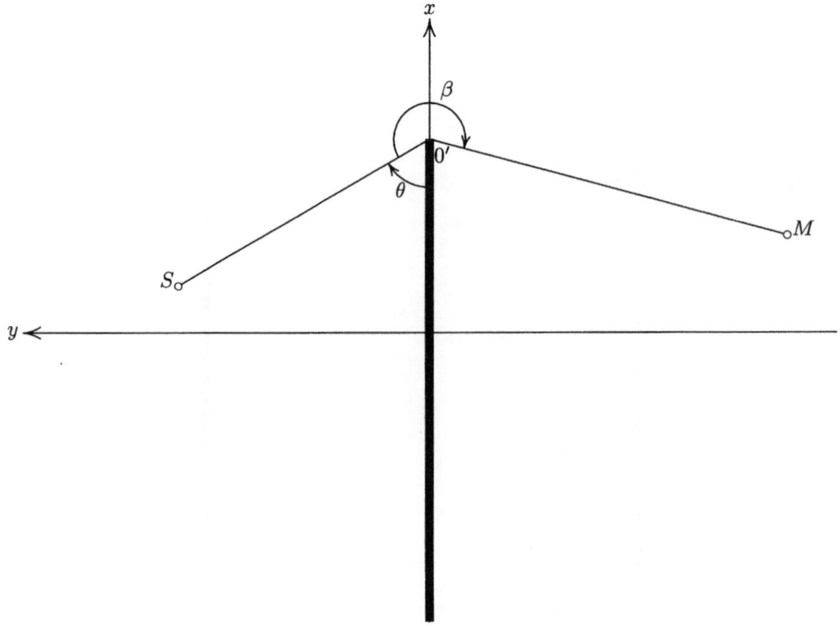

Fig. 4.17. Determination of the angle α.

where index j corresponds to the four cases (S and S', Σ_1 and Σ_2). $D_j = d(s, M)$ and $\Delta_j = d(s, O_j) + d(O_j, M)$ with $s = S$ or S', $O_j = O'$ or O'' depending on the screen, and α_j is defined similarly to α.

An experimental study has been carried out in an anechoic room. Figure 4.18 shows the positions of source and receivers.

The curves in Fig. 4.19 show the efficiency of the screen for one position of the source and three receiver heights (lines D_1, D_2 and D_3). The efficiency of the screen is defined by

$$\text{EFF} = 20 \log \left| \frac{p(M)}{2p_{OS}(M)} \right|$$

EFF represents the comparison between the pressure obtained with the screen and the pressure obtained when there is no screen. The full lines correspond to the solution of the integral equation, the dashed lines correspond to the approximation (4.28) and the circles represent the measurements.

(b) Other kinds of approximations. G.T.D. of course provides 'high frequency' approximations. Other kinds of approximations can be found. They are often based on the Kirchhoff–Fresnel approximation. In [27], many references can be found along with a comparison of several approximations for one geometry. It must be noticed, however, that for the case of a screen on an absorbing plane, the curves are not compared with an exact solution.

Let us end this section with the curves proposed by Maekawa [28], established from several experiments in the case of an infinite half-plane. The author provides a typical attenuation curve versus the number $N = 2\delta/\lambda$. λ is the wavelength. $\delta = d(O, S) + d(O, M) - d(S, M)$ where S is the source and O the end of the screen. Although quite elementary, these curves provide a rough idea of the efficiency of the screen. Kirchhoff approximations or others give similar results within 3 dB.

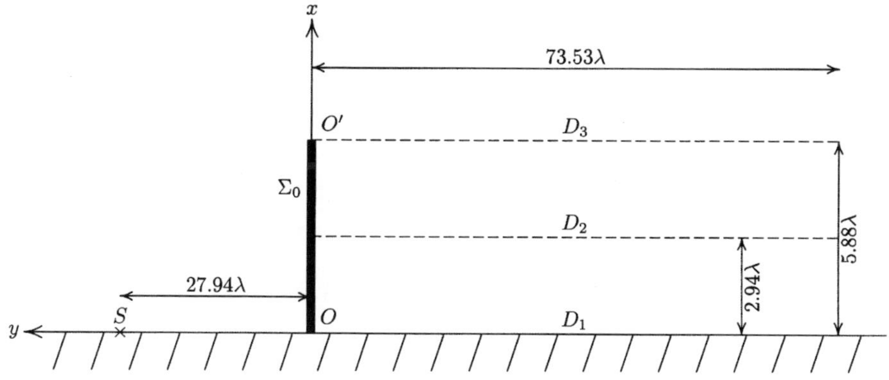

Fig. 4.18. Experiment conducted in an anechoic room.

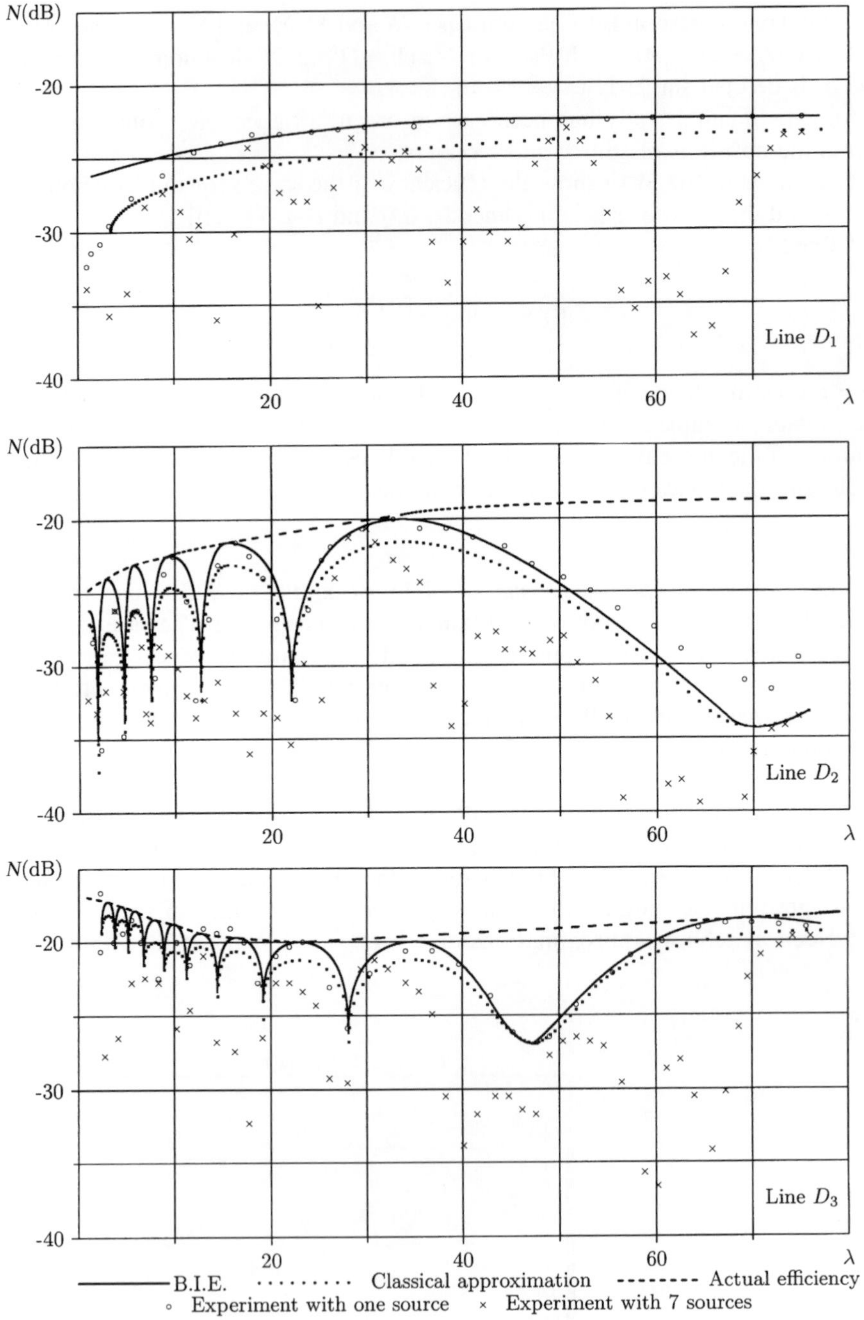

Fig. 4.19. Efficiency of the screen versus distance between screen and receiver [22].

Diffraction by thick screens
High frequency approximations can again be obtained by using G.T.D. Comparisons with experimental results show that these approximations are quite satisfactory. The boundary integral equation methods can provide an accurate prediction of the sound field for any shape or thickness of the screen. However, they are used mainly at low and middle frequencies since computer time and storage increase with frequency. For the particular case of a wedge, results can be found in [29] and [30] for example.

4.3. Sound Propagation in an Inhomogeneous Medium

4.3.1. Introduction

The main application of this paragraph is wave propagation in air and in water.

The study of sound propagation in these media is quite complicated. For propagation in air, two phenomena can become important: (1) gradients of temperature and wind, which imply a varying sound speed; (2) turbulence effects, which imply a random model. The acoustical effects of the inversion of a gradient of temperature are qualitatively well known. They appear, for example, on summer evenings; it is a kind of guided wave phenomenon, the rays propagate within a finite depth layer and their energy is reinforced. Figure 4.20 presents an example of temperature profile as a function of the height above the ground.

The propagation phenomena in water are at least as complex as in air. Because the ocean is non-homogeneous (particles in suspension, changes of temperature, etc.) the density and the sound speed of the medium are functions of space. Because of swell, the surface is in random motion. The model must also take into account

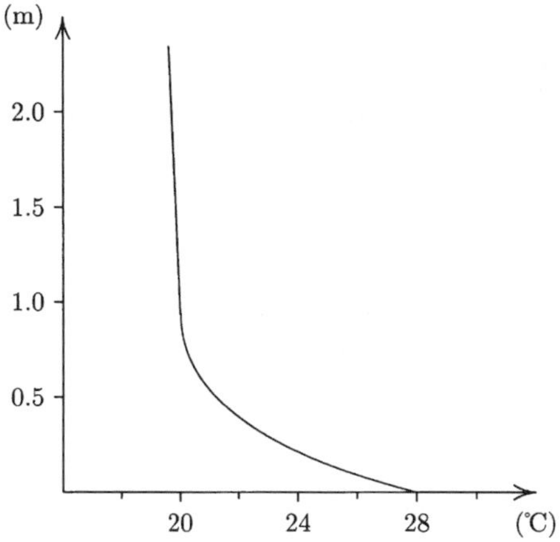

Fig. 4.20. Example of temperature profile in air.

the bottom of the ocean (its geometrical and acoustical or mechanical properties). Also, in real-life problems, all kinds of obstacles are present (from small particles to fish and submarines).

To predict the sound field in such a medium, it is necessary to use simplified models taking into account only the main phenomena of interest. The following sub-sections present several techniques which provide a description of the propagation in an inhomogeneous medium. Their limitations are reviewed and some numerical examples are presented. Random models are not included here: some basic ideas along with references can be found in [31]. A more recent article [32] presents a detailed survey of the models mostly used for underwater propagation. For a quite extensive presentation of the computational aspects of underwater sound propagation, the reader is also referred to the book by Jensen *et al.* [33].

These models can generally be applied to propagation in air. In the following, the propagation medium is characterized by its refractive index (inverse of sound speed) which is a function of depth only. All the techniques can be applied to both media, air and water. The surface of the ocean is assumed to be a plane surface which does not depend on time. It is described by a homogeneous Dirichlet condition. The bottom of the ocean (water–sediment interface) can be characterized by a Neumann condition (perfectly reflecting surface), and an impedance condition (absorbing surface). However, a much better prediction is obtained if the propagation in the sediments is also taken into account, with conditions of continuity (pressure and velocity, or displacement and stress).

All these simplified assumptions provide a first description of the propagation phenomena in an inhomogeneous medium.

Section 4.3.2 is devoted to plane wave propagation. Section 4.3.3 is devoted to cylindrical and spherical wave propagation. In what follows, the index $n(z)$ is defined by $n(z) = c_0/c(z)$. $c(z)$ is the sound speed at depth z and $c_0 = c(z_0)$ is a reference value.

4.3.2. Plane wave propagation

The plane wave case corresponds to simple problems but the results obtained can be used to find the solutions of the spherical wave case or their asymptotic behaviour.

Layered medium

In some cases, the propagation medium can be modelled as a multi-layered medium (Fig. 4.21). Within each layer j, the index $n(z)$ is assumed to be a constant n_j. Each layer j is characterized by an impedance Z_j and a thickness d_j.

When a plane wave impinges on the interface I_p with an angle of incidence θ_{p+1}, a reflected wave is emitted in layer $(p+1)$ and a transmitted wave is emitted in layer p. This transmitted wave then impinges on interface I_{p-1} and two waves (one reflected and one transmitted) are created, and so on. At interface I_1, there will finally appear a transmitted wave in medium 1, with an angle θ_1. At each interface,

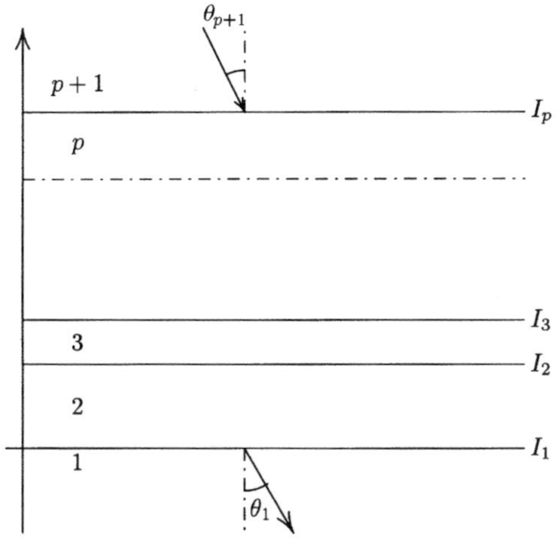

Fig. 4.21. Layered medium. Media (1) and $(p + 1)$ are semi-infinite.

the angles θ_j are given by Snell's law:

$$\frac{\sin \theta_j}{\sin \theta_{j+1}} = \frac{n_{j+1}}{n_j} \tag{4.29}$$

The reflection and transmission coefficients can be expressed as functions of impedances Z_j by using the conditions of continuity of pressure and normal velocity at each interface I_j $(Z_j = \rho_j c_j / \cos \theta_j)$.

For example, if A_{p+1} is the amplitude of the incident wave and A_1 the amplitude of the transmitted wave in layer 1, then [7]

$$\frac{A_1}{A_{p+1}} = \prod_{j=1}^{p} (Z_{in}^{(j)} + Z_j)/(Z_{in}^{(j)} + Z_{j+1})\, e^{\iota \varphi_j} \tag{4.30}$$

where $\varphi_j = k_j d_j \cos \theta_j$ and $Z_{in}^{(j)}$ is the impedance at interface I_j:

$$Z_{in}^{(j)} = Z_j \frac{Z_{in}^{(j-1)} - \iota Z_j \tan \varphi_j}{Z^j - \iota Z_{in}^{(j-1)} \tan \varphi_j}$$

Infinite medium characterized by a varying index. Exact solutions

Let $n(z)$, a continuous function of depth, be the index of the propagation medium. For some particular functions $n(z)$, exact representations of the sound field can be obtained. They are based on special functions, such as Airy functions, hypergeometric functions, etc. Their advantage is mainly to provide an asymptotic behaviour of the sound field. In [7], L. Brekhovskikh presents a list of such functions $n(z)$ for which there exist exact solutions. The author considers the following two-dimensional problem.

The propagation is characterized by a function $k(z)$, $(k(z) = \omega/c(z))$. The inhomogeneities of the medium are assumed to be concentrated in a layer and constants k_0 and k_1 are the limits of $k(z)$ when z tends to $(-\infty)$ and to $(+\infty)$ respectively. An incident plane wave, of amplitude 1, propagates with an angle θ_0 from the z-axis:

$$p_{inc}(x, z) = \exp[\iota(k_0 x \sin\theta_0 - k_0 z \cos\theta_0)]$$

Because of the limits k_0 and k_1 of $k(z)$, there is a reflected wave in the region $(z \longrightarrow -\infty)$, which can be written as

$$p_{ref}(x, z) = V \exp[\iota(k_0 x \sin\theta_0 + k_0 z \cos\theta_0)]$$

and a transmitted wave in the region $(z \longrightarrow +\infty)$, written as

$$p_t(x, z) = W \exp[\iota(k_1 x \sin\theta_1 - k_1 z \cos\theta_1)]$$

where θ_1 is defined by $k_0 \sin\theta_0 = k_1 \sin\theta_1$. V and W are the reflection and transmission coefficients.

In the layer, $p(x, z)$ must be a solution of the propagation equation

$$(\Delta + k^2(z))p(x, z) = 0 \tag{4.31}$$

It is expressed as $p(x, z) = A(z)\exp(\iota\xi x)$. Then A is the solution of the following equation

$$A''(z) + (k^2(z) - \xi^2)A(z) = 0 \tag{4.32}$$

If $k(z)$ corresponds to an Epstein profile, that is if n is such that

$$n^2(z) = 1 - N\mathrm{e}^{mz}/(1 + \mathrm{e}^{mz}) - 4M\mathrm{e}^{mz}/(1 + \mathrm{e}^{mz})^2$$

where N, M and m are constants, $A(z)$ is a hypergeometric function. Figure 4.22 shows two examples of function $(1 - n^2(z))$.

If $k(z)$ is such that $n^2(z) = 1 \pm az$, with a equal to a constant, $A(z)$ is an Airy function.

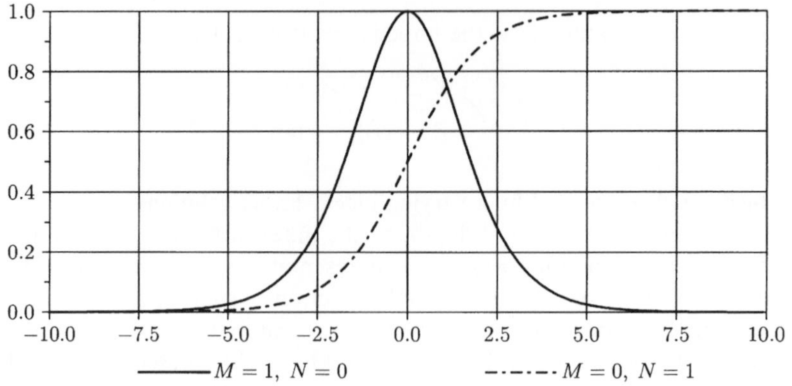

Fig. 4.22. Functions $(1 - n^2(z))$ for $N = 0$, $M = 1$ and $M = 0$, $N = 1$.

The expressions for V and W can be deduced from the conditions of continuity of the sound field at the boundaries of the layer. In [7], some examples present the behaviour of V and W as functions of frequency, of the angle of incidence and of the width of the domain in which k varies.

Infinite medium characterized by a varying index. W.K.B. method

In most cases, the solution of (4.31) cannot be obtained in a closed form, but the W.K.B. method can be used to find its asymptotic behaviour for $((\lambda/\Lambda) \ll 1)$, where λ is the acoustic wavelength and Λ the characteristic length of the index and of the solution.

The basic features of the method are presented in chapter 5, section 5.4.1. Let us consider the simple problem of determining the function $p(x, z)$ solution of a Helmholtz equation, where $n(z)$ is a slowly varying function compared with the wavelength:

$$(\Delta + k_0^2 n^2(z))p(x, z) = 0$$

$k_0 = \omega/c_0$, c_0 is a reference value of the sound speed. p is then written as

$$p(x, z) = \mathrm{e}^{\iota k_0 x \sin\theta} \mathrm{e}^{\iota k_0 M(z)}$$

where $M(z)$ is a series in $(k_0)^{-1}$.

Substituting this representation into the Helmholtz equation, the left-hand side becomes a series and the right-hand side is still zero. By equating each term of the series to zero, it is possible to find the coefficients of $M(z)$. Using only the first terms, p can be approximated by

$$\begin{aligned} p(x, z) \simeq{} & (n^2 - \sin^2\theta)^{-1/4} \exp(\iota k_0 x \sin\theta) \\ & \times \left\{ C_1 \exp\left(\iota k_0 \int_{z_0}^{z} \sqrt{n^2 - \sin^2\theta}\, dz \right) \right. \\ & \left. + C_2 \exp\left(-\iota k_0 \int_{z_0}^{z} \sqrt{n^2 - \sin^2\theta}\, dz \right) \right\} \end{aligned} \tag{4.33}$$

p is expressed as the sum of two waves propagating in opposite directions, with no interaction. It must be noted that (4.33) cannot be used if $n^2(z)$ is close to $\sin^2\theta$. In this last case, it is still possible to find another representation of p, by using the first terms of the Taylor expansion of $n^2(z)$ for example (see chapter 5).

This kind of representation (4.33) can also be used as the initial data of an iterative algorithm (see [34] for example). Some applications of the W.K.B. method can be found in [35] where several examples of propagation in an inhomogeneous medium are studied, and in [36] for underwater sound propagation.

4.3.3. Propagation of cylindrical and spherical waves

The books by J. Keller [31] and by Jensen *et al.* [33] present a very good survey of the methods used in underwater acoustics. The same methods can also be applied in air. If the index is a function of depth z only and if the surface and bottom of the ocean are planes ($z = \text{constant}$) described by 'local reaction' conditions, it is possible to obtain a representation of the sound pressure by using a spatial Fourier transform (as in Section 4.1.2), or a series of modes.

At high frequency, the W.K.B. method and ray methods also provide approximations of the sound pressure. A ray method is used, for example, in [37] to describe the sound propagation above the (perfectly reflecting) surface of a lake, with a wind profile. A comparison between experimental and theoretical results shows that the method is quite satisfactory for source and receiver above the surface. For receiver heights equal to 5 and 12 m and a horizontal distance source-receiver equal to 750 m, the attenuation is about 60 dB for a wind speed equal to 4 m/s. The difference of sound levels (15 dB) for a receiver in the wind direction and a receiver in the opposite direction is also correctly predicted.

When n is a function of both radial distance and depth and when the surface or the bottom of the ocean are not parallel planes, two main methods are left: G.T.D. and parabolic approximation.

Geometrical theory of diffraction

With G.T.D., it is possible to take into account the reflection phenomena on the boundaries and the refraction phenomena caused by the inhomogeneities of

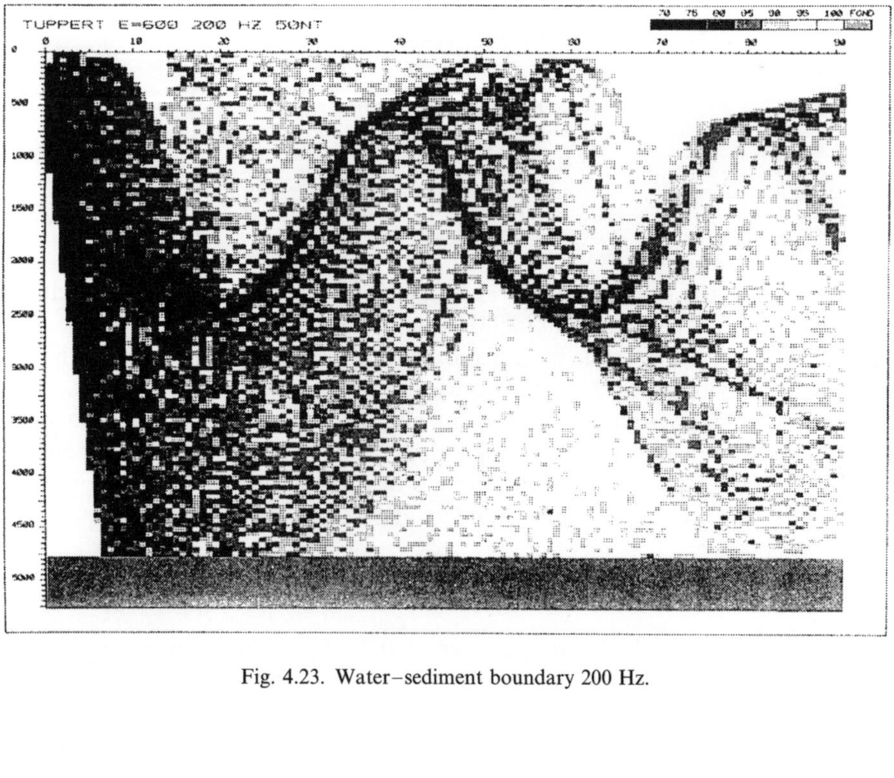

Fig. 4.23. Water–sediment boundary 200 Hz.

the medium. The general procedure is presented in chapter 5, section 5.5.3. The sound field is expressed as a sum of sound fields evaluated along incident, reflected and diffracted rays as well as complex rays. Because the equations are quite difficult to solve (like the transport equation which gives the amplitude of the field on the ray) other methods are also used: a method using normal modes (in depth) and horizontal rays [31] or the parabolic approximation (especially at low frequency).

Parabolic approximation
This method consists in replacing the Helmholtz equation by a parabolic equation, which is easier to solve numerically. Details on the procedure and the advantages of the method are presented in chapter 5, section 5.6, along with references.

The parabolic equation can be solved by a split-step Fourier method, based on an FFT-method as in [31]. One of the examples presented in [31] corresponds to the ocean medium with a bilinear speed profile. The sound speed decreases from the surface down to $h = 1200$ m and then increases from $h = 1200$ m to the bottom of the ocean ($h = 4800$ m). Figures which present the ray pattern for horizontal distances up to 150 km and for frequencies between 25 and 200 Hz clearly point out the shadow zones caused by the curvature of the rays.

The parabolic equation can also be solved by techniques based on the finite difference method (see [33] for example and Figs 4.23 and 4.24 [38]).

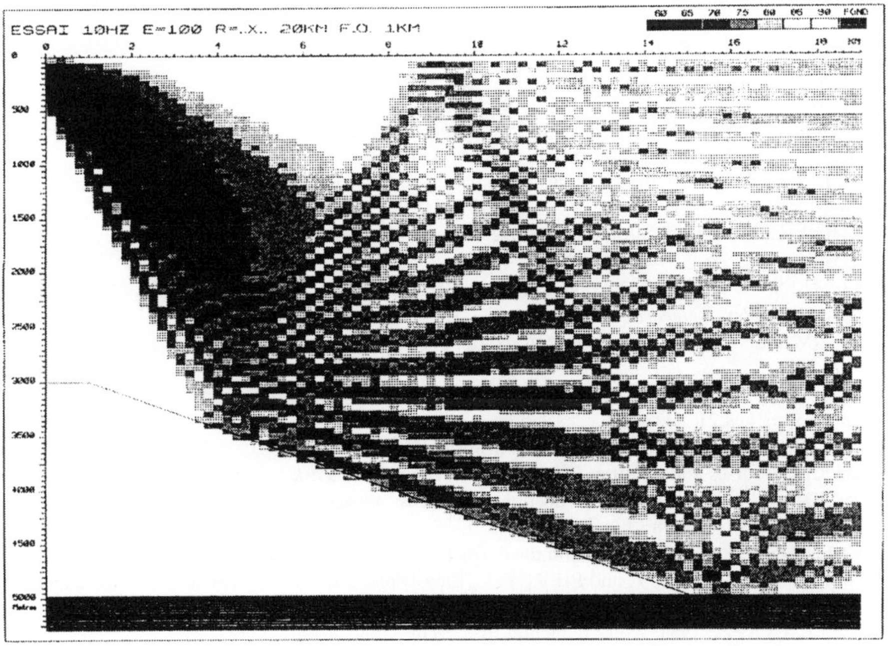

Fig. 4.24. Oblique bottom 10 Hz.

Bibliography

[1] ERDELYI, A., 1956. *Asymptotic expansions.* Dover Publications, New York.

[2] BRIQUET, M. and FILIPPI, P., 1977. Diffraction of a spherical wave by an absorbing plane. *J. Acoust. Soc. Am.* 61(3), 640–646.

[3] FILIPPI, P., 1983. Extended sources radiation and Laplace type representation: application to wave propagation above and within layered media. *J. Sound Vib.* 91(1), 65–84.

[4] LUKE, Y., 1975. *Mathematical functions and their approximations.* Academic Press Inc., New York.

[5] ABRAMOWITZ, M. and STEGUN, I., 1972. *Handbook of mathematical functions.* Dover Publications, New York.

[6] INGARD, U., 1951. On the reflection of a spherical sound wave from an infinite plane. *J. Acoust. Soc. Am.* 23(3), 329–335.

[7] BREKHOVSKIKH, L.M., 1980. *Waves in layered media.* Academic Press, New York.

[8] DICKINSON, P. and DOAK, P., 1970. Measurements of the normal acoustic impedance of ground surfaces. *J. Sound Vib.* 13(3), 309–322.

[9] HAZEBROUCK, R., 1981. Mesure de l'impédance acoustique et du coefficient d'absorption d'un revêtement par une méthode de filtrage temporel. *Rev. Acoust.* 56, 4–10.

[10] BERENGIER, M., 1983. *Propagation acoustique au voisinage du sol.* Internal report, Laboratoire Central des Ponts et Chaussées, Nantes.

[11] HABAULT, D. and CORSAIN, G., 1985. Identification of the acoustical properties of a ground surface. *J. Sound Vib.* 100(2), 169–180.

[12] LEGEAY, V. and SEZNEC, R., 1983. Sur la détermination des caractéristiques acoustiques des matériaux absorbants. *Acustica* 53(4), 171–192.

[13] DELANY, M. and BAZLEY, E., 1970. Acoustical properties of fibrous absorbant materials. *Appl. Acoust.* 3(2), 105–116.

[14] HABAULT, D., 1985. Sound propagation above an inhomogeneous plane: boundary integral equation methods. *J. Sound Vib.* 100(1), 55–67.

[15] DURNIN, J. and BERTONI, H.L., 1981. Acoustic propagation over ground having inhomogeneous surface impedance. *J. Acoust. Soc. Am.* 70(3), 852–859.

[16] HEINS, A. and FESCHBACH, H., 1954. On the coupling of two half-planes. *Proc. of Symposia in Appl. Math. V*, pp. 75–87. McGraw-Hill, New York.

[17] MORSE, P. and FESCHBACH, H., 1953. *Methods of theoretical physics.* McGraw-Hill, New York.

[18] BOWMAN, J., SENIOR, T. and USLENGHI, P., 1969. *Electromagnetic and acoustic scattering by simple shapes.* North-Holland, Amsterdam.

[19] KELLER, J., 1955. Diffraction by a convex cylinder. *Electromagnetic wave theory symposium,* pp. 312–321.

[20] LUDWIG, D., 1966. Uniform asymptotic expansions at a caustic. *Commun. Pure Appl. Math.* 19, 215–250.

[21] COMBES, P., 1978. Introduction à la théorie géométrique de la diffraction et aux coefficients de diffraction. *Rev. Cethedec* 55, 77–104.

[22] DAUMAS, A., 1978. Etude de la diffraction par un écran mince disposé sur le sol. *Acustica* 40(4), 213–222.

[23] FILIPPI, P. and DUMERY, G., 1969. Etude théorique et numérique de la difraction par un écran mince. *Acustica* 21, 343–350.

[24] FILIPPI, P., 1983. *Integral equations in acoustics in theoretical acoustics and numerical techniques.* C.I.S.M. Courses and Lectures 277. Springer-Verlag, New York.

[25] MAEKAWA, Z., 1965. Noise reduction by screens. *Memoirs of the Faculty of Engineering* (11). Kobe University, Japan.

[26] CLEMMOW, P.C., 1950. A note on the diffraction of a cylindrical wave. *Q. J. Mech. Appl. Math.* 3.

[27] ISEI, T., EMBLETON, T. and PIERCY, J., 1980. Noise reduction by barriers on finite impedance ground. *J. Acoust. Soc. Am.* 67(1), 46–58.

[28] MAEKAWA, Z., 1968. Noise reduction by screens. *Appl. Acoust.* 1, 157–173.

[29] PIERCE, A., 1981. *Acoustics: an introduction to its physical principles and applications.* McGraw-Hill, New York.

[30] AMBAUD, P. and BERGASSOLI, A., 1972. Le problème du dièdre en acoustique. *Acustica* 27, 291–298.

[31] KELLER, J., 1977. *Wave propagation and underwater acoustics.* Lecture Notes in Physics 70 Springer-Verlag, New York.

[32] BUCKINGHAM, M.J., 1992. Ocean acoustic propagation models. *J. Acoust.* 3, 223–287.

[33] JENSEN, F.B., KUPERMAN, W.A., PORTER, M.B., and SCHMIDT, H., 1994. *Computational ocean acoustics.* American Institute of Physics, New York.

[34] DE SANTO, J., 1979. *Ocean Acoustics.* Topics in Current Physics 8. Springer-Verlag, New York.

[35] BAHAR, E., 1967. Generalized WKB method with applications to problems of propagation in nonhomogeneous media. *J. Math. Phys.* 8(9), 1746–1753.

[36] HENRICK, R.F., BRANNAN, J.R., WARNER, D.B. and FORNEY, G.P., 1983. The uniform WKB modal approach to pulsed and broadband propagation. *J. Acoust. Soc. Am.* 74(5), 1464–1473.

[37] SIMONET, M., 1979. Sur l'estimation de la gêne maximale provoquée par un bruiteur aux distances moyennes. Thèse de 3ème Cycle en Acoustique, Université d'Aix-Marseille I, France.

[38] GRANDVUILLEMIN, B., 1985. Application de l'approximation parabolique à l'Acoustique sous-marine. Thèse de 3ème Cycle en Acoustique, Université d'Aix-Marseille I, France.

CHAPTER 5

Analytic Expansions and Approximation Methods

Dominique Habault

Introduction

This chapter presents a general survey of the main approximation methods used in acoustics. The methods used to solve acoustics problems can be very roughly divided into two groups:

- The methods here called 'purely numerical', such as finite element or boundary integral equation methods. They consist in solving the Helmholtz equation or an equivalent equation in a straightforward way. They can be used with no particular assumptions.
- The analytical and asymptotic methods such as the method of steepest descent or geometrical theory of diffraction which provide approximate expressions of the solution or simpler equations (parabolic approximation) which are then solved by a numerical procedure. They are based on assumptions such as low or high frequency, large distance, etc.

The methods presented in this chapter belong to the second group.

Most of them provide asymptotic expansions of the solution and can be a good tool for studying the respective influence of the parameters of a problem. This is quite interesting because of the following observation. Complicated problems are always difficult to solve satisfactorily even with powerful numerical means and it is useless to include a complete description in all details. Thus before using a numerical procedure, it is necessary to analyse all the aspects of the phenomena and their relative influence on the behaviour of the solution in order to reduce the initial problem to as simple a model as possible.

These approximation methods can be used in computer programs in several ways:

- approximate expressions can be used as input data of an iterative algorithm;
- simple problems for which approximate solutions are known can be used as 'benchmarks' to validate numerical algorithms;

- approximate expressions can be used to provide *a priori* estimates of the solution and the behaviour can be introduced in test functions to speed up the convergence of an algorithm.

All these remarks point out that even though 'real-life' problems are too complex to be solved by some of the methods presented in this chapter, these methods must not be ignored, even nowadays with regularly increasing computer power.

Each method is applied here to the Helmholtz equation, for a time-harmonic signal ($\exp(-\iota\omega t)$). Most of the approximation methods consist in expressing the solution as a series, with a large or small parameter. This series can be convergent. In most cases, it is an asymptotic series. Let us recall that the series $\sum_{n=0}^{N} a_n u_n(x)$ is an asymptotic expansion of $u(x)$ of order N when x tends to x_0, if

$$u(x) = \sum_{n=0}^{N} a_n u_n(x) + o(u_N) \qquad \text{when } x \longrightarrow x_0$$

that is the first neglected term u_{N+1} is small compared with the last included term u_N. From a mathematical point of view, the relation $u_{N+1} = o(u_N)$ when x tends to x_0 means that for any $\varepsilon > 0$, there exists a neighbourhood D_ε of x_0 such that $|u_{N+1}| \leqslant \varepsilon |u_N|$ for every x in D_ε [1].

The term 'asymptotic' means that the sum of the first N terms is a better approximation when x is closer to x_0. From a numerical point of view, this means that for x fixed, taking more terms of the series into account does not necessarily improve the approximation of $u(x)$, as is the case for a convergent series. In [2], G. Arfken presents the asymptotic behaviour of the exponential integral $E_1(x)$ when x tends to infinity. A numerical example shows that, for x equal to 5, the sum of the first N terms provides a correct approximation for $N = 6$ and then diverges for $N > 6$.

Let us also emphasize that it is difficult to know *a priori* the exact limitations of an approximation. For example, 'high frequency approximation' means that the accuracy of the approximation increases when frequency increases but it may be also satisfactory at middle or even low frequency. An approximation method is considered to be satisfactory if predicted results are close to measured results. In that sense, comparisons between calculations and measurements provide a better practical knowledge of the limitations of a method.

Section 5.1 is devoted to some methods which provide asymptotic expansions from integral representations. Section 5.2 presents the Kirchhoff approximation for diffraction by a hole or a plane screen; it corresponds to the geometrical optics approximation. Section 5.3 shows how to use a Neumann series to obtain an approximation of the solution of an integral equation. Section 5.4 presents approximation techniques applied to propagation in a slowly varying medium. Section 5.5 is devoted to image and ray methods and to the geometrical theory of diffraction (G.T.D.). G.T.D. extends the geometrical optics laws to take into account diffraction phenomena. It applies to wave propagation in inhomogeneous media and wave diffraction by obstacles. The parabolic approximation method is presented in Section 5.6. It applies to wave propagation in inhomogeneous media.

It consists in replacing the Helmholtz equation by a parabolic equation which is then solved numerically. It is mainly a numerical method but it is based on some assumptions such as narrow- or wide-angle aperture, large distances, etc. Finally, a description of the Wiener–Hopf method is included in Section 5.7. Strictly speaking, the Wiener–Hopf method is not an approximation method but in most cases only provides approximations of the solution.

To get a more detailed knowledge of the methods presented here, the reader will find references at the end of the chapter. It must be noted that most of these methods come from other fields of physics (optics, electromagnetism, etc.).

5.1. Asymptotic Expansions Obtained from Integral Expressions

For propagation and diffraction problems, the sound pressure can sometimes be obtained as an integral expression by using a spatial Fourier transform or a Green's representation, for example. As seen in chapter 4, section 4.1, in some typical cases (infinite plane boundaries) it is easier to find the Fourier transform of the solution; the solution is then expressed as an inverse Fourier transform of a known function.

For certain types of propagation problems, the sound pressure can be expressed by using integrals of the kind

$$I(x) = \int_a^b g(t) \exp(xh(t))\, dt$$

where g and h are two real or complex functions. a and b are finite or infinite, and x is a 'large' parameter.

This kind of integral can be obtained, for example, by using Fourier or Laplace transforms. In acoustics, the 'large' parameter is often a distance R or the product kR of the wavenumber and the distance.

The methods which provide the asymptotic expansions of this kind of integral are presented in detail in [1], see also [6]. The last three sections present a brief survey of the main results.

5.1.1. Elementary kernels

Let $p(M)$ be the solution of

$$\begin{cases} (\Delta + k^2)p(M) = f(M) & \text{in } \mathbb{R}^3 \\ \text{Sommerfeld conditions} \end{cases}$$

f represents the sources. M is a point of the 3-dimensional space $\mathbb{R}^3$, with spherical coordinates (R, φ, θ). Let $G(M, M')$ be the elementary kernel which satisfies

$$\begin{cases} (\Delta + k^2)G(M, M') = \delta_M(M') & \text{in } \mathbb{R}^3 \\ \text{Sommerfeld conditions} \end{cases}$$

$p(M)$ can then be expressed as

$$p(M) = \int f(M')G(M, M')d\sigma(M') \tag{5.1}$$

where $M' = (R', \varphi', \theta')$ is another point of $\mathbb{R}^3$. The integration domain is $\mathbb{R}^3$ but obviously reduces to the support of the sources (domain where $f \neq 0$). The series expansion of G provides an asymptotic expansion of p. Indeed [3]

$$G(M, M') = -\frac{\iota k}{2\pi} \sum_{n=0}^{\infty} (2n+1) \sum_{m=0}^{n} \varepsilon_n \frac{(n-m)!}{(n+m)!} \cos m(\varphi - \varphi')P_n^m(\cos\theta)$$

$$\times P_n^m(\cos\theta') \left| \begin{array}{ll} j_n(kR')h_n(kR) & \text{if } R > R' \\ j_n(kR)h_n(kR') & \text{if } R' > R \end{array} \right. \tag{5.2}$$

P_n^m is the Legendre function of degree n and of order m. j_n and h_n are the spherical Bessel functions of first and third kind respectively and of order n. h_n is written for $h_n^{(1)} = j_n + \iota y_n$.

Using:

$$h_n(kR) = \frac{e^{\iota kR}}{\iota^{n+1}kR} \sum_{p=0}^{n} \frac{\Gamma(n+1+p)}{p!\Gamma(n+1-p)} \left(\frac{-1}{2\iota kR}\right)^p$$

and the expansion

$$\exp[-\iota kR'(\sin\theta \sin\theta' \cos(\varphi - \varphi') + \cos\theta \cos\theta')]$$

$$= \sum_{n=0}^{\infty} (2n+1) \sum_{m=0}^{n} \varepsilon_n \frac{(n-m)!}{(n+m)!} \cos(m(\varphi - \varphi'))P_n^m(\cos\theta)P_n^m(\cos\theta')j_n(kR')$$

the first term of the asymptotic series of G (for $R > R'$) can easily be obtained:

$$G(M, M') = -\frac{e^{\iota kR}}{2\pi R} \exp(-\iota kR'(\sin\theta \sin\theta' \cos(\varphi - \varphi') + \cos\theta \cos\theta')) + \mathbb{O}(R^{-2}) \tag{5.3}$$

To find the asymptotic behaviour of $p(M)$ when R tends to infinity, the procedure is the following:

- the expansion of G (5.2) is substituted into the integral expression (5.1);
- two integrals are obtained, one on $[0, R] \times [0, 2\pi] \times [-\pi/2, \pi/2]$, the other one on $[R, \infty] \times [0, 2\pi] \times [-\pi/2, \pi/2]$;
- since R tends to infinity, the second integral is neglected and the approximation (5.3) is introduced in the first integral;
- the integral terms are expressed as a Fourier transform of f.

This leads, for instance, to

$$p(M) = -\frac{e^{\iota kR}}{2\pi R} \hat{f}(k \sin\theta \cos\varphi, k \sin\theta \sin\varphi, k \cos\theta) + \mathbb{O}(R)^{-2} \tag{5.4}$$

where $\hat{f}$ is the spatial Fourier transform of $f(x, y, z)$ defined by

$$\hat{f}(\xi, \eta, \zeta) = \int_{-\infty}^{+\infty} \int_{-\infty}^{+\infty} \int_{-\infty}^{+\infty} e^{\iota(x\xi + y\eta + z\zeta)} f(x, y, z)\, dx\, dy\, dz$$

The asymptotic expansion of G then provides, when introduced in relation (5.2), an expression of the asymptotic field radiated (at large distance) by a source distribution f. If f is known, it is generally easy to find its Fourier transform $\hat{f}$. The same method can of course provide higher order approximations, if necessary.

The particular case of layer potentials. The same method still applies. For example, let $\psi(M)$ be a simple layer potential, with a density μ which depends only on the radial coordinate ρ:

$$\psi(M) = \int_\Sigma \mu(\rho(M')) \frac{e^{\iota kR(M, M')}}{4\pi R(M, M')} d\sigma(M')$$

M' is a point of the plane Σ: $(z = 0)$. Then $\psi(M)$ can be approximated by [4]

$$\psi(M) \simeq \frac{e^{\iota kR(O, M)}}{4\pi R(O, M)} \left[\hat{\mu}(k \sin\theta) + \frac{1}{2\iota kR(O, M)} \left(\frac{\partial \hat{\mu}(k \sin\theta)}{\partial\theta} \cot\theta + \frac{\partial^2 \hat{\mu}(k \sin\theta)}{\partial^2\theta} \right) \right] \tag{5.5}$$

θ is the angular coordinate of M, measured from the normal to the surface Σ.

In the previous chapters, it has been shown that, for several kinds of propagation problems, the sound pressure can be expressed as a sum of layer potentials. In [4] and [5], two examples of this kind of expansion are presented. The first one is applied to the prediction of sound propagation above the ground, the second to the sound radiation of a plate immersed in a fluid. Similar results have also been obtained in two dimensions by using the expansion of the Green's kernel $H_0(kR)$ in cylindrical harmonics.

5.1.2. Integration by parts, Watson's lemma

The asymptotic expansion of some integrals can be simply obtained by a method of integration by parts. Let us consider the following example:

$$I(x) = \int_0^{+\infty} \frac{\exp(-xt)}{(1+t)} dt \tag{5.6}$$

Integrating N times by parts leads to:

$$I(x) = \sum_{n=1}^{N} (-1)^n \frac{(n-1)!}{x^n} + (-1)^N \frac{N!}{x^N} \int_0^{+\infty} \frac{\exp(-xt)}{(1+t)^{N+1}} dt$$

It can be shown that the integral on the right-hand side behaves as $(1/x)$ when x tends to infinity. The sum of the N first terms is then an asymptotic expansion of $I(x)$ when x tends to infinity. Another way to obtain this asymptotic expansion of $I(x)$ is to replace $(1+t)^{-1}$ by its convergent series:

$$\frac{1}{1+t} = \sum_{n=0}^{N} (-1)^n t^n \qquad \text{for } |t| < 1 \tag{5.7}$$

Although this series is not convergent for all t real and positive, by introducing (5.7) in expression (5.6) and by integrating term by term, one obtains an asymptotic expansion of $I(x)$ for large x.

This result is more general, following Watson's lemma:

Watson's lemma. Let $I(x) = \int_0^A t^m f(t) \exp(-xt)\, dt$ where $f(t)$ is an analytic function on $[0, A]$ such that $f(0) \neq 0$. A is real and can be infinite. m is real and greater than (-1). Let us assume that $I(x)$ exists for at least one value x_0 of x. If $f(t)$ can be written as

$$f(t) = \sum_{n \geqslant 0} a_n t^n$$

with the series convergent for $|t| < t_0$, then $I(x)$ has the following asymptotic expansion:

$$I(x) \sim \sum_{n \geqslant 0} \frac{(m+n)!a_n}{x^{m+n+1}}$$

when x tends to infinity. The '$\sim$' sign is used instead of the equals sign because the series is asymptotic (often not convergent).

Remark [7]. The lemma still holds if f is finite on $[0, A]$ or if f is infinite only at some points of $[0, A]$. Essentially, $I(x)$ must exist for some value x_0.

5.1.3. The method of stationary phase

This method applies to rapidly oscillating integrals such as

$$I(x) = \int_a^b g(t) \exp(\iota x h(t))\, dt$$

a, b and x are real, x is 'large'. h is a real function, twice continuously differentiable. g is a continuous, real function. $I(x)$ can also be written

$$I(x) = \int_a^b \frac{g(t)}{\iota x h'(t)} \frac{d}{dt} [\exp(\iota x h(t))] dt$$

Then by integrating by parts, one obtains

$$I(x) = \frac{1}{\iota x} \left[\frac{g(t) e^{\iota x h(t)}}{h'(t)} \right]_a^b - \frac{1}{\iota x} \int_a^b \left(\frac{g(t)}{h'(t)} \right)' e^{\iota x h(t)}\, dt$$

It can be shown that if $h'(t)$ never equals zero on $[a, b]$, this is an asymptotic expansion in $(1/x)$ for x large. However, if there exists a point t_0 on $[a, b]$ such that $h'(t_0) = 0$ (t_0 is called a stationary point), the method of stationary phase shows that the first term of the asymptotic expansion is of order $(1/x)^{-1/2}$. The book by F.W.J. Olver [8] presents this method in detail, along with several examples of applications and a brief history.

In the following, the main steps used to evaluate the first term of the expansion are described without mathematical proof.

It is assumed that t_0 such that $(a < t_0 < b)$ is the only stationary point on $[a, b]$ and that $h''(t_0) \neq 0$. With the following change of variable

$$h(t) = h(t_0) + \frac{1}{2}\, h''(t_0)u^2$$

$I(x)$ can be expressed as:

$$I(x) = h''(t_0)e^{\iota xh(t_0)} \int_{u(a)}^{u(b)} \frac{g(t(u))}{h'(t(u))}\, u \; \exp\,(\iota xu^2h''(t_0)/2)\; du$$

Then u tends to zero when t tends to t_0. The main idea of the method is that, when x tends to infinity, the major contribution for $I(x)$ is given by the behaviour of the expression under the integral sign around the stationary point ($t \simeq t_0$ or $u \simeq 0$). Indeed, because of the exponential term (and as far as the behaviour of g does not cancel this exponential behaviour), the function under the integral sign has a behaviour similar to the curve shown in Fig. 5.1 ($\Re(e^{\iota xt^2})$). The integrations on the domains with rapid oscillations almost cancel one another. Only the integration on the neighbourhood of the stationary point provides a significant term. $I(x)$ is then approximated by

$$h''(t_0)e^{\iota xh(t_0)} \int_{u_{-1}}^{u_2} \frac{g(t(u))}{h'(t(u))}\, u \; \exp\,(\iota xu^2h''(t_0)/2)\; du$$

where u_1 and u_2 are the real positive roots of $(u(t_0 - \varepsilon))^2$ and $(u(t_0 + \varepsilon))^2$. Because of the definition of u, these two squared terms are real and positive.

The function ug/h' is approximated by its limit when u tends to zero. By using also $h'(t) = h'(t) - h'(t_0) \simeq (t - t_0)h''(t_0)$, $I(x)$ is written as

$$I(x) \simeq g(t_0)e^{\iota xh(t_0)} \int_{u_1}^{u_2} \exp\,(\iota xu^2h''(t_0)/2)\; du$$

By using the same argument as previously, the integration domain is now extended to $]-\infty, +\infty[$. It can be shown [6] that this corresponds to adding terms of order $(1/x)$ which are then small compared with a term in $x^{-1/2}$. Finally,

$$I(x) = g(t_0)e^{\iota xh(t_0)} \int_{-\infty}^{\infty} \exp\,(\iota xu^2h''(t_0)/2)\; du + \mathcal{O}\left(\frac{1}{x}\right)$$

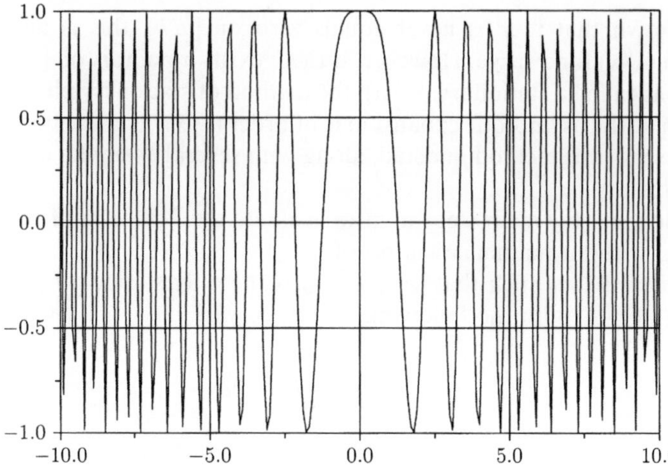

Fig. 5.1. $\Re(e^{\iota x t^2})$ for $x = 1$

By introducing another change of variable and the result

$$\int_{-\infty}^{\infty} \exp(-t^2)\, dt = \sqrt{\pi}$$

$I(x)$ is finally obtained as

$$I(x) \simeq g(t_0) e^{\iota x h(t_0)} \left(\frac{2\pi}{x|h''(t_0)|} \right)^{1/2} e^{\pm \iota\pi/4} \equiv \frac{a_0}{x^{1/2}} \tag{5.8}$$

when x tends to infinity. The $\pm$ sign corresponds to $h''(t_0)$ positive or negative.

Remarks:

- If there are several stationary points on $[a, b]$, their contributions must be added. This result can be seen immediately by dividing $[a, b]$ into subintervals which include only one stationary point.
- If a (or b) is a stationary point itself, its contribution must be divided by 2.
- If $a = -\infty$ and $b = +\infty$, $I(x)$ has an expansion of the form $\sum_{n \geqslant 0}(a_n/x^{n+1/2})$ where the first term a_0 is defined by (5.8). If a or b is finite, $I(x)$ has an expansion of the kind $a_0/x^{1/2} + \mathcal{O}(1/x)$; the finite ends give terms of order $(1/x)$.

5.1.4. The method of steepest descent

The method of stationary phase is a particular case of the method of steepest descent which is used to evaluate the asymptotic expansions, for $|x|$ large, of integrals of the type

$$I(x) = \int_{\mathscr{C}} g(z) \exp(xf(z))\, dz \tag{5.9}$$

where $\mathscr{C}$ is an integration contour in the complex plane. The parameter x can be complex but is assumed to be real here, for simplicity.

The results of the method of steepest descent have been proved rigorously. They are based on the theory of analytic functions. The aim of this section is only to provide a general presentation of the method and how it can be used. The reader will find in [6] a very good description of the method for simple cases and an application to a classical diffraction problem in acoustics.

In the following, f is assumed to be analytic (then twice continuously differentiable). The first step of the method is to draw a map of f. More precisely, if $z = x + \iota y$ and $f(z) = u(x, y) + \iota v(x, y)$, the first step is to represent the values of u and v (and in particular the curves $u =$ constant and $v =$ constant) in the complex plane (x, y). Let us consider, for example, the function $f(z) = z^2$. The curves $u =$ constant and $v =$constant correspond respectively to the equations $x^2 - y^2 = C'$ and $xy = C$ (see Fig. 5.2). The dotted lines and the full lines respectively correspond to $u = C'$ and $v = C$.

Because f is analytic, u and v satisfy the Cauchy–Riemann equations. Then $\nabla u \nabla v = 0$, that is the curves $u =$ constant and $v =$ constant are orthogonal. Furthermore, if there exists a stationary point $z_0 = x_0 + \iota y_0$ (such that $f'(z_0) = 0$ and $f''(z_0) \neq 0$), then $\partial u(x_0, y_0)/\partial x = \partial u(x_0, y_0)/\partial y = 0$ but $u(x_0, y_0)$ is not a global extremum. Indeed, depending on the way chosen in the plane (x, y) to go to $u(x_0, y_0)$, its value can be a maximum or a minimum. In the example in Fig. 5.2, $f(z) = z^2$ has a stationary point (or saddle point) at $z = 0$. The arrows show the direction of increasing u. On the contour $x = y$, $u(0)$ is a maximum. On $x = -y$, $u(0)$ is a minimum.

The general idea of the method is to change the initial contour $\mathscr{C}$ into a contour passing through z_0 and such that $u(x_0, y_0)$ is a maximum on this new contour. Indeed, because of the exponential behaviour of the function under the integral sign in $I(x)$ (and as far as g does not cancel this behaviour), the main contribution for $I(x)$, when x tends to infinity, is given by the neighbourhood of z_0. Further from z_0 on the contour, the function under the integral sign exponentially decreases and the decrease is faster when x is larger.

The second step is to find a contour, called the steepest descent path. This contour is by definition orthogonal to the curve $u(x, y) = u(x_0, y_0)$ around z_0 and then corresponds to a curve $v(x, y) = v(x_0, y_0)$. In Fig. 5.2, the steepest descent path coincides with the curve $x = -y$. Let us call this path $\mathscr{C}_1$. Then,

$$I(x) = \int_{\mathscr{C}_1} g(z) \exp[xf(z)]\, dz + F(x) \tag{5.10}$$

where $F(x)$ includes, if necessary, the contributions of the poles and/or branch points of f and g. $F(x)$ is a sum of residues and/or of branch integrals. For example, $F(x) \equiv 0$, if f and g are uniquely defined and if g has no poles in the domain between $\mathscr{C}$ and $\mathscr{C}_1$. This is a classical result of the theory of complex functions.

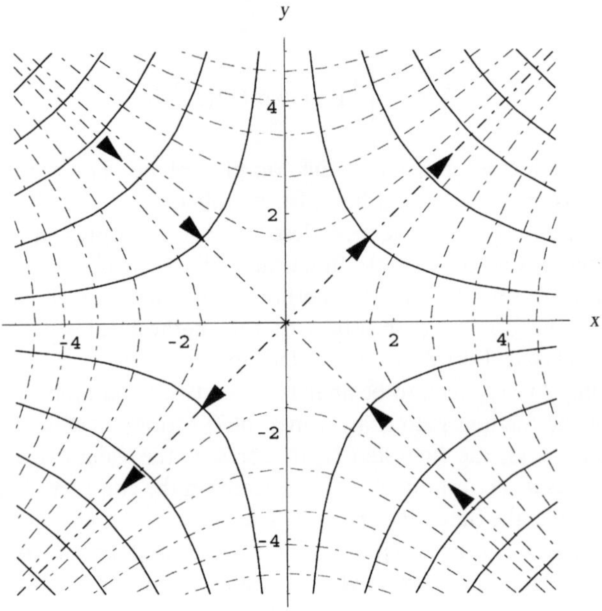

Fig. 5.2. $f(z) = z^2$

Let us assume for simplicity that F is identically zero. The following change of variable is used:

$$f(z) - f(z_0) = -t^2$$

Because u has a maximum on $\mathscr{C}_1$, t is real. Then,

$$I(x) = \exp(xf(z_0)) \int_{-\infty}^{+\infty} \varphi(t) \exp(-st^2)\, dt \tag{5.11}$$

with $\varphi(t)\, dt = g(z)\, dz$ and $f'(z)\, dz = -2t\, dt$. The integration domain is $]-\infty, +\infty[$ if the integration contour coincides exactly with the steepest descent path. It is restricted to a finite interval if they partly coincide.

The explicit expression of function φ is often difficult to obtain. However, since the main contribution for $I(x)$ comes from the neighbourhood of z_0 ($t \simeq 0$), only the behaviour of φ and its derivatives around 0 is needed. Then,

$$I(x) \simeq \exp(xf(z_0)) \int_{-\infty}^{\infty} [\varphi(0) + t\varphi'(0) + \cdots] \exp(-st^2)\, dt$$

and

$$I(x) \simeq \frac{\exp(xf(z_0))}{\sqrt{x}} \left[\varphi(0) + \frac{1}{4x}\varphi'(0) + \mathcal{O}\left(\frac{1}{x^2}\right)\right] \quad \text{when } x \rightarrow \infty \tag{5.12}$$

The terms $\varphi^{(n)}(0)$ are evaluated from the Lagrange theorem [6, 7] by using Taylor's expansions of f and g around z_0. In particular,

$$\varphi(0) = \left(-\frac{2}{f''(z_0)} \right)^{1/2} g(z_0) \tag{5.13}$$

5.2. Kirchhoff Approximation

The Kirchhoff approximation is a kind of 'geometrical optics' approximation; it is then valid at high frequency. It provides an approximate expression of the solution of the Helmholtz equation with constant coefficients. It was first used to describe the diffraction by a thin screen of finite dimensions [9].

In three-dimensional space, the screen corresponds to the domain Σ of the plane $(z = 0)$. The sound sources are located in the half-space $(z < 0)$. By analogy with optics, the side $(z < 0)$ of the screen is called the illuminated side. Let p_0 be the incident field (field in the absence of the screen) and $G(M, P) = -e^{\iota kr(M, P)}/(4\pi r(M, P))$ the classical Green's kernel. The Green's representation of the total field can be written as (see chapter 3, section 3.2)

$$p(M) = p_0(M) - \int_{\Sigma_+ \cup \Sigma_-} [p(P)\partial_{n(P)}G(M, P) - G(M, P)\partial_{n(P)}p(P)]\, d\Sigma \tag{5.14}$$

Σ_+ and Σ_- denote the two sides of the screen $(z > 0)$ and $(z < 0)$ respectively. $\vec{n}$ is the unit vector normal to Σ and interior to the propagation domain. The functions p and $\partial_{n(P)}p(P)$ which appear under the integral sign are then approximated by:

$$p = p_0 \quad \text{and} \quad \partial_n p = \partial_n p_0 \quad \text{on } \textstyle\sum_-$$
$$p = 0 \quad \text{and} \quad \partial_n p = 0 \quad \text{on } \textstyle\sum_+$$

This seems 'reasonable' under the geometrical optics approximation, but it is not rigorous since for a Helmholtz equation (elliptic of degree 2) it is not possible to fix both p and $\partial_n p$ on the same surface. The sound field p at any point of the space is then obtained by simply integrating the right-hand side of (5.14). There is no longer an integral equation to solve.

The Kirchhoff approximation also applies to the case of a thin plane screen of infinite dimensions, perfectly rigid, with a hole of dimensions large compared with the acoustic wavelength.

Let the infinite screen coincide with plane $(z = 0)$. It is characterized by the homogeneous Neumann condition. Let D denote the surface of the hole. If a sound wave is emitted on the $(z < 0)$ side, the sound pressure and its normal derivative are equal to zero on the screen and equal to the incident pressure and its normal derivative on the surface D. More precisely, if p_0 denotes the incident pressure:

$$p = 0 \quad \text{and} \quad \partial_n p = 0 \quad \text{on the screen, for } z > 0$$
$$p = p_0 \quad \text{and} \quad \partial_n p = \partial_n p_0 \quad \text{on } D, \text{ for } z > 0$$

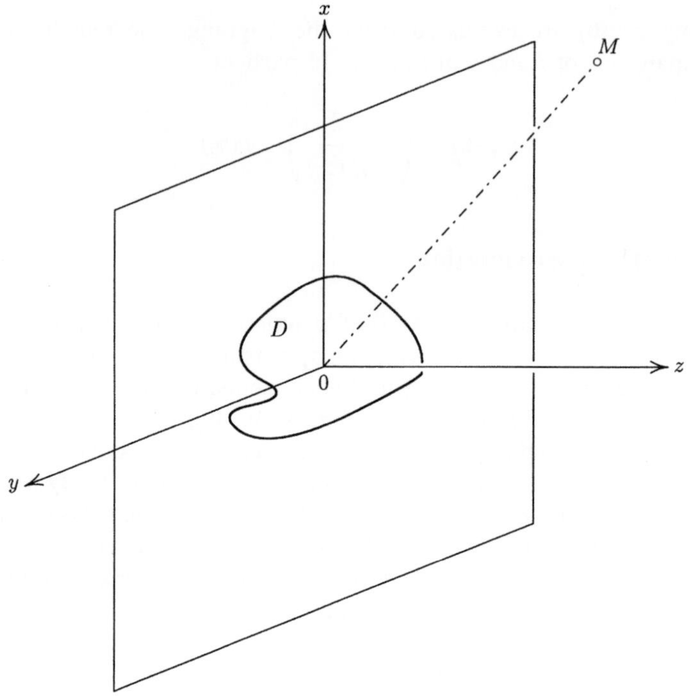

Fig. 5.3. Aperture in an infinite screen.

For the reasons presented previously, any representation of p which is based on these assumptions cannot be a solution of the Helmholtz equation. The domain of validity of this approximation is not precisely defined. It corresponds *a priori* to the 'geometrical optics' domain: high frequency, large dimensions of the hole compared with the wavelength, and, in the case of Fig. 5.3, observation point M such that the axis OM is close to the z-axis (incidence close to the normal).

5.3. Neumann series

The aim of the method presented in this paragraph is to express the solution of the Helmholtz equation as a series. In practice, only the first term or the first two terms are used. The series is called the Neumann series.

Let us consider the general case of a function ψ, solution of an integral equation:

$$\psi(x) = \psi_0(x) - K\psi(x) \tag{5.15}$$

for all x in a domain Γ. K is an integral operator on Γ.

Any kind of diffraction problem described by a Helmholtz equation with constant coefficients can be replaced by an integral problem of this type, by using a Green's formula for example. ψ could represent the sound field emitted in the presence of an obstacle, ψ_0 would be the incident field and Γ the boundary of the obstacle.

Instead of solving integral equation (5.15), one constructs a sequence of functions $\psi^{(p)}, p \geqslant 0$, defined by:

$$\psi^{(0)}(x) = \psi_0(x)$$
$$\psi^{(p)}(x) = (\psi_0 - K\psi^{(p-1)})(x) \qquad \text{for } p \geqslant 1$$

Each function $\psi^{(p)}$ is an approximation of ψ. If there is any convergence $\psi^{(p)}$ is a better approximation than $\psi^{(p-1)}$.

With this method, it is possible to avoid solving the integral equation. Each $\psi^{(p)}$ is the sum of the first p terms of a series. Indeed, the integral equation (5.15) can be written $(Id + K)\psi = \psi_0$, where Id is the identity operator. If the series is convergent, ψ can be written as

$$\psi = (Id + K)^{-1}\psi_0 = (Id - K + K^2 - \cdots)\psi_0 = \sum_{j=0}^{\infty} (-1)^j K^j \psi_0 \qquad (5.16)$$

with $K^0 = Id$. This series is called a Neumann series. For the case of the Helmholtz equation, it is proved that there exists a wavenumber k_0 such that the series converges for any k less than k_0. The first terms of the series then provide an approximation valid at low frequency, which is really interesting if only the first term or the first two terms are needed.

5.4. W.K.B. Method. Born and Rytov Approximations

5.4.1. W.K.B. method

The W.K.B. (or W.K.B.J.) method is named after the works of G. Wentzel, K. Kramers, L. Brillouin and H. Jeffreys. It is presented here in outline, on a simple case. A much more detailed presentation as well as references can be found in the classic book [10].

The W.K.B. method consists in developing an approximation of the solution of a Helmholtz equation with varying coefficients, under two assumptions:

- $k \gg 1$ ('geometrical optics' approximation);
- the index $n(z)$ and the solution are functions which vary slowly over a distance equal to one wavelength.

The index $n(M)$ is the function equal to the ratio of the sound speeds $c(M_0)/c(M)$ where $c(M_0)$ is chosen as a reference sound speed.

The W.K.B. method belongs to the group of multi-scale methods.

Let us consider the one-dimensional Helmholtz equation:

$$\frac{d^2\psi}{dz^2}(z) + k_0^2 n^2(z)\psi(z) = 0 \qquad (5.17)$$

where $k_0 = \omega/c_0$, $c_0 = c(z_0)$ is the sound speed at a reference point z_0 and $n(z) = c_0/c(z)$. The solution ψ is first expressed as

$$\psi(z) = \exp\,(\iota k_0 M(z))$$

where $M(z)$ is a series in $(k_0)^{-m}$. For example, M can be chosen as in [10]:

$$M(z) = \int_{z_0}^{z} \sum_{m=0}^{\infty} \frac{A_m(z)}{k_0^m}\, dz$$

It is then easier to determine the coefficients of the series. They are obtained by introducing this representation of ψ into the Helmholtz equation and by equating term by term. The first coefficients are

$$A_0(z) = \pm\, n(z)$$

$$A_1(z) = \iota \frac{d}{dz}\,(\ln (n(z))^{1/2})$$

$$A_2(z) = \pm \tfrac{1}{2} n(z)^{-1/2} \frac{d^2}{dz^2}\,(n(z)^{-1/2})$$

Using the first three terms, ψ is approximated by

$$\psi(z) \simeq n(z)^{-1/2} \exp\left[\pm \iota k_0 \int_{z_0}^{z} (1 + \varepsilon(z)) n(z)\, dz\right]$$

with

$$\varepsilon(z) = \frac{1}{2k_0^2}\, n(z)^{-3/2} \frac{d^2}{dz^2}\,(n(z)^{-1/2})$$

ψ is often approximated by the first terms only:

$$\psi(z) \simeq n(z)^{-1/2} \exp\left[\pm \iota k_0 \int_{z_0}^{z} n(z)\, dz\right] \tag{5.18}$$

The right-hand side term corresponds to two waves travelling in opposite directions. This approximation cannot be used in the neighbourhood of a point z such that $n(z)$ tends to zero. Such points are called 'turning points'. If z_0 is a turning point, then the representation (5.18) is used for z far from z_0. Closer to z_0, the Helmholtz equation is transformed by using changes of variable and a limited expansion of $n(z)$ around z_0. In [10], the approximation of the solution of the transformed Helmholtz equation is obtained as an Airy function [11]. ψ is then approximated by two expressions: one for z close to z_0 and the other one for z far from z_0. Obviously, it must be checked that they match in between.

5.4.2. Born approximation and Rytov approximation

These approximations are used to describe the solution of a Helmholtz equation in a medium with a slowly varying index. Let ε denote the 'small' parameter which describes the variation of the index. For the sound speed profile shown in Fig. 5.4, ε can be chosen as $\varepsilon = a$.

Let us present these approximations in the one-dimensional case:

$$\left(\frac{d^2}{dx^2} + k_0^2 n^2(x, \varepsilon)\right)\psi(x, \varepsilon) = 0$$

First, ψ is written as

$$\psi(x, \varepsilon) = \exp(\iota k(\varepsilon)x)$$

where $k(\varepsilon)$ is a series in ε^n:

$$k(\varepsilon) = \sum_{j=0}^{\infty} k_j \varepsilon^j$$

Such an approximation is called the Rytov approximation:

$$\psi(x, \varepsilon) = \exp\left(\iota x \sum_{j=0}^{\infty} k_j \varepsilon^j\right) \tag{5.19}$$

The Born approximation is then obtained by replacing in (5.19) the exponential by its series and re-ordering the terms in increasing powers of ε. Then:

$$\psi(x, \varepsilon) = e^{\iota k_0 x} \sum_{p=0}^{n} \varepsilon^p \sum_{q=0}^{p} \frac{(\iota x)^q}{q!} \sum_{j_1 + \cdots + j_q = p} k_{j_1} \ldots k_{j_q} \tag{5.20}$$

As an example, a comparison of these two methods can be found in [12] where J. Keller shows that the error obtained with the first n terms of each series is of order $(\varepsilon x)^{n+1}$ for the Born formula and $(\varepsilon^{n+1} x)$ for the Rytov formula.

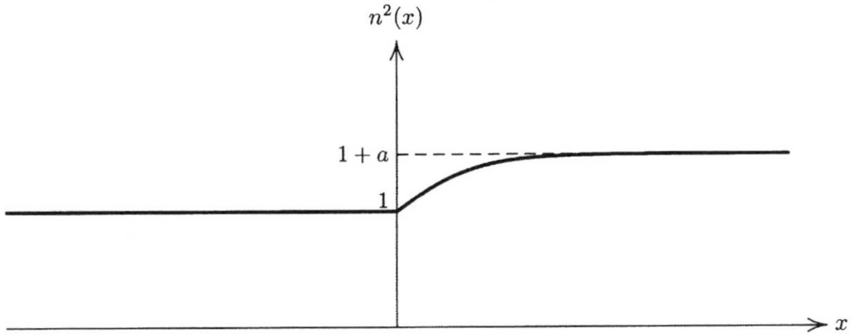

Fig. 5.4. Index $n^2(x) = 1$ for $x < 0$ and $= (1 + a\tanh(x))$ for $x > 0$.

In [13], the authors compare both approximations with exact solutions obtained for Epstein profiles (in the case of Fig. 5.4, particularly). They show that the first order approximations coincide.

5.5 Image method, ray method, geometrical theory of diffraction

5.5.1. Image method

A particular case: exact solution.   The image method is based on the following remark. Let $p(M)$ be the sound pressure emitted by a source S at a point M above a plane Σ. Let S' be symmetrical to S with respect to Σ. If Σ is characterized by a homogeneous Dirichlet condition ($p = 0$) the pressure $p(M)$ is exactly equal to:

$$p(M) = p_S(M) - p_{S'}(M)$$

where $p_S(M)$ and $p_{S'}(M)$ are the sound pressures emitted at M by S and S' respectively.

If Σ is characterized by a homogeneous Neumann condition ($\partial_n p = 0$ on Σ), $p(M)$ is then equal to

$$p(M) = p_S(M) + p_{S'}(M)$$

The general case.   The image method *a priori* applies to any problem of propagation between two or more plane surfaces. It is an approximation method and its limitations are specified at the end of the paragraph.

The simplest applications correspond to propagation between two parallel planes (waveguide), four or six parallel surfaces (room acoustics).

Each surface is characterized by a reflection coefficient. Let S be the point source and M the observation point. S_m, ($m = 1, \ldots, \infty$) denotes the images of S relative to the boundaries. Each ray 'emitted' by the image source S_m is considered as a ray emitted by S and reflected n times by the boundaries (see Fig. 5.5). n obviously depends on the index m. The sound pressure p, solution of a Helmholtz equation with constant coefficients, is written as

$$p(M) = -\frac{e^{\iota k R(S, M)}}{4\pi R(S, M)} + \sum_{m=1}^{\infty} \left(\prod_{j=1}^{n} Q_j(\theta_{mj}) \right) \frac{e^{\iota k r_m}}{4\pi r_m} \qquad (5.21)$$

where $r_m = R(S_m, M)$ is the arclength on ray m emitted by S_m, $n =$ number of reflections on ray m, $\theta_{mj} =$ angle of incidence of ray m at jth reflection, $Q_j(\theta_{mj}) =$ reflection coefficient for ray m at jth reflection.

Application to two parallel planes.   In three-dimensional space, let the planes ($z = 0$) and ($z = h$) denote the boundaries of the domain. A harmonic signal ($e^{-\iota \omega t}$)

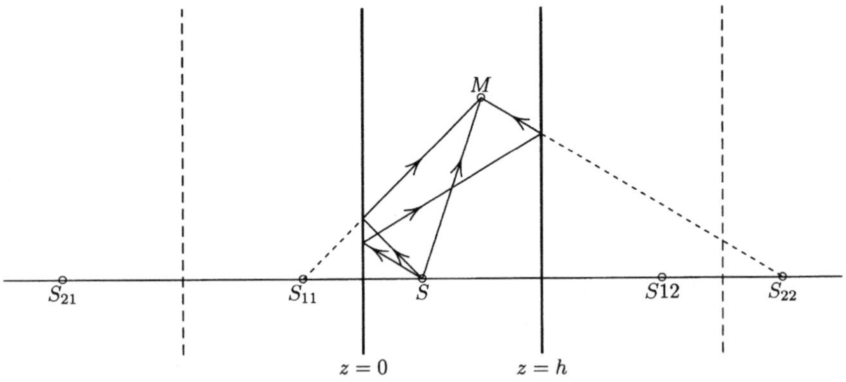

Fig. 5.5. The case of two parallel planes.

is emitted by an omnidirectional, point source $S = (0, 0, 0 < z_0 < h)$ (see Fig. 5.5). The sound pressure p is the solution of the following system:

$$\begin{cases} (\Delta + k^2)p(x, y, z) = \delta(x)\delta(y)\delta(z - z_0) \qquad 0 < z < h \\ \left(\dfrac{\partial}{\partial \vec{n}} + \dfrac{\imath k}{\zeta_1}\right) p(x, y, z) = 0 \qquad \text{on } z = 0 \\ \left(\dfrac{\partial}{\partial \vec{n}} + \dfrac{\imath k}{\zeta_2}\right) p(x, y, z) = 0 \qquad \text{on } z = h \\ \text{Sommerfeld conditions} \end{cases}$$

ζ_1 and ζ_2 are the reduced specific normal impedances which characterize planes $(z = 0)$ and $(z = h)$ respectively. They can be equal to zero or infinity to include Dirichlet or Neumann conditions.

The first step is to define the set of the sources S_{mj}, images of S, with their coordinates. In this simple case, they are given by

$$S_{m1} = \begin{cases} (0, 0, -mh + z_0) & \text{for even } m \\ (0, 0, -(m-1)h - z_0) & \text{for odd } m \end{cases}$$

$$S_{m2} = \begin{cases} (0, 0, mh + z_0) & \text{for even } m \\ (0, 0, (m+1)h - z_0) & \text{for odd } m \end{cases}$$

Let A_1 and A_2 denote the plane wave reflection coefficients of planes $(z = 0)$ and $(z = h)$ respectively:

$$A_j = \frac{\zeta_j \cos\theta - 1}{\zeta_j \cos\theta + 1} \qquad \text{with } j = 1, 2$$

θ is the angle of incidence from the normal vector to each plane. The sound pressure p is then

$$p(M) = -\frac{e^{\iota kR(S, M)}}{4\pi R(S, M)} - \sum_{m=1}^{\infty} \sum_{j=1}^{2} Q_{mj} \frac{e^{\iota kR(S_{mj}, M)}}{4\pi R(S_{mj}, M)}$$

Q_{mj} is the reflection coefficient on ray mj which comes from S_{mj}. It is equal to

$$Q_{2m,j} = A_1^m A_2^m \qquad \text{if } j = 1, 2$$

$$Q_{2m+1,j} = \left| \begin{array}{ll} A_1^{m+1} A_2^m & \text{if } j = 1 \\ A_1^m A_2^{m+1} & \text{if } j = 2 \end{array} \right.$$

Limitations of the method. The approximation obtained by the image method is particularly interesting at short distance from the source, in the region where the incident field is almost dominating and only the first sources must be taken into account. Far from the source, the series is not so efficient from a numerical point of view since it becomes a sum of terms with almost equal phases (see [14], for example).

It must also be emphasized that the series is not always convergent. In the case of two parallel planes described by a homogeneous Neumann condition, it becomes

$$p(M) = -\frac{e^{\iota kR(S, M)}}{4\pi R(S, M)} - \sum_{m=1}^{\infty} \sum_{j=1}^{2} \frac{e^{\iota kR(S_{mj}, M)}}{4\pi R(S_{mj}, M)}$$

which is not convergent because when m tends to infinity, the distance $R(S_{mj}, M)$ is of order mh.

5.5.2. Ray method

This method is also based on the laws of geometrical optics, that is it is a high frequency approximation. For the case of plane obstacles, it is similar to the image method. The ray method consists in representing the sound field by using direct, reflected and refracted rays. It is based on the classical laws (Fig. 5.6):

- in a homogeneous medium, direct ray paths are straight lines;
- when an incident ray impinges on a surface, with an angle of incidence θ, a reflected ray is emitted, located in the plane defined by the incident ray and the normal to the surface, with an angle $(-\theta)$;
- when an incident ray impinges at the boundary between two media (1) and (2) described by indices n_1 and n_2, with an angle of incidence θ_1, a refracted ray is emitted with an angle θ_2 defined by Snell's law:

$$\frac{\sin \theta_1}{\sin \theta_2} = \frac{n_1}{n_2}$$

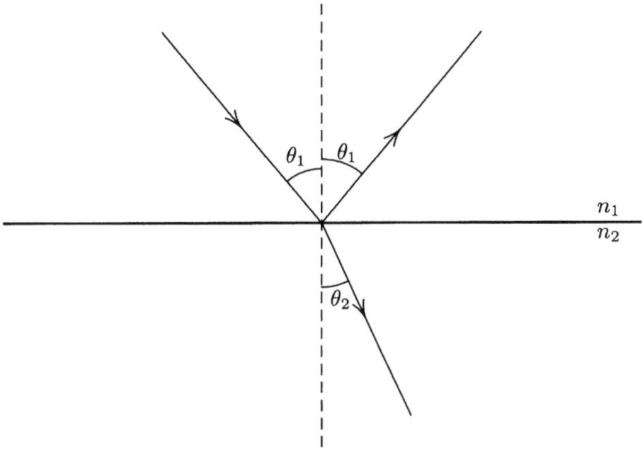

Fig. 5.6. Plane wave on a plane boundary.

The amplitude of a reflected ray at a point P on a surface is the product of the amplitude of the incident ray and the reflection coefficient which describes the surface at P. Similarly, the amplitude of a refracted ray at a point P on a surface is the product of the amplitude of the incident ray and the transmission coefficient which characterizes the surface.

The method consists in taking into account a large number of rays R_m, which come from the source with an energy E_m (the number of rays, their initial directions and the values E_m depend on the characteristics of the source). To evaluate the sound pressure at a point M, only the rays which pass close to M are taken into account. The amplitude on each of these rays must be calculated. It depends on the trajectory.

In room acoustics, the ray method is easier to use than the image method, in particular if the room has more than six faces, if obstacles must be taken into account, and if the faces are not flat.

No more details are given here on this method since it is a particular case of the geometrical theory of diffraction which is presented in the next paragraph.

5.5.3. Geometrical theory of diffraction

The geometrical theory of diffraction was developed by J. Keller [14]. It generalizes the laws of geometrical optics to include the phenomenon of diffraction and describe the propagation around obstacles in a homogeneous or inhomogeneous medium. It leads to high frequency approximations (wavelength $\lambda \ll 1$).

For this purpose, new types of rays (diffracted and curved rays) are introduced.

The sound field is expressed as a sum of sound fields, each one corresponding to one ray:

$$p(M) = \sum_{j \geqslant 0} A^{(j)}(M) e^{\iota k \varphi^{(j)}(M)} \tag{5.22}$$

where $M = (x_1, x_2, x_3)$ is a point of the propagation medium. The amplitude $A^{(j)}$ and phase $\varphi^{(j)}$ correspond to each ray j which passes close to M. To compute $p(M)$, one must evaluate:

- the trajectories of all the rays emitted by S and passing close to M;
- the amplitude and phase on each ray.

On each ray j, the sound field ψ is written as an expansion in $(1/k)^m$:

$$\psi(M) = e^{\iota k \varphi(M)} \sum_{m=0}^{\infty} \frac{A_m(M)}{(\iota k)^m} \tag{5.23}$$

In general, ψ is approximated by the first term of the expansion.

The main features of the method are presented in the next paragraph, for the most general case of propagation in inhomogeneous media. A few remarks are added for propagation in a homogeneous medium, which is only a particular case and is presented in more detail in chapter 4 for diffraction problems.

Propagation in an inhomogeneous medium. Each field ψ is the solution of a Helmholtz equation:

$$(\Delta + k^2 n^2(x_1, x_2, x_3))\psi(x_1, x_2, x_3) = F$$

F represents the source. $n(x) = n(x_1, x_2, x_3)$ is the index of the medium. Let us assume that F is zero; the source term can equivalently be introduced in the boundary conditions.

Replacing ψ by its expansion (5.23) leads to the following equations:

$$(\nabla\varphi)^2 = n^2 \tag{5.24}$$

$$2\nabla\varphi.\nabla A_m + A_m.\Delta\varphi = -\Delta A_{m-1} \text{ for } m \geqslant 0, \qquad \text{with } A_{-1} = 0 \tag{5.25}$$

The eikonal equation (5.24) provides the phase φ. The coefficients A_m are then evaluated recursively by solving equations (5.25).

One more system of equations is needed to determine the trajectories of the rays, which are orthogonal to the surfaces $\varphi =$ constant. Introducing a system of coordinates (s, u, v) where s denotes the arclength along the ray, the equations which express the orthogonality of the rays and the surfaces of constant phase are [14]

$$n \frac{d}{ds}\left(n \frac{dx_i}{ds}\right) = \frac{1}{2} \nabla(n^2) \qquad \text{for } i = 1, 2, 3 \tag{5.26}$$

From these equations, parametrical representations of the ray trajectories are deduced.

In [14], Keller gives the general expressions of the phase φ and the coefficients A_m. In particular,

$$\varphi(s, u, v) = \varphi(s_0, u, v) + \int_{s_0}^{s} n(x(s', u, v))\, ds' \tag{5.27}$$

and

$$A_0(s) = \left| \frac{n(s_0)J(s_0)}{n(s)J(s)} \right|^{1/2} A_0(s_0) \tag{5.28}$$

where J is the Jacobian:

$$\left| \frac{\partial(x_1, x_2, x_3)}{\partial(s, u, v)} \right|$$

$s_0 = s(x_0)$, x_0 is the initial point on the ray path. For an incident ray, x_0 can be the source.

Remark: In a homogeneous medium, $n(x) \equiv 1$. From (5.26) and (5.27), it is easy to check that the rays are straight lines (s is proportional to the radial distance r) and that the phase is proportional to the distance r.

The initial values $\varphi(s_0)$ and $A_0(s_0)$ are deduced from the initial conditions on each ray. These conditions depend on the type of the ray (incident, reflected, refracted). To take into account the boundaries of the medium and its inhomogeneity, one must include rays reflected by the boundaries and diffracted. In an inhomogeneous medium, the phase φ may be complex; this corresponds to evanescent rays.

Propagation in a homogeneous medium. As seen in the previous section, in a homogeneous medium ($n(x) \equiv 1$), the ray trajectories are straight lines. They are then easily determined. Rays can be incident, reflected, refracted (in the case of obstacles in which rays can penetrate) and diffracted.

The calculation of phase and amplitude on each ray follows the laws presented in chapter 4 for the particular case of a convex cylinder. A description of the general procedure and some applications can be found in [15] and [16].

One of the drawbacks of the G.T.D. is that the sound field obtained is infinite on caustics. However, several kinds of improvements have been proposed to avoid this problem (see [17] for example).

5.6. Parabolic approximation

After being used in electromagnetism or plasma physics, the parabolic approximation is now extensively used in acoustics, especially to describe the sound propagation in the sea or in the atmosphere. The main idea is to replace the Helmholtz equation with varying index (elliptic equation) by a parabolic equation which is more convenient from a numerical point of view. The parabolic equation obtained is not unique, it depends on the assumptions made on the propagation medium and the acoustical characteristics. It is then solved by techniques based on FFT or on finite difference methods.

It must be pointed out that the parabolic equation is only valid at large distance from the source; it is then *a priori* necessary to evaluate the near-field in another way. A description of all these aspects can be found in [14] and [18], along with references. Many articles are still published on this subject (see [19] to [22], for example).

5.6.1. Replacing the Helmholtz equation by a parabolic equation

To transform the Helmholtz equation into a parabolic equation, two types of methods have been developed. The method presented here is the most extensively used now and the most general. It is based on the approximation of operators.

Let p denote the sound pressure solution of

$$(\Delta + k_0^2 n^2(r, z))p(r, z) = f(r, z) \tag{5.29}$$

z is depth, r is the horizontal distance. Sound speed c depends on these two coordinates and the refractive index $n(r, z)$ is defined by $n = c_0/c$, where c_0 is a reference sound speed. $k_0 = \omega/c_0$.

The basic assumptions of the parabolic approximation are

- large distance between source and receiver ($k_0 r \gg 1$);
- depth of the sound channel is small compared with the propagation distance or small aperture angle θ, i.e.

$$|\theta| \simeq (|z - z_0|/r) \ll 1 \tag{5.30}$$

It is first assumed that p, far from the source, can be expressed as the product of a Hankel function $H_0^{(1)}(k_0 r)$ and a function $\psi(r, z)$ which varies slowly with r:

$$p(r, z) = \psi(r, z) H_0^{(1)}(k_0 r)$$

Since $k_0 r \gg 1$, $H_0^{(1)}$ can be approximated by its asymptotic behaviour:

$$p(r, z) \simeq \psi(r, z)\sqrt{\frac{2}{\imath k_0 r}}\, e^{\imath k_0 r}$$

By introducing the expression in (5.29), one obtains

$$\frac{\partial^2 \psi}{\partial r^2} + 2\imath k_0 \frac{\partial \psi}{\partial r} + \frac{\partial^2 \psi}{\partial z^2} + k_0^2(n^2 - 1)\psi = 0 \tag{5.31}$$

The source term is omitted at this stage and is actually taken into account in the initial conditions (see Section 5.6.3). Let P and Q denote the following operators:

$$P = \frac{\partial}{\partial r} \quad \text{and} \quad Q = \left(n^2 + \frac{1}{k_0^2}\frac{\partial^2}{\partial z^2}\right)^{1/2}$$

Then equation (5.31) can be written

$$(P^2 + 2\imath k_0 P + k_0^2(Q^2 - 1))\psi = 0 \tag{5.32}$$

or

$$(P + \iota k_0 - \iota k_0 Q)(P + \iota k_0 + \iota k_0 Q)\psi - \iota k_0(PQ - QP)\psi = 0$$

If the index n is a slowly varying function of r, the $(PQ - QP)$ term can be neglected. In particular, this term is zero if n depends only on z. By taking into account only 'outgoing' waves, it is possible to replace equation (5.31) by

$$(P + \iota k_0 - \iota k_0 Q)\psi = 0$$

Let us now define

$$q = n^2 - 1 + \frac{1}{k_0^2}\frac{\partial^2}{\partial z^2} \qquad \text{then} \qquad Q = \sqrt{1+q}$$

If $|q| < 1$, Q can be formally approximated by the first terms of its Taylor series

$$\sqrt{1+q} \simeq 1 + q/2 - q^2/8 + \cdots$$

Taking into account only the first two terms, equation (5.32) becomes

$$2\frac{\partial\psi}{\partial r} - \iota k_0\left((n^2 - 1)\psi + \frac{1}{k_0^2}\frac{\partial^2\psi}{\partial z^2}\right) = 0 \tag{5.33}$$

which is the most classical parabolic equation used to describe the sound propagation [14]. This equation is a better approximation if q is small, that is if the index is close to 1 and the aperture angles are small. However, it must be noticed that this way of using operators also leads to parabolic equations with a wider range of validity. To obtain such equations, it is possible to take into account more terms in the Taylor series of Q or to chose other approximations of Q (Padé approximants, for example). These approximations lead to equations which are more complicated but which are still valid for large aperture angles. Numerous studies are devoted to this aspect of parabolic approximation, and the reader is highly recommended to refer to them (in the *Journal of the Acoustical Society of America*, for example).

Because of the assumption of 'outgoing wave' in (5.32), let us emphasize that equation (5.33) cannot take into account backscattering effects. For example, if there is an obstacle at $r = R_0$ and if the equation is solved up to $r = R_1 < R_0$, the result does not take into account the presence of the obstacle.

5.6.2. Solution techniques

There are two main methods for solving the parabolic equation.

- One is called the 'split-step Fourier' method [14]; the z-Fourier transform of the field is first computed and then the inverse Fourier transform.
- The other is based on finite difference methods [18, 19]. In the plane (r, z), the propagation domain is discretized by drawing lines $r =$ constant and $z =$ constant. The points of the grid are then (lr, mz), $l = 0, \ldots, N$ and $m = 0, \ldots, M$. At each

step l, a linear system of order M is solved. For more details on the calculation of the coefficients and the properties of the matrices, see [18].

5.6.3. A word about starters

The parabolic equation is valid only at large distances from the source (let us say $r > r_0$). On the other hand, a parabolic equation cannot be solved without an initial condition; here it must be the sound field at $r = r_0$. To find the initial data, it is possible to use the methods based on rays or modes. It is, however, far too complicated. The first and simplest technique used has been to approximate the sound field at r_0 by a Gaussian law. This representation corresponds to a small angle of aperture. A great number of studies have also been published on the improvement of the initial condition.

5.6.4. Error estimates

The errors made on the computation of the sound field are first caused by the use of the parabolic approximation. In [14], F. Tappert gives for a simple case a comparison between an exact solution and the solution of the parabolic equation. The errors are also caused by the technique used to solve the parabolic equation.

Finally it must be noted that, because of the mathematical properties of the parabolic equations, the error increases with distance.

5.7. Wiener–Hopf method

Strictly speaking, the Wiener–Hopf method [23] is not an approximation method. From a theoretical point of view, it leads to an exact expression of the solution. But, except for very simple cases, the expressions are not adapted to numerical or analytical computations. A method such as the method of stationary phase is then used to obtain an asymptotic behaviour (see [24], for example).

The Wiener–Hopf method is based on partial Fourier transforms. It is quite suitable for propagation problems with discontinuous boundary conditions such as sound propagation above an inhomogeneous surface (see chapter 4, section 4.1.3).

5.7.1. Description

The general procedure is as follows:

- Using partial Fourier transforms, the initial system of equations is replaced by an equation of the following kind:

$$\hat{\ell}(\xi)\hat{p}_+(\xi) + \hat{p}_-(\xi) = \hat{h}(\xi) \qquad \text{for all real } \xi \tag{5.34}$$

  The functions $\hat{p}_+$ and $\hat{p}_-$ are unknown. The notations $\hat{p}_+$ and $\hat{p}_-$ correspond to the properties of these functions: $\hat{p}_+$ and $\hat{p}_-$ are analytic functions in the upper half-plane ($\xi + \iota\tau, \tau > 0$) and in the lower half-plane ($\xi + \iota\tau, \tau < 0$) respectively. They are both continuous for $\tau = 0$. $\hat{\ell}$ and $\hat{h}$ depend on the data. Equation (5.34) is called a Wiener–Hopf equation; the way to find it is presented in detail in Section 5.7.2.

- $\hat{\ell}$ is then written as a product $\hat{\ell}_-\hat{\ell}_+$ where $\hat{\ell}_-$ and $\hat{\ell}_+$ have no zeros and are analytic in the lower ($\tau < 0$) and the upper ($\tau > 0$) half-plane respectively. This leads to:

$$\hat{\ell}_+(\xi)\hat{p}_+(\xi) + \frac{\hat{p}_-(\xi)}{\hat{\ell}_-(\xi)} = \frac{\hat{h}(\xi)}{\hat{\ell}_-(\xi)} \tag{5.35}$$

- $(\hat{h}/\hat{\ell}_-)$ is in turn written as a sum $(\hat{q}_+ + \hat{q}_-)$, $\hat{q}_+$ and $\hat{q}_-$ must have the same properties as $\hat{p}_+$ and $\hat{p}_-$. Then

$$\hat{\ell}_+(\xi)\hat{p}_+(\xi) - \hat{q}_+(\xi) = -\frac{\hat{p}_-(\xi)}{\hat{\ell}_-(\xi)} + \hat{q}_-(\xi) \tag{5.36}$$

The left-hand side of this equation is analytic in the upper half-plane, the right-hand side is analytic in the lower half-plane. Both functions are equal and continuous for $\tau = 0$.

The theorem of analytic continuation [25] implies that they can be extended to two functions equal in the whole complex plane. This leads to

$$\hat{\ell}_+(\xi)\hat{p}_+(\xi) - \hat{q}_+(\xi) = \hat{E}(\xi) = -\frac{\hat{p}_-(\xi)}{\hat{\ell}_-(\xi)} + \hat{q}_-(\xi) \tag{5.37}$$

where $\hat{E}(\xi)$ is an analytic function which is deduced by examining the asymptotic behaviour of the right-hand side and left-hand side terms. The unknown functions $\hat{p}_+$ and $\hat{p}_-$ are determined from the first and the second equality respectively.

5.7.2. Obtaining a Wiener–Hopf equation

Several procedures can be used to replace the initial system of equations by a Wiener–Hopf equation. Three of them are presented here. The first two correspond to the following two-dimensional problem. A point source is located in the half-plane ($y, z > 0$) at $S = (0, z_0)$. The boundary of the half-plane is characterized by a homogeneous Dirichlet condition on ($y < 0, z = 0$) and a homogeneous Neumann condition on ($y > 0, z = 0$). The signal is harmonic ($\exp(-\iota\omega t)$). The sound pressure p is the solution of

$$\begin{cases} (\Delta + k^2)p(y, z) = \delta(y)\delta(z - z_0) \quad \text{for } z > 0 \\ p(y, z) = 0 \quad \text{for } y < 0 \text{ and } z = 0 \\ \dfrac{\partial p(y, z)}{\partial z} = 0 \quad \text{for } y > 0 \text{ and } z = 0 \\ \text{Sommerfeld conditions} \end{cases} \tag{5.38}$$

(a) One of the methods consists in writing an integral equation equivalent to the differential system (5.38) and using partial Fourier transforms. Let G denote the Green's function which satisfies the homogeneous Neumann condition on $(z=0)$. The Green's representation applied to p and G leads to

$$p(y,0) = G(y,0;\ 0,z_0) + \int_{-\infty}^{0} \frac{\partial}{\partial z'} p(y', z'=0)\, G(y',0;\ y,0)\, dy' \tag{5.39}$$

for all y. Let us define the following transforms:

$$\hat{p}_+(\xi, z) = \int_0^{+\infty} p(y,z) e^{\iota \xi y}\, dy; \qquad \hat{p}_-(\xi, z) = \int_{-\infty}^{0} p(y,z) e^{\iota \xi y}\, dy \tag{5.40}$$

and

$$\widehat{\partial_z p}_+(\xi, z) = \int_0^{+\infty} \frac{\partial p(y,z)}{\partial z} e^{\iota \xi y}\, dy; \qquad \widehat{\partial_z p}_-(\xi, z) = \int_{-\infty}^{0} \frac{\partial p(y,z)}{\partial z} e^{\iota \xi y}\, dy \tag{5.41}$$

The Paley–Wiener theorem applies and shows that function $\hat{p}_+(\xi, z)$ (resp. $\hat{p}_-(\xi, z)$) can be extended to a function $\hat{p}_+(\xi + \iota\tau, z)$ (resp. $\hat{p}_-(\xi + \iota\tau, z)$), analytic for $\tau > 0$ (resp. for $\tau < 0$) and continuous for $\tau \geqslant 0$ (resp. $\tau \leqslant 0$). Functions $\widehat{\partial_z p}_+(\xi, z)$ and $\widehat{\partial_z p}_-(\xi, z)$ obviously have the same properties.

The y-Fourier transform applied to equation (5.39) leads to

$$\iota K \hat{p}_+(\xi, 0) = e^{\iota K z_0} + \widehat{\partial_z p}_-(\xi, 0)$$

with $K^2 = k^2 - \xi^2$, $\Im(K) > 0$, since the y-Fourier transform of the kernel $G(S, M)$ is: $\exp(\iota K\,|\,z - z_0)/\iota K$.

The equation is of the same kind as (5.34) with $\hat{\ell}(\xi) = 2\iota K$. It must be noticed that the decomposition of $\hat{\ell}$ in $\hat{\ell}_- \hat{\ell}_+$ is readily obtained (it is not unique):

$$\hat{\ell}_+(\xi) = 2\iota\sqrt{k+\xi} \qquad \text{and } \hat{\ell}_-(\xi) = \sqrt{k-\xi}$$

(b) Another method consists in applying the Fourier transform to the differential system (5.38). Let

$$\hat{p}(\xi, z) = \int_{-\infty}^{+\infty} p(y,z) e^{\iota \xi y}\, dy$$

$\hat{p}$ is the solution of

$$\left(\frac{d^2}{dz^2} + K^2\right) \hat{p}(\xi, z) = \delta(z - z_0) \qquad \text{for } z > 0$$

with K defined as in the previous section. Then

$$\hat{p}(\xi, z) = \frac{e^{\iota K |z - z_0|}}{2\iota K} + \hat{A}(\xi)\, \frac{e^{\iota K |z + z_0|}}{2\iota K} \qquad \text{for } z \geqslant 0$$

where $\hat{A}$ is the plane wave reflection coefficient, to be deduced from the boundary conditions. For functions $\hat{p}_\pm$ and $\partial_z p_\pm$, one obtains:

$$\hat{p}_+(\xi, 0) + \hat{p}_-(\xi, 0) = \hat{p}_+(\xi, 0) = (1 + \hat{A}(\xi))\,\frac{e^{\iota K z_0}}{2\iota K}$$

$$\widehat{\partial_z p}_+(\xi, 0) + \widehat{\partial_z p}_-(\xi, 0) = \widehat{\partial_z p}_-(\xi, 0) = (-1 + \hat{A}(\xi))\,\frac{e^{\iota K z_0}}{2\iota K}$$

and by eliminating $\hat{A}$:

$$\iota K \hat{p}_+(\xi, 0) = e^{\iota K z_0} + \widehat{\partial_z p}_-(\xi, 0)$$

which is the same equation as previously.

(c) The third method generalizes the previous one. It is presented here for the same problem in a 3-dimensional space.

Let $p(x, y, z)$ be the pressure, solution of

$$\begin{cases} (\Delta + k^2)p(x, y, z) = 0 & \text{for } z > 0 \\ p(x, y, z) = f(x, y) & \text{at } (x, y < 0, z = 0) \\ \dfrac{\partial}{\partial z} p(x, y, z) = g(x, y) & \text{at } (x, y > 0, z = 0) \\ \text{Sommerfeld conditions} & \end{cases} \tag{5.42}$$

p satisfies non-homogeneous boundary conditions. f and g are square integrable, known functions. Let us define the following Fourier transforms:

$$\hat{h}_+(\xi, z) = \int_0^{\infty} \left(\int_{-\infty}^{+\infty} h(x, y) e^{\iota \xi_1 x}\, dx \right) e^{\iota \xi_2 y}\, dy$$

$$\hat{h}_-(\xi, z) = \int_{-\infty}^{0} \left(\int_{-\infty}^{+\infty} h(x, y) e^{\iota \xi_1 x}\, dx \right) e^{\iota \xi_2 y}\, dy$$

with $\xi = (\xi_1, \xi_2)$ and $\tau = (\tau_1, \tau_2)$.

Let $\tilde{f}$ be any continuation of f that is such that

$$\tilde{f}(x, y) = f(x, y) \qquad \text{if } y < 0$$

The function $(p(x, y, 0) - \tilde{f}(x, y))$ is zero for y negative. Let $\hat{E}_+(\xi)$ denote its Fourier transform, $\hat{E}_+(\xi)$ can be extended to $\hat{E}_+(\xi_1, \xi_2 + \iota\tau_2)$, analytic for $\tau_2 > 0$ and continuous for $\tau_2 \geqslant 0$. Similarly, if $\tilde{g}$ is any continuation of g, that is:

$$\tilde{g}(x, y) = g(x, y) \qquad \text{if } y > 0$$

$\hat{F}_-$, the Fourier transform of $(\partial p(x, y, 0)/\partial z - \tilde{g}(x, y))$, can be extended to an analytic function in the lower half-plane $(\tau_2 < 0)$ and continuous for $(\tau_2 \leqslant 0)$. $\hat{p}(\xi, z)$

is such that

$$\left(\frac{d^2}{dz^2}+K^2\right)\hat{p}(\xi, z)=0 \qquad \text{then} \qquad \hat{p}(\xi, z)=\hat{A}(\xi)e^{\iota Kz}$$

The boundary conditions lead to

$$\hat{A}-\hat{\hat{f}}=\hat{E}_+ \qquad \text{and} \qquad \iota K\hat{A}-\hat{\hat{g}}=\hat{F}_-$$

By eliminating $\hat{A}$, one obtains

$$\iota K(\hat{E}_+ + \hat{\hat{f}}) - \hat{\hat{g}} = \hat{F}_-$$

where $\hat{E}_+$ and $\hat{F}_-$ are the unknowns.

The three methods presented in this section provide for the same problem Wiener–Hopf equations which are identical or equivalent.

5.7.3. The decomposition theorem

The main difficulty of the Wiener–Hopf method is the decomposition of $\hat{\ell}$ and $\hat{h}$ in a product $\hat{\ell}_+\,\hat{\ell}_-$ or a sum $\hat{h}_+ + \hat{h}_-$.

If the decomposition of $\hat{\ell}$ is not obvious as in the previous example, it can be deduced from Cauchy integrals in the complex plane.

Theorem 5.1 ([23]). Let $f(\alpha)$ be a function of $\alpha=\xi+\iota\tau$, analytic in the strip $\tau_- < \tau < \tau_+$, such that

$$|f(\xi+\iota\tau)| < c\,|\xi|^{-p}, p>0, \qquad \text{for } |\xi| \to \infty$$

the inequality is uniformly true for all τ in the strip $\tau_- + \varepsilon \leqslant \tau \leqslant \tau_+ - \varepsilon, \varepsilon > 0$. Then, for $\tau_- < c < \tau < d < \tau_+$,

$$f(\alpha) = f_+(\alpha) + f_-(\alpha)$$

where f_+ and f_- are analytic respectively for $\tau > \tau_-$ and $\tau < \tau_+$. They are given by

$$f_+(\alpha) = \frac{1}{2\iota\pi}\int_{-\infty+\iota c}^{+\infty+\iota c}\frac{f(x)}{x-\alpha}\,dx; \qquad f_-(\alpha) = \frac{-1}{2\iota\pi}\int_{-\infty+\iota d}^{+\infty+\iota d}\frac{f(x)}{x-\alpha}\,dx$$

The decomposition of $\hat{h}$ into a sum $(\hat{h}_+ + \hat{h}_-)$ is based on the same theorem, using a logarithmic function.

5.7.4. The main difficulties of the method

The function $\hat{\ell}$ depends on the operators which correspond to the boundary conditions. For example, a Neumann condition and an 'impedance' condition lead to

$$\hat{\ell}(\xi) = \frac{\iota K}{(\iota K + \alpha)}$$

where α is a complex number. It is then not easy to obtain the factorization of $\hat{\ell}(\xi)$.

Furthermore, in the Wiener–Hopf equation (5.34), $\hat{h}$ depends on the right-hand side members of the differential system. The decomposition of $\hat{h}$ is then simple if the incident wave is a plane wave and becomes much more complicated otherwise. These difficulties can sometimes be partly overcome. It is then possible to obtain the solution of the Wiener–Hopf equation as an integral, and asymptotic expressions are deduced through methods like the method of stationary phase.

Bibliography

[1] ERDELYI, A., 1956. *Asymptotic expansions.* Dover Publications, New York.

[2] ARFKEN, G., 1966. *Mathematical methods for physicists.* Academic Press, New York.

[3] MORSE, P. and FESHBACH, H., 1953. *Methods of theoretical physics.* McGraw-Hill, New York.

[4] HABAULT, D., 1980. Diffraction of a spherical wave by different models of ground: approximate formulas. *J. Sound Vib.* 68(3), 413–425.

[5] FILIPPI, P., 1985. Sound radiation by baffled plates and related boundary integral equations. *J. Sound Vib.*, 100(1), 69–81.

[6] LEPPINGTON, F., 1992. Asymptotic evaluations of integrals. in *Modern Methods in Analytical Acoustics*, Springer-Verlag, New York.

[7] JEFFREYS, H. and JEFFREYS, B., 1972. *Methods of mathematical physics*, 3rd edn. Cambridge University Press, Cambridge.

[8] OLVER, F.W.J., 1974. *Asymptotics and special functions.* Computer Science and Applied Mathematics. Academic Press, New York.

[9] BOUKWAMP, C.J., 1954. Diffraction theory. *Rep. Prog. Phys.* 17, 35–100.

[10] BREKHOVSKIKH, L.M., 1960. *Waves in layered media.* Academic Press, New York.

[11] ABRAMOWITZ, M. and STEGUN, I.A., 1964. *Handbook of mathematical functions.* Appl. Math. Ser. 55. National Bureau of Standards, Washington D.C.

[12] KELLER, J.B., 1969. Accuracy and validity of the Born and Rytov approximations. *J. Opt. Soc. Am.* 59(1), 1003–1004.

[13] HADDEN, W. and MINTZER, D., 1978. Test of the Born and Rytov approximations using the Epstein problem. *J. Acoust. Soc. Am.* 63(5), 1279–1286.

[14] KELLER, J.B., 1977. *Wave propagation and underwater acoustics.* Lecture Notes Physics. 70. Springer-Verlag, New York.

[15] LEVY, B.R. and KELLER, J.B., 1959. Diffraction by a smooth object. *Commun. Pure Appl. Math.* 12, 159–209.

[16] COMBES, P., 1978. Introduction à la théorie géométrique de la diffraction et aux coefficients de diffraction. *Rev. Cethedec* 55, 77–104.

[17] LUDWIG, D., 1966. Uniform asymptotic expansions at a caustic. *Commun. Pure Appl. Math.* 19, 215–250.

[18] JENSEN, F.B., KUPERMAN, W.A., PORTER, M.B. and SCHMIDT, H., 1994. *Computational ocean acoustics.* American Institute of Physics, New York.

[19] LEE, D., BOTSEAS, G. and PAPADAKIS, J.S., 1981. Finite-difference solution to the parabolic wave equation. *J. Acoust. Soc. Am.* 70(3), 795–800.

[20] McDANIEL, S.T. and LEE, D., 1982. A finite-difference treatment of interface conditions for the parabolic wave equation: the horizontal interface. *J. Acoust. Soc. Am.* 71(4), 855–858.

[21] BAMBERGER, A., ENQUIST, B., HALPERN, L. and JOLY, P., 1988. Higher-order paraxial wave equation approximations in heterogeneous media. *SIAM J. Appl. Math.* 48, 129–154.

[22] COLLINS, M.D., 1990. Benchmark calculations for higher-order parabolic equations. *J. Acoust. Soc. Am.* 87(4), 1535–1538.

[23] NOBLE, B., 1958. *Methods based on the Wiener-Hopf technique for the solution of partial differential equations.* International Series of Monographs in Pure and Applied Mathematics. Pergamon Press, Oxford.

[24] HEINS, A. and FESHBACH, H., 1954. On the coupling of two half-planes. *Proc. of Symposia in Appl. Math. V*, McGraw-Hill, New York.

[25] CARTAN, H., 1961. *Théorie élémentaire des fonctions analytiques d'une ou plusieurs variables complexes.* Hermann, Paris.

CHAPTER 6

Boundary Integral Equation Methods – Numerical Techniques

Dominique Habault

Introduction

This chapter is devoted to the description of the main numerical aspects of the integral equation methods.

The theoretical aspect of these methods is presented in chapter 3: writing of an integral equation, existence and uniqueness of the solution, and so on. Several applications are described in chapter 4.

The main advantage of these methods in acoustics is that they can deal with propagation problems in domains with any kind of geometry. However, there is an essential limitation: the propagation medium must be homogeneous. This restriction comes from the fact that the integral equations are based on the use of a Green's function which must be at least in a suitable form for numerical computations. Up to now, such functions are known only for homogeneous media (Helmholtz equation with constant coefficients).

From a theoretical point of view, these methods can be used for any frequency band. From a numerical point of view, they are mainly used at low frequency since the computation time and storage needed increase with frequency. Then at higher frequency, specific 'high frequency' methods such as G.T.D. are often preferred.

In this chapter, we will not describe again how to obtain an integral equation. However, let us point out how essential it is to check that the integral equation is equivalent to the initial differential system, that is that they both lead to the same solution and that the conditions for existence and uniqueness of the solution are the same.

Section 6.1 is devoted to the methods used to solve integral equations. They consist in replacing the integral equation by an algebraic linear system of order N. To do so, the boundary of the propagation domain is divided into sub-intervals and the function, solution of the integral equation, is approximated by a linear combination of known functions (called 'approximation functions'). The coefficients of the combination are the unknowns of the linear system.

Section 6.2 is devoted to the particular problem of eigenvalues. Interior and exterior problems are defined in chapter 3. The 'interior' term is used to characterize a problem of propagation in a closed domain. For this kind of problem, there exist eigenvalues (eigenfrequencies). They cannot be obtained by separation methods if the geometry of the domain is too complicated. The integral equation methods are a good tool for finding these eigenfrequencies which are obtained by looking for the zeros of the determinant of the linear system. For exterior problems, the propagation domain extends up to infinity and there are no (real) eigenfrequencies. Nevertheless, as explained in chapter 3, the same integral equation can correspond to an interior problem and an exterior problem. Then, it has eigenfrequencies of the interior problem. To avoid this, one of the simplest and most rigorous methods consists in representing the solution of the differential problem as a linear combination of simple and double potentials with complex coefficients.

Section 6.3 presents a brief survey of problems of singularity. The problems of the singularity of the Green's kernel are examined in chapter 3. Generally speaking, discontinuities of the boundary or the boundary conditions imply a singular behaviour of the solution around these discontinuities. From an analytical study of this asymptotic behaviour, it is possible to introduce some *a priori* information in the system to be solved and then to improve the convergence of the algorithms.

Although finite element methods (F.E.M.) are not presented here, it is interesting to compare the advantages and drawbacks of these methods and boundary integral equation methods (B.I.E.M.).

The main advantage of B.I.E.M. is that the initial differential system written in a domain Ω is replaced by an equation or a system of equations on the boundary of Ω. The integral equation is then written on a domain of dimension n ($n = 1$ or 2) instead of a domain of dimension $(n + 1)$. This property is essential from a numerical point of view. For the same reason, there is no difficulty in solving problems of propagation in infinite media, while with F.E.M. the need to mesh the whole domain leads to a problem of a boundary at infinity (it is solved by using infinite finite elements, matching the numerical solution with asymptotic expressions, etc.).

On the other hand, F.E.M. apply to propagation problems in inhomogeneous media.

These properties of both methods are deduced from the following observation. The formulation of an integral equation is based on an *a priori* partial knowledge of the solution (representation of the pressure as a layer potential by using a known Green's function). This use of a known function leads to a simplified formulation on the boundary. In contrast, no *a priori* knowledge is required for F.E.M. and then they can be used to solve the Helmholtz equation with varying coefficients.

From a purely numerical point of view, both methods are similar and are based on the mesh of a domain, the use of approximation functions and the solution of a linear system. But for B.I.E.M. there is no use of special techniques to number the nodes or the elements: the matrix of the linear system is generally a full matrix, but it must be noted that the terms of the diagonal are larger than the others.

Finally, let us point out that F.E.M. and B.I.E.M. can be used jointly to solve problems such as sound radiation by a vibrating structure immersed in a fluid. If the structure is quite complex, it is described by using finite elements and the effect of the fluid is taken into account by an integral equation on the surface of the structure.

We end this introduction with a quite philosophical remark. The increasing complexity of the problems along with the increasing power of computers lead to an extensive use of F.E.M. and B.I.E.M. However, this does not mean that analytic expansions and approximations are no longer useful. On the contrary, they cannot be ignored. First, they provide tests of software on simple cases. They can also lead to a better choice of the approximation functions. Before solving a quite complex problem, it is very often useful to consider, as a first step, a simplified version for which partial analytical expansions can be obtained or simple computations can be carried out. Such a step can consist, for example, in changing the values of some parameters to evaluate their respective influence and then identifying the hypotheses to add to simplify the initial problem, depending on the aim of the study (frequency bands, accuracy required, and so on).

6.1 Techniques of Solution of Integral Equations

The methods presented in this section transform the integral equation into an algebraic linear system. There are mainly two types of methods: the collocation method and the Galerkin method. A third has been proposed which is a mixture of the first two.

Any integral equation can be expressed in the general form

$$Kp(P) = f(P), \qquad \forall P \in \Gamma \tag{6.1}$$

Γ is the boundary of a domain Ω of $\mathbb{R}^n$ (mainly $n = 2$ or 3). p is the unknown function. f is known and defined on Γ. The integral operator K is defined by

$$Kp(P) \equiv \varepsilon p(P) + \int_\Gamma p(P')G(P, P')\, d\sigma(P')$$

P and P' are points of Γ, G is any kernel (Green's kernel, its derivative). ε is a constant and can be equal to zero.

6.1.1. Collocation method

With the collocation method, the solution is expressed as

$$p(P) = \sum_{\ell=1}^{N} \nu_\ell \gamma_\ell(P) \tag{6.2}$$

Functions γ_ℓ, $(\ell = 1, ..., N)$ are the approximation (or test) functions. Coefficients ν_ℓ, $(\ell = 1, ..., N)$ are the unknowns. The collocation method consists in choosing a set of N points P_j of Γ. The integral equation is then written at these points P_j and this leads to the following linear system:

$$\sum_{\ell=1}^{N} \nu_\ell K\gamma_\ell(P_j) = f(P_j) \qquad \text{for } j = 1, ..., N$$

This system is solved to obtain the coefficients ν_ℓ and then the solution p on Γ. Points P_j are called collocation points.

Choice of the collocation points and the approximation functions

This choice obviously depends on the kind of problem to be solved and on the *a priori* information on the behaviour of the solution. This information can be obtained from physical or mathematical considerations.

The functions γ_ℓ are often chosen as spline functions (i.e. often piecewise polynomial functions). In acoustics, applications of the collocation method (see [1] and [2] for example) show that very often piecewise constant functions give quite satisfactory results.

The numerical procedure is as follows:

- Γ is divided into N sub-intervals Γ_j, such that the Γ_j are disjoint and that their sum is equal to Γ. For simplicity, the Γ_j can be chosen of the same length or area, unless special attention must be given to some particular parts of Γ. Let P_j be a point of Γ_j; it is often chosen as the centre. In $\mathbb{R}$, numerical tests show that the length of Γ_j must be of order of a sixth of the acoustic wavelength.
- p is approximated by constants:

$$p(P) = p(P_j) = \mu_j \qquad \text{for } P \in \Gamma_j, j = 1, ..., N$$

- The integral equation is written at the N points P_j, leading to the linear system of order N: $A\mu = B$. μ is the vector of the unknowns. B is the vector given by $B_j = f(P_j), j = 1, ..., N$. The matrix A given later in an example depends on all the data of the problem (frequency, geometry of the domain, boundary conditions, ...) except on the source which appears only in vector B.

Example. Let $p(y, z)$ be the pressure, solution of the two-dimensional problem:

$$\begin{cases} (\Delta + k^2)p(y, z) = \delta(y)\delta(z - z_0) & \text{if } z > 0 \\ \dfrac{\partial p(y, z)}{\partial \vec{n}} = 0 & \text{if } z = 0 \text{ and } a < y < b \\ \left(\dfrac{\partial}{\partial \vec{n}} + \dfrac{\iota k}{\zeta}\right) p(y, z) = 0 & \text{if } z = 0 \text{ and } y < a \text{ or } y > b \\ \text{Sommerfeld conditions} & \end{cases} \tag{6.3}$$

A harmonic signal ($\exp(-\iota\omega t)$) is emitted by an omnidirectional point source S, located at $(0, z_0)$.

This example describes the sound propagation above an inhomogeneous ground, made of a constant-width strip characterized by a Neumann condition and two absorbing half-planes characterized by the same normal impedance ζ (a complex constant, see chapter 4, section 4.1.3). If the source is cylindrical and its axis parallel to the axis of the strip, the three-dimensional problem can be replaced by the two-dimensional problem (6.3). $\vec{n}$ is the normal to the plane $(z=0)$.

Let G be the Green's kernel which satisfies the same impedance condition on $(z=0)$:

$$\begin{cases} (\Delta + k^2)G(S, M) = \delta_S(M) \qquad \text{if } M = (y, z > 0) \\ \left(\dfrac{\partial}{\partial \vec{n}} + \dfrac{\iota k}{\zeta}\right) G(S, M) = 0 \qquad \text{if } z = 0 \\ \text{Sommerfeld conditions} \end{cases}$$

The Green's formula applied to p and G leads to the following representation:

$$p(M) = G(S, M) + \frac{\iota k}{\zeta} \int_a^b p(P')G(P', M)\, d\sigma(P') \tag{6.4}$$

M is a point of the half-plane $(z > 0)$ and P' is a point of the interval $[a, b]$. An integral equation is obtained by letting M tend to a point P_0 of $[a, b]$:

$$p(P_0) = G(S, P_0) + \frac{\iota k}{\zeta} \int_a^b p(P')G(P', P_0)\, d\sigma(P') \tag{6.5}$$

The unknown is the value of the sound pressure on $[a, b]$. Once this equation is solved, the sound pressure can be calculated anywhere in the half-plane $(z > 0)$ by using the integral representation (6.4).

Equation (6.5) is solved by the simplest collocation method:

- The interval $[a, b]$ is divided into N sub-intervals $\Gamma_j = [a_j, a_{j+1}[$ of the same length, such that $a_1 = a$ and $a_{N+1} = b$. P_j is the middle of Γ_j. N is chosen such that $|a_{j+1} - a_j| \simeq \lambda/6$.
- The pressure p on $[a, b]$ is approximated by a function constant on each sub-interval: $p(P) = p(P_j) = \mu_j, j = 1, ..., N$.
- The integral equation is written at points P_j. This leads to the linear system $A\mu = B$, where A is a $N \times N$ matrix given by

$$A_{mn} = \delta_{mn} - \frac{\iota k}{\zeta} \int_{a_n}^{a_{n+1}} G(P, P_m)\, d\sigma(P), \qquad m = 1, ..., N;\; n = 1, ..., N$$

δ_{mn} is the Kronecker symbol. B is the vector given by

$$B_m = G(S, P_m), \qquad m = 1, ..., N$$

The next step is to compute A and B and then to solve the linear system. The pressure at any point M of the half-plane $(z > 0)$ can be written

$$p(M) = G(S, M) - \frac{\iota k}{\zeta} \sum_{m=1}^{N} \mu_m \int_{a_m}^{a_{m+1}} G(P', M)\, d\sigma(P') \qquad (6.6)$$

An example of this result is presented in Fig. 4.8 of chapter 4 where the solid curve is obtained from (6.6).

Integration on an infinite domain

The boundary Γ may be infinite (straight line, plane, ...). Let us consider an integral equation of the following type:

$$p(P_0) = f(P_0) + \int_{\Gamma} p(P')G(P_0, P')\, d\sigma(P')$$

Γ denotes the boundary $(z = 0)$ and f is a known function (the incident field, for example). P' and P_0 are two points of Γ.

The integral can be divided into two integrals: one on $[-A, A]$ and the other on $]-\infty, -A[\cup]A, +\infty[$, where A is a large positive number. With the collocation method, the first integral is divided into a sum of N integrals which are evaluated numerically. The other one is evaluated numerically or analytically, using the asymptotic behaviour of G. In some cases, if A is large enough, it can be neglected.

In the previous example, there appears an integration on an infinite domain if, instead of G, use is made of the Green's kernel $D(S, M)$ which satisfies the homogeneous Neumann condition on $(z = 0)$. Another integral equation is then obtained (it is equivalent to (6.5)):

$$p(P_0) = D(S, P_0) - \frac{\iota k}{\zeta}\left(\int_{-\infty}^{a} + \int_{b}^{+\infty}\right) p(P')D(P_0, P')\, d\sigma(P') \qquad (6.7)$$

The unknown is the value of p on $]-\infty, a[\cup]b, +\infty[$. The integration domain is divided into $]-\infty, A[$ and $]A, a[$ for the first part and $]b, B[$ and $]B, +\infty[$ for the second. A is a negative number and B a positive number. The intervals of finite length are divided into sub-intervals Γ_j to apply the collocation method. On the intervals $]-\infty, A[$ and $]B, +\infty[$, the integration can be made by using asymptotic behaviours. If the values of $|A|$ and B are greater than several wavelengths, both integrals are ignored. But it must be emphasized that such a sharp truncation can lead to numerical oscillations around A and B.

6.1.2. Galerkin method

The Galerkin method applied to equation (6.1) consists in choosing an approximation space for p. p is written as previously (6.2) where the functions γ_m are a basis of this space. The coefficients ν_m are determined by the equation

$$\langle Kp, \gamma_n \rangle = \langle f, \gamma_n \rangle, \qquad n = 1, ..., N \qquad (6.8)$$

where $\langle , \rangle$ represents the scalar product defined in the approximation space. This leads to the following linear system:

$$\sum_{m=1}^{N} \nu_m \langle K\gamma_m, \gamma_n \rangle = \langle f, \gamma_n \rangle, \qquad n = 1, ..., N \tag{6.9}$$

The numerical procedure is similar to the one developed for the collocation method: evaluation of a matrix A and a vector B before solving the linear system. The matrix A is given by

$$A_{mn} = \langle K\gamma_m, \gamma_n \rangle, \qquad m = 1, ..., N;\ n = 1, ..., N$$

The scalar product $\langle , \rangle$ is generally defined as an integral. This means that the evaluation of A leads to the computation of integrals of order $(d+1)$ for the Galerkin method and integrals of order d for the collocation method. The collocation method then leads to simpler computations. However, Wendland [3] shows that when spline functions are chosen as approximation functions, it is necessary to use spline functions of order $(2m+1)$ for the collocation method and order m for the Galerkin method to obtain the same rate of convergence. The choice of the method depends on the type of problem, the accuracy required and the computer power. Generally speaking, in acoustics, the simplest collocation method often gives satisfactory results.

6.1.3. Method of Galerkin-collocation

This method has been proposed by Wendland among others. It consists in separating the kernel of the integral equation into a singular part and a remaining part (which is regular): $G = G_s + G_r$. The equation which includes the singular part is solved by a Galerkin method. The other one is solved by a collocation method. An example of this technique is presented in [4] by Atkinson and de Hoog. The aim of the method is to obtain a good accuracy for the integration of the singular part, because the influence of this singular part is greater than that of the regular part, which can then be computed more roughly. Let us finally point out that the collocation method can be seen as a particular case of the Galerkin method where the second integration is computed through a one-point integration formula, that is by approximating the integrals by

$$\int_{\alpha}^{\beta} f(t)\, dt \approx (\beta - \alpha) f(\tau) \qquad \text{with } \tau \text{ a point of } [\alpha, \beta]$$

6.2. Eigenvalue Problems

6.2.1. Interior problems

When the domain of propagation is bounded (i.e. it does not extend to infinity), the system of differential equations has eigenvalues, the wavenumbers k for which there is no existence or uniqueness of the solution. Since the integral equation deduced from this system is equivalent to it, it has the same eigenvalues which are then solutions of

$$|\det A| = 0$$

if A is the matrix of the linear system deduced from the integral equation. For numerical reasons, the eigenvalues k are the wavenumbers for which $|\det A|$ has a minimum value. To point out the efficiency of the method, let us chose a two-dimensional problem of sound propagation inside a disc, for which an exact representation of the solution is known. The eigenvalues obtained from this exact representation are compared with the values obtained from the integral equation. This example has been studied by Cassot [5] (see also [6]).

It is convenient to use polar coordinates. The disc D with radius a is centred at 0. A point source $S = (R, 0)$ is located inside D. The boundary Γ of D is described by a homogeneous Neumann condition. The sound pressure inside the disc is the solution of

$$\begin{cases} (\Delta + k^2)p(M) = \delta_S(M) & \text{inside } D \\ \dfrac{\partial p(M)}{\partial \vec{n}} = 0 & \text{on } \Gamma \end{cases}$$

$\vec{n}$ is the outward unit vector normal to Γ.

Exact solution. An exact expression for p can be obtained by using the separation method:

$$p(M) = -\frac{\iota}{4} H_0^{(1)}(kd(S, M)) + \frac{\iota}{4} \sum_{m=-\infty}^{+\infty} \frac{H'_m(ka)}{J'_m(ka)} J_m(kR) J_m(kr) e^{\iota m\theta} \tag{6.10}$$

where $M = (r, \theta)$ is a point of the disc. J_m and H_m are respectively the Bessel and Hankel functions of order m. $J'_m(ka) \equiv J'_m(z)$ for $z = ka$.

The eigenvalues are the eigenwavenumbers k such that $J'_m(ka) = 0$.

Numerical solution. The pressure p can also be expressed by using a simple layer potential:

$$p(M) = -\frac{\iota}{4} H_0^{(1)}(kd(S, M)) + \frac{\iota}{4} \int_\Gamma \mu(P) H_0(kd(M, P))\, d\sigma(P)$$

P is a point of Γ. The integral equation is obtained by writing the boundary condition:

$$-\frac{\mu(P_0)}{2} - \frac{\iota}{4} \int_0^{2\pi} \mu(P) \frac{\partial H_0(kd(P_0, P))}{\partial r_0} a\, d\theta = -\frac{\iota}{4} \frac{\partial H_0(kd(P_0, S))}{\partial r_0}$$

$P=(a,\theta)$ and $P_0=(r_0\rightarrow a,\theta_0)$ are two points of Γ. The unknown is the function μ on Γ.

The linear system is obtained by a collocation method and μ is approximated by a piecewise constant function. The boundary Γ is divided into N sub-intervals Γ_j. This leads to $A\mu=B$ with

- the vector $(\mu_j), j=1,\ldots,N$ is the unknown $(\mu_j=\mu(P_j))$
- B_j given by

$$B_j=\iota k H_1(k\rho_j)\cos(\vec{\rho}_j,\vec{n}_j),\qquad \text{with } \vec{\rho}_j=\overrightarrow{SP_j} \text{ and } \rho_j=\|\vec{\rho}_j\|$$

- $A_{j\ell}$ given by

$$A_{j\ell}=-\iota k H_1(kd_{\ell j})\cos(\vec{d}_{\ell j},\vec{n}_j)L(\Gamma_j)\qquad \text{if } \ell\neq j, \ell=1,\ldots,N;\ j=1,\ldots,N$$

and

$$A_{\ell\ell}=2-\frac{\iota\pi L(\Gamma_\ell)}{4a}\{S_0(z)H_1(z)-H_0(z)S_1(z)\}$$

$$\text{with } z=k\frac{L(\Gamma_\ell)}{2} \text{ and } \ell=1,\ldots,N$$

$$\vec{n}_j=\vec{n}(P_j);\qquad \vec{d}_{\ell j}=\overrightarrow{P_\ell P_j};\qquad d_{\ell j}=\|\vec{d}_{\ell j}\|;\qquad L(\Gamma_j)=\text{length of }\Gamma_j$$

S_0 and S_1 are the Struve functions [7].

The problem has a symmetry but this symmetry is ignored for demonstration purposes. Table 6.1 presents a comparison between the first eigenvalues k_e obtained from the exact representation and the eigenvalues k_a obtained from the integral equation (i.e. such that $|\det A|$ is minimum).

Table 6.1.

	$N=20$		$N=40$	
k_e	k_a	$\frac{\Delta k}{k}$	k_a	$\frac{\Delta k}{k}$
1.841 18	1.841 05	7.06(−5)	1.841 165	0.815(−5)
3.054 24	3.053 93	1.015(−4)	3.054 19	1.64(−5)
3.831 71	3.831 38	8.61(−5)	3.831 75	1.04(−5)
4.201 19	4.200 52	1.57(−4)	4.201 10	2.14(−5)
5.317 55	5.316 50	1.97(−4)	5.317 38	3.13(−5)
5.331 44	5.330 83	1.14(−4)	5.331 39	0.938(−5)
6.415 62	6.414 04	2.46(−4)	6.415 79	2.72(−5)
6.706 13	6.705 16	1.44(−4)	6.706 00	1.94(−5)
7.015 59	7.014 62	1.38(−4)	7.015 42	2.42(−5)

For $N = 20$, the relative error $(\Delta k/k)$ on the eigenvalues is less than 2.5×10^{-4}. For $N = 40$, this error is less than 3.2×10^{-5}. These results are quite satisfactory for engineering purposes.

6.2.2. Exterior problems

To avoid the spurious eigenvalues which appear when solving an exterior problem by an integral equation method, several methods have been proposed (see [8], [9], [10] for example). Among the simplest and most rigorous methods is the one presented in [9]. It consists in representing the solution as a linear combination of a simple and a double layer potential, with a complex coefficient. The example presented here shows the efficiency of this method.

An incident plane wave, travelling in the plane (O, x, y) is represented by a function $p_0(x) = \exp(\iota kx)$. It is diffracted by a cylindrical obstacle of radius a and boundary Γ. Γ is described by a Dirichlet condition. Let $p(M)$ denote the pressure at any point M, outside the obstacle. $p(M)$ can be written as

$$p(M) = p_0(M) - \int_\Gamma \frac{\partial p(P')}{\partial \vec{n}(P')} G(M, P')\, d\sigma(P')$$

P' is a point of Γ and $G(M, P') = -\iota H_0^{(1)}(kd(M, P'))/4$. An integral equation is obtained by deriving this expression and letting M tend to a point P of Γ:

$$\frac{1}{2}\frac{\partial p(P)}{\partial \vec{n}(P)} + \int_\Gamma \frac{\partial p(P')}{\partial \vec{n}(P')} \frac{\partial G(P, P')}{\partial \vec{n}(P)}\, d\sigma(P') = \frac{\partial p_0(P)}{\partial \vec{n}(P)}, \qquad \forall P \in \Gamma \tag{6.11}$$

This integral equation has eigenvalues which are the eigenvalues of the corresponding interior problem with a homogeneous Neumann condition, studied in the previous section.

To avoid this difficulty, the solution p is expressed as a sum of a simple and a double layer potential, with a complex coefficient:

$$p(M) = p_0(M) + \int_\Gamma \mu(P') \left(\frac{\partial G(M, P')}{\partial \vec{n}(P')} - \iota G(M, P') \right) d\sigma(P')$$

for any point M exterior to the obstacle. The integral equation is obtained from the Dirichlet condition:

$$-\frac{\mu(P)}{2} + \int_\Gamma \mu(P') \left(\frac{\partial G(P, P')}{\partial \vec{n}(P')} - \iota G(M, P') \right) d\sigma(P') = -p_0(P) \tag{6.12}$$

for any point P of Γ.

For comparison, both integral equations (6.11) and (6.12) have been solved by a collocation method.

Figure 6.1 presents the behaviour of $|\det A|$ versus ka, where A is the matrix of the linear system obtained from (6.11) and (6.12). Clearly, the linear system

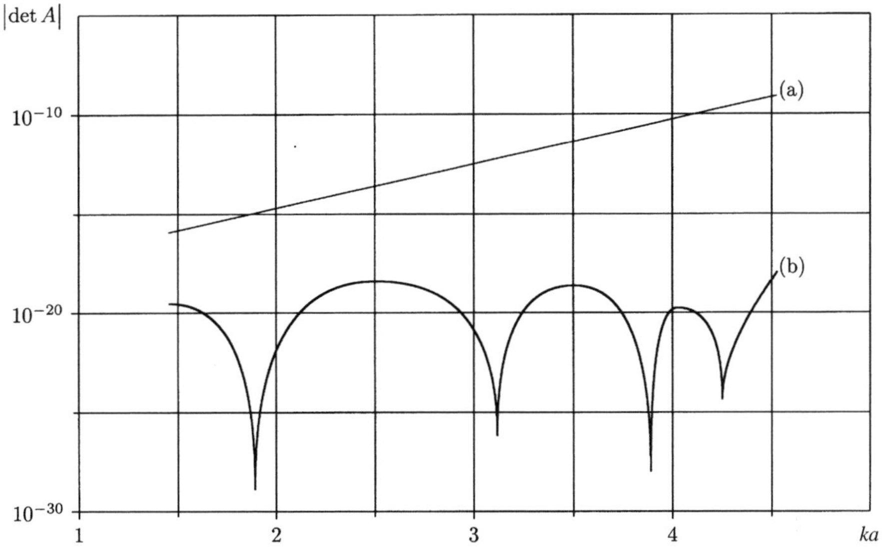

Fig. 6.1. | det A | versus ka: (a) equation (6.12), (b) equation (6.11).

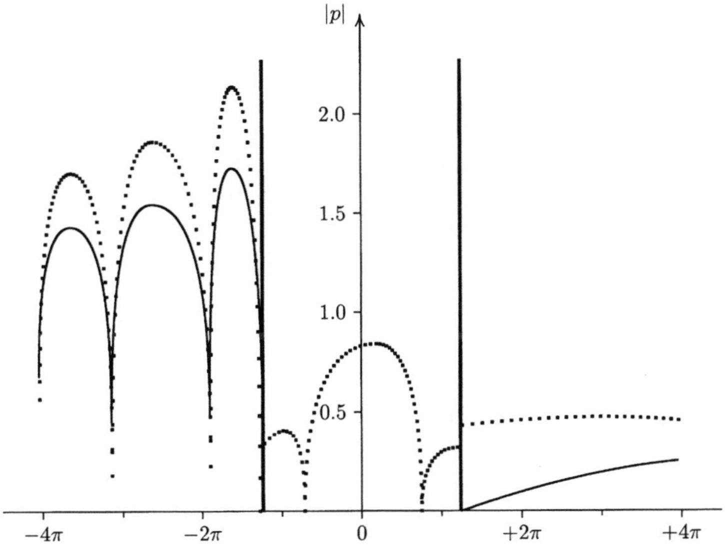

Fig. 6.2. Modulus of the sound pressure: (—) exact curve and curve deduced from (6.12), (·······) curve deduced from (6.11).

deduced from (6.12) does not have real eigenvalues. In contrast, the determinant of A deduced from (6.11) has minima which correspond to the wavenumbers such that $J_n'(ka) = 0$, the eigenvalues of the interior problem with a Neumann condition.

Figure 6.2 presents the total pressure, computed for the eigenvalue k such that $ka = 3.8317$. The dotted curve corresponds to the pressure obtained from the Green's representation (6.11). The solid curve corresponds to the representation (6.12). It is quite similar to the curve obtained from the exact representation in a series of Bessel functions for the exterior problem. The curves are drawn versus kx, and x is the distance from the centre of the obstacle to the observation point.

6.3. Singularities

6.3.1. Singularity of the kernel

All the integral equations deduced from the Helmholtz equation correspond to a Green's kernel which is singular: $G(M, P)$ tends to infinity when distance $r(M, P)$ tends to zero. For this reason, all the numerical computations must be carried out with care. For the collocation method, this difficulty appears when computing the diagonal terms of matrix A.

The integrals corresponding to a simple layer potential are well defined because the kernel is integrable. For example, in two dimensions, the Hankel function $H_0(kr)$ behaves as $\ln r$. For numerical computations, the kernel can then be approximated by the first terms of its asymptotic behaviour when r tends to zero. This leads to an integral which can be analytically evaluated.

The integrals corresponding to a double layer potential are defined as Cauchy principal values (see chapter 3). A presentation of the numerical procedure and some examples can be found in [10].

The integrals corresponding to a derivative of a double layer potential can only be defined as the finite part of an integral following Hadamard's works. This leads to the difficulty of an intrinsic definition of the limit. The theory of pseudo-differential operators is a good tool for defining this limit and proving that the techniques used in the case of diffraction by a thin screen, for example [11], are rigorous.

6.3.2. Domains with corners, polygons

The singular behaviour of the solutions of boundary value problems in polygonal domains has been studied by Grisvard [12], among others. For example, let Ω denote a polygonal domain in the plane $(0, y, z)$ and let $p(y, z)$ be the solution of the system:

$$\begin{cases} Ep = 0 & \text{inside } \Omega \\ B_j p = g_j & \text{on } \Gamma_j, j = 1, \ldots, N \end{cases}$$

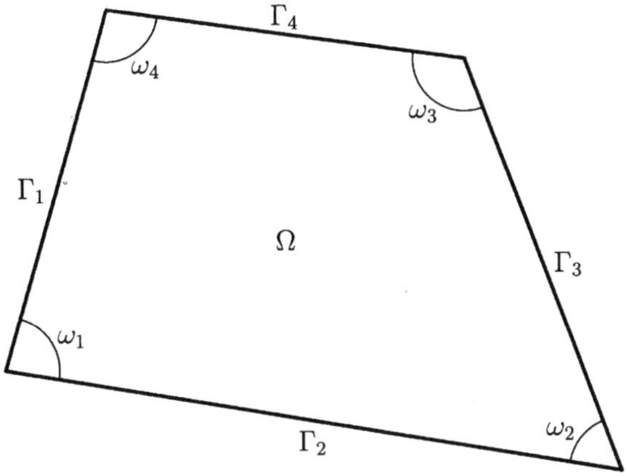

Fig. 6.3. Domain Ω.

where the sum of the intervals Γ_j is equal to the boundary Γ of Ω. E is a differential operator, of degree 2, with constant coefficients. The operators B_j are defined on the boundary and correspond to the boundary conditions. The g_j are functions defined and known on Γ_j.

The existence and uniqueness of the solution p are proved. Around each corner, p has a singular behaviour and is written $p = p_s + p_r$ where p_s is a singular function which is explicitly given and p_r is a regular rest.

The number and the behaviour of the singular functions depend on the values of the angles ω_j and on the type of the boundary conditions. Similar results have also been obtained in three dimensions.

Discontinuous boundary conditions, cracks. Problems described by partial differential equations and mixed discontinuous boundary conditions have been studied by Eskin [13]. For example, let p be the solution of the following two-dimensional problem in the half-plane ($z \geqslant 0$):

$$\begin{cases} Ep(y,z) = 0 & \text{if } z > 0 \\ B_1 p(y,z) = f(y) & \text{if } y < 0 \text{ and } z = 0 \\ B_2 p(y,z) = g(y) & \text{if } y > 0 \text{ and } z = 0 \\ \text{conditions at infinity} & \end{cases}$$

where E is a differential operator. B_1 and B_2 are boundary operators. The functions f and g are defined respectively on ($y < 0, z = 0$) and ($y > 0, z = 0$).

This problem is similar to those presented in chapter 5 where the Wiener–Hopf method is described. The proofs of the existence and the uniqueness are partly based on the Wiener–Hopf method. The discontinuity of the boundary conditions implies that p has a singular behaviour around (0, 0) which can be obtained explicitly.

Example: Let ($y < 0, z = 0$) be described by a Neumann condition and ($y > 0, z = 0$) be described by an impedance condition. It can be shown [14] that the solution p around $(0, 0)$ behaves as

$$p(y, 0) = \mathcal{O}(y \log y) + \text{constant}, \quad \text{when } y \to 0$$

The method proposed by Eskin to obtain the behaviour of the solution around such discontinuities is quite general. It applies, for example, to propagation problems and problems with cracks, when the conditions on both sides of the crack are different. Let us also note that Achenbach ([15] for example) has developed applications of integral equations and Wiener–Hopf methods to mixed boundary value problems such as diffraction of elastic waves by a slit.

It is always interesting to obtain *a priori* information on the solution such as its singular behaviour. Introducing this information into the numerical solution technique leads to more efficient algorithms.

Bibliography

[1] DAUMAS, A., 1978. Etude de la diffraction par un écran mince posé sur le sol. *Acustica* 40(4), 213–222.

[2] EXTREMET, G., 1970. Propagation du son dans une enceinte fermée. *Acustica* 23, 307–314.

[3] WENDLAND, W., 1983. Boundary element methods and their asymptotic convergence. *Theoretical acoustics and numerical techniques.* C.I.S.M. Courses and Lectures 277. Springer-Verlag, New York.

[4] ATKINSON, K. and DE HOOG, F., 1984. The numerical solution of Laplace's equation on a wedge. *I.M.A. J. Num. Anal.* 4, 19–41.

[5] CASSOT, F. and EXTREMET, G., 1972. Détermination numérique du champ sonore et des fréquences propres dans une enceinte circulaire par la méthode de discrétisation. *Acustica* 27, 238–245.

[6] FILIPPI, P., 1983. Integral equations in acoustics. *Theoretical Acoustics and Numerical Techniques.* C.I.S.M. Courses and Lectures 277. Springer-Verlag, New York.

[7] ABRAMOWITZ, M. and STEGUN, I.A., 1964. *Handbook of mathematical functions.* Appl. Math. Ser. 55, National Bureau of Standards, Washington D.C.

[8] SCHENK, H., 1968. Improved integral formulation for acoustic radiation problems. *J. Acoust. Soc. Am.* 44, 41–58.

[9] FILIPPI, P.J.T., 1977. Layer potentials and acoustic diffraction. *J. Sound Vib.* 54, 473–500.

[10] SAYHI, M.N., OUSSET, Y. and VERCHERY, G., 1981. Solution of radiation problems by collocation of integral formulations in terms of single and double layer potentials. *J. Sound Vib.* 74(2), 187–204.

[11] DURAND, M., 1983. Boundary value problems analysis and pseudo-differential operators in acoustics. *Theoretical Acoustics and Numerical Techniques.* C.I.S.M. Courses and Lectures 277. Springer-Verlag, New York.

[12] GRISVARD, P., 1985. *Elliptic problems in non smooth domains.* Adv. Publ. Prog. Pitman, Boston.

[13] ESKIN, G., 1981. Boundary value problems for elliptic pseudo-differential equations. *Transl. Math. Monographs* 52. American Mathematical Society, Providence.

[14] HABAULT, D., 1984. Etude de l'influence des sols sur la propagation sonore. Thèse de Doctorat d'Etat, Université d'Aix-Marseille I, France.

[15] ACHENBACH, J.D., 1984. *Wave propagation in elastic solids.* Elsevier Science Publishers, Amsterdam.

CHAPTER 7

Introduction to Guided Waves

*Aimé Bergassoli**

Introduction

This chapter presents an introduction to the study of guided waves. Our aim is to present some basic results in order to help the reader to get a better understanding of more specialized publications.

We have tried to give more importance to the explanation of the physical phenomena than to numerical calculations, sometimes purely academic. This chapter has been written for both engineering and research purposes.

Only modal methods are described here. They are suitable for numerical computations. The straightforward study of the propagation of guided plane waves with more classical techniques such as electrical analogies and variational methods applied to discontinuities is still the most practical method.

Finally, let us point out that finite element methods are a good tool when the previous methods cannot be used.

7.1. Definitions and General Remarks

7.1.1. Guided waves

If the sound field emitted by a simple source has larger values at an observation point than it would have at the same point in an isotropic, homogeneous, infinite medium, this means that there is a guiding phenomenon along the path. This very general definition includes reflection by obstacles, diffraction, focusing effects in the propagation medium, and so on (of course, a more restrictive definition could be chosen).

From this, it can be seen that any kind of propagation in a realistic medium is more or less guided. Guiding effects have been observed for a long time, even before a theoretical description was proposed: short waves in electromagnetism, acoustic ducts.

Guided wave phenomena are quite similar in electromagnetism and in acoustics in fluids and solids (surface of the earth). Actually, many theoretical results were

---

*Chapter translated by Dominique Habault and Paul Filippi.

obtained during World War II, especially for the study of large distance radio communications and radar applications. At the same period, a better knowledge of underwater sound propagation was obtained because of applications to target detection. The phenomenon of waves guided by the surface of the earth was also extensively studied when the first nuclear explosion tests were carried out underground.

Let us point out that artificial guides (ducts) often have simple shapes, with finite length. They are used to produce or guide sounds (musical instruments, stethoscope) but more often they are used to carry gas, smokes or liquids from one point to another. The acoustical phenomenon is then added: it can be a nuisance (sound radiation from the walls or the ends of the guide) or it can be used for measurements (detection of cracks or defects).

Let us also briefly note that a tube of varying section and irregular shape (along its axis or/and in the section) can be as efficient as a circular cylinder with a straight axis. However, in this more complicated case, separation methods cannot apply, even in the direction of the axis. It must be noted that even a slowly varying section produces backward reflections. But if the angular frequency ω is small enough, the plane wave is filtered everywhere so that, in a finite length tube, there is a standing wave system which can include energy loss by the guide ends. If the tube is quite long, the stationary regime will appear after some time. In real cases, only a transient regime is present. The stationary regime corresponds mainly to academic problems (room acoustics, machinery noise).

All these remarks will appear again in the problems studied in the following sections. But it is always essential to determine how the chosen model fits the problem.

On the contrary, if the angular frequency ω is large, all the reflections which can be modelled by a random model will produce a backward flow. If the emitted signal is quite complicated, this leads to many difficulties and restrictions.

Finally, when a simple model cannot be used, the solution must be computed through a finite element method. As far as possible, one must try to mix both methods: separation methods for simple parts of ducts and finite element methods for more complicated volumes. When numerical methods are used, it is essential to obtain estimates of the errors. Furthermore, without an enormous number of runs with different values of the parameters (which is too much work and quite unrealistic), numerical computations alone seldom provide a good aid to a better understanding of the observed physical phenomena and their possible consequences.

This is the reason why the study of simple guides is essential. It must also be noted that many natural guides (surface of the earth, shallow water, etc.) as well as artificial guides (rigid tubes) can be described as simple guides. Even a slow variation of the axis or planes of symmetry does not change the results.

In the following, the behaviour of the sound pressure is assumed to be harmonic with time and is described by the term $(\exp(-\iota\omega t))$.

7.1.2. Boundaries

Only results related to wave guides are presented here.

Preamble

The first problem to solve is the total reflection of a wave by an interface. Except in particular cases such as quasi-rigid tubes, when flexural waves cannot be excited, the reflection coefficient depends on the angle of incidence. Indeed, the reflected field is rapidly attenuated since the incident energy is absorbed in the pores of the material. These losses are much larger than the classical losses caused by viscosity and thermoconduction in the fluid. The guiding phenomenon no longer exists. Let us consider the interface between two media characterized by different physical properties. Both media may have properties varying with width. If they vary rapidly (with the wavelength λ), the interface is sharp; if they vary slowly, it is a 'soft' interface. For particular conditions, media with varying properties can correspond to a total reflection case: deep ocean, atmosphere, ionosphere, deep layers of the earth.

We first consider the case of two infinite half-planes. The notion of an 'infinite half-space' is a good model of a deep layer: deep enough that any wave inside is sufficiently attenuated from one side of the layer to the other.

The sharp, plane interface between two fluids

Both fluids are characterized by a density ρ_i and a sound speed c_i, $i = 1, 2$. Any source radiation can formally be decomposed into plane waves (propagative, evanescent, inhomogeneous). Let us then consider an incident plane wave on the boundary ($z = 0$). The incident, reflected and transmitted fields can be written as

$$\phi_i(x, z) = e^{\iota k_1(x \sin \theta_1 - z \cos \theta_1)}; \qquad \phi_r(x, z) = \mathcal{R} e^{\iota k_1(x \sin \theta_1 + z \cos \theta_1)};$$
$$\phi_t(x, z) = \mathcal{T} e^{\iota k_2(x \sin \theta_2 - z \cos \theta_2)}$$

with $k_j = \omega/c_j$, $j = 1, 2$. The functions ϕ are in general complex. The angles θ_j are the angles measured from the normal direction to the surface. $\mathcal{R}$ and $\mathcal{T}$ are the reflection and transmission coefficients respectively.

The boundary conditions on ($z = 0$) are the conditions of continuity of the pressure and the normal velocity:

$$\rho_1\phi_1 = \rho_2\phi_2 \qquad \text{and} \qquad \frac{\partial \phi_1}{\partial z} = \frac{\partial \phi_2}{\partial z}$$

with $\phi_1 = \phi_i + \phi_r$ and $\phi_2 = \phi_t$.

This leads to the Snell–Descartes law and $\mathcal{R}$ and $\mathcal{T}$ are given by

$$\mathcal{R} = \frac{\rho_1 c_1 \cos \theta_1 - \rho_2 c_2 \cos \theta_2}{\rho_1 c_1 \cos \theta_1 + \rho_2 c_2 \cos \theta_2} = \frac{(\rho_2/\rho_1) \cos \theta_1 - \sqrt{(c_1/c_2)^2 - \sin^2 \theta_1}}{(\rho_2/\rho_1) \cos \theta_1 + \sqrt{(c_1/c_2)^2 - \sin^2 \theta_1}} \tag{7.1}$$

$$\mathcal{T} = \frac{2(\rho_2/\rho_1) \cos \theta_1}{(\rho_2/\rho_1) \cos \theta_1 + \sqrt{(c_1/c_2)^2 - \sin^2 \theta_1}} \tag{7.2}$$

The relations still hold for complex parameters. By analogy with electromagnetism (E.M.), the ratio c_1/c_2 is called the index.

In the particular case of an air/water boundary, $c_1 \simeq 345\ \mathrm{m\,s^{-1}}$, $c_2 \simeq 1500\ \mathrm{m\,s^{-1}}$, $\rho_1 = 1.29$, $\rho_2 = 10^3$, formulae (7.1) and (7.2) show that if the source is in water, the interface is perfectly reflecting for any real θ. If the source is in air, a plane wave cannot propagate in water if the angle is greater than 13°.

From these results, it is possible to check that the law of energy conservation is satisfied. In both media, the intensity I_j of the plane wave is equal to

$$\frac{p_j^2}{2\rho_j c_j} = \frac{(\omega\rho_j)^2 \mid \phi_j \mid^2}{2\rho_j c_j}, \qquad j = 1, 2$$

The normal components of the energy vector are continuous.

Expression (7.1) shows that the reflection is total for $\theta \geqslant \arcsin(c_1/c_2)$. In this case, the amplitude of the transmitted wave tends to zero when z tends to $(-\infty)$. In this medium, for the evanescent waves, the surfaces of equal amplitudes are parallel to the interface and the surfaces of equal phases (wave surfaces) are oblique, following Snell's law.

It must also be noticed that $\mathcal{R}$ can be zero for an angle called Brewster's angle. This is a fundamental solution for radio waves to penetrate the ionosphere. The condition is

$$0 < \frac{(\rho_2/\rho_1)^2 - (c_1/c_2)^2}{(\rho_2/\rho_1)^2 - 1} < 1$$

which is always true if $c_1 > c_2$ and always false if not. For the air/water interface, this angle does not exist. It can exist for the water/solid interface.

Sharp interface between a fluid and a solid

The elastic properties of the solid medium are characterized by Lamé coefficients. This approximation is correct for metals but not so much for rocky grounds and even less for sediments. Let $\phi_1(x, z)$ denote the potential in the fluid and $\phi_2(x, z)$ and $\psi_2(x, z)$ the scalar and vector potentials in the solid. Let us introduce:

$$u_1 = -\nabla\phi_1; \qquad u_2 = -\nabla\phi_2 + \nabla \times \vec{\psi}_2$$

In the case of a harmonic plane wave, the continuity of the stress components (τ_{zz} = pressure in the fluid, and $\tau_{zx} = 0$ for no tangential coupling) and the continuity of the normal velocity u_z leads to

$$\lambda_1 \Delta\phi_1 = \lambda_2 \Delta\phi_2 + 2\mu_2 \left(\frac{\partial^2 \phi_2}{\partial z^2} + \frac{\partial^2 \psi_2}{\partial z \partial x} \right)$$

$$2\frac{\partial^2 \phi_1}{\partial x \partial z} + \frac{\partial^2 \psi_2}{\partial x^2} - \frac{\partial^2 \phi_1}{\partial z^2} = 0$$

$$\frac{\partial \phi_1}{\partial z} = \frac{\partial \phi_2}{\partial z} + \frac{\partial \psi_2}{\partial x}$$

In the solid, the velocities $c_{2,L}$ and $c_{2,T}$ of the longitudinal and transverse waves are given by $\rho_2 c_{2,L}^2 = \lambda_2 + 2\mu_2$ and $\rho_2 c_{2,T}^2 = \mu_2$. If θ_2 and γ_2 are the angles of refraction for both waves, the reflection coefficient in the fluid can be written as

$$\mathcal{R} = \frac{\sin^2(2\gamma_2)(\rho_2 c_{2,T}/\cos\gamma_2 - \rho_2 c_{2,L}/\cos\theta_2) + \rho_2 c_{2,L}/\cos\theta_2 - \rho_1 c_{1,L}/\cos\theta_1}{\sin^2(2\gamma_2)(\rho_2 c_{2,T}/\cos\gamma_2 - \rho_2 c_{2,L}/\cos\theta_2) + \rho_2 c_{2,L}/\cos\theta_2 + \rho_1 c_{1,L}/\cos\theta_1} \tag{7.3}$$

By using Snell's law, it is possible to express $\mathcal{R}$ as a function of one angle only. For a source in the solid, the same method can be used to evaluate the reflection coefficient of the longitudinal and transverse waves in the solid. Polarization phenomena are then observed.

This model of two infinite half-spaces is quite academic but can often be used to describe deep layers (deep compared with the wavelength λ) if there is enough attenuation. No attenuation term is explicitly introduced here but it must be taken into account to describe realistic cases.

Let us remark that the case studied in the previous section is a particular case of the interface fluid/solid with $\mu_2 = 0$.

Two particular values of the angles must be considered: $\arcsin(c_1/c_{2,L})$ and $\arcsin(c_1/c_{2,T})$. For a metallic medium, $c_{2,L}$ is about 6000 m s^{-1} and $c_{2,T}$ about 3000 m s^{-1}, then $\gamma_2 < \theta_2$. In the particular case of water/solid:

$$\arcsin(c_1/c_{2,L}) \simeq 14^\circ > \arcsin(c_1/c_{2,T}) \simeq 30^\circ$$

The water/solid interface is perfectly reflecting for $\theta_1 > 30^\circ$ and (almost) perfectly reflecting for any value of θ_1.

For seismic applications, it is interesting to study the case of a longitudinal wave travelling in the solid and transformed into a transverse wave after reflection on the boundary. These surface waves are similar to Rayleigh waves which exist at the solid/vacuum boundary. Such surface waves are 'free' solutions of the Helmholtz equation. In other words, following Brekhovskikh [3], there exists an angle such that the denominator of $\mathcal{R}$ and $\mathcal{T}$ (such as in (7.3)) is equal to zero. Since their numerators are not zero for this angle, the coefficients tend to infinity. Formally, there can be reflection and transmission in the absence of excitation.

This surface wave which is the sum of a longitudinal and a transverse wave (with an elliptic trajectory) can easily be observed in the far field, in earthquakes. It is a wave guided by the interface.

It must be noted that the velocity of Rayleigh waves, V_R, is smaller than $c_{2,T}$ and that their phase velocity, parallel to the boundary, is $c_{2,T}/\sin\gamma_2$. It is then possible to solve the equation for this variable. γ_2 necessarily has the form $(\pi/2 - \alpha)$, which means that the amplitude of the surface wave exponentially decreases with depth. This is the reason why it is called a surface wave and it satisfies the law of conservation of energy.

This wave exists if $(c_{2,L}/c_{1,L}) < \sqrt{2}/2$ which is generally true. Furthermore, it can be shown that a solution exists also in the fluid, close to the boundary; its velocity is smaller than $c_{1,L}$. This type of wave does not exist at the boundary between two fluids since it is based on the existence of transverse waves. For plates immersed in a fluid, surface waves also exist around boundaries; they are called Lamb waves.

Soft interface

In this section, the sound field is evaluated from a ray method but the quantitative study must be based on a wave approach. Anyhow in some cases, the analysis based on a ray method appears to be very similar to that based on plane wave decomposition. Soft interfaces are encountered in media such as the ionosphere (E.M.), troposphere (E.M. and infra-sound) and ocean (sound waves).

Let us consider an (almost) infinite layer in which the wave propagation is characterized by parameters slowly varying in the direction Oz perpendicular to the boundaries of the layer. This layer is assumed to be suitably modelled as a multi-layered medium, with constant physical properties in each sub-layer.

If the velocity of an acoustic wave varies with the depth z, it is easy to imagine that the ray path will behave as shown in Fig. 7.1. A plane wave has, in the reference medium, the velocity c. Its wavevector $\vec{k}$ makes an angle φ with the boundary Ox. φ is the complementary of the angle θ previously used; this is a notation classically used in E.M. and in underwater acoustics. Snell's law then gives

$$\frac{\cos\varphi}{c} = \cdots \frac{\cos\varphi_j}{c_j} = \frac{\cos\varphi_{j+1}}{c_{j+1}} = \frac{1}{c_0}$$

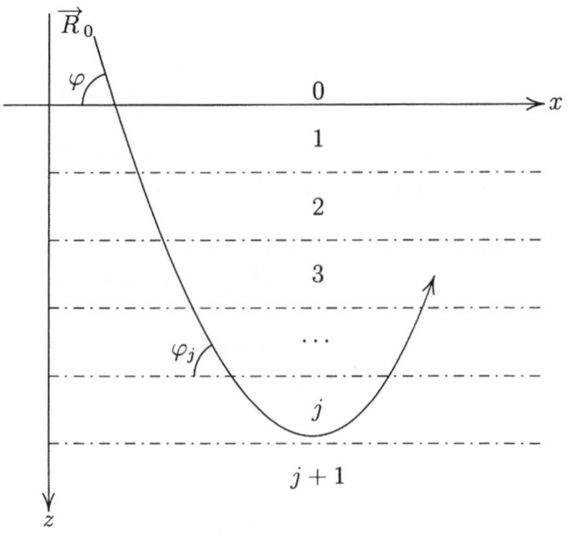

Fig. 7.1. Ray curving.

The last term corresponds to a layer in which $\varphi = 0$ if it exists. Let us define the index by $n_j = c/c_j$ or $k_j = n_j k$. The velocity potential in each sub-layer can be written

$$\begin{aligned}\phi(x, z) &= \phi_j \exp[\iota k_j(x \cos \varphi_j + z \sin \phi_j)] \\ &= \phi_j \exp[\iota\omega(x \cos \varphi_j + z \sin \varphi_j)/c_j] \\ &= \phi_j \exp(\iota kx \cos \varphi) \exp\left[\iota \frac{\omega}{c}\left(\frac{c}{c_j} z \sin \varphi_j\right)\right]\end{aligned}$$

If q_j is equal to $n_j \sin \varphi_j$, Snell's law implies $q_j^2 = n_j^2 - \cos^2 \varphi$. q_j has been introduced in E.M. by Booker. It is generally complex. If β denotes the total complex phase $\beta = kx(\cos \varphi + q_j z)$ and if φ_j is assumed to be a constant for any z, $\phi(x, z) = \phi \exp(\iota\beta)$ and there necessarily exists a reflection plane with $|\mathfrak{R}| = 1$.

The variations of β in the multi-layered medium are given by

$$\Delta\beta = k \sum_{j=1}^{n} q_j \delta z$$

for n sub-layers. Let δz tend to zero and the number of layers tend to infinity, then $\Delta\beta = k \int_{z_1}^{z_2} q\, dz$ depends only on q and

$$\phi(x, z) = \phi \exp\left[-\iota k(x \cos \varphi + \int_0^z q\, dz)\right]$$

This formula is called a phase integral. It means that the phase term is cumulative. A more detailed study shows that the pressure can be written

$$p(x, z) = p_{ref}\, q^{1/2} \exp\left[\iota k(x \cos \varphi + \int_0^z q\, dz)\right]$$

where k and p_{ref} are evaluated close to the surface.

This solution is correct in the optics approximation: it is called W.K.B. (after Wentzel, Kramers, Brillouin). It is valid if the sound speed varies slowly over a wavelength along the path. This can be seen, for example, from a comparison with an exact solution. Then this method cannot be used in the case of a sharp interface.

The problem is, as before, to compute the reflection coefficient and incidentally the level at which $\varphi(z)$ is zero.

Since q is in general complex, it is necessary to find the right zero of q, denoted q_0, among the zeros of $q(z)$ in the complex plane and to evaluate the phase integral on a contour, which can be difficult to determine if q_0 is not an isolated point. q_0 is called a reflection point. This technique is an extension of the ray theory for complex indices.

In order to use the geometrical approximation, let us assume that $n^2(z)$, $\cos \varphi$ and $q(z)$ are real and that n^2 varies slowly with z and is equal to 1 in the reference medium. At the level z_0 for which φ becomes zero and the z-direction of the ray changes, the reflected potential can be written

$$\phi_r(x, z) = \phi \exp\left[\iota kx \cos\varphi - \int_{z_0}^{0} q\, dz\right]$$

This leads to, $\mathfrak{R} = \exp[2\iota k \int_0^{z_0} q\, dz]$, which corresponds to total reflection with a phase change.

The exact theory shows that $\mathfrak{R}$ is given by $\mathfrak{R} = \iota \exp[2\iota k \int_0^{z_0} q\, dz]$. This additional rotation term ι is in general small compared with the integral. It can be neglected if the integral is evaluated in distance and time, that is for the study of a wavefront propagation.

The following example is quite classical:

$$n^2(z) = \begin{cases} 1 & \text{for } z < z_1 \\ 1 - \alpha(z - z_1) & \text{for } z > z_1 \end{cases}$$

α is real, positive. Then

$$q = \sqrt{\sin^2\varphi - \alpha(z - z_1)}$$

The reflection level is such that

$$z_0 = z_1 + \frac{\sin^2\varphi}{\alpha} \qquad \text{and} \qquad \int_{z_1}^{z_0} q\, dz = \frac{2}{3}\frac{\sin^3\varphi}{\alpha}$$

This leads to

$$\mathfrak{R} = \iota \exp\left(\frac{4}{3}\iota k \frac{\sin^3\varphi}{\alpha}\right)$$

if the factor ι is introduced, even though its existence has not been proved up to here. To do so, the solutions are assumed to be of the following form:

$$\phi(x, z) = \phi(z) \exp(\iota\gamma(z)) \exp(\iota kx \cos\varphi)$$

When substituting this expression into the propagation equation, derivatives γ' and γ'' appear. They are assumed to vary slowly compared with a wavelength. This leads to a correct W.K.B. approximation. If $z_1 = 0$ the equation for ϕ is

$$\frac{\partial^2\phi}{\partial z^2} + k^2[n^2(z) - \cos^2\varphi]\phi = 0$$

If ζ is defined as $\zeta = \sin^2\varphi - \alpha z$, the equation becomes

$$\frac{\partial^2\phi}{\partial z^2} + k^2\zeta\phi = 0$$

This is a classical equation [1, 2, 3]. Its solution is expressed in terms of Bessel functions of order 1/3, also called Airy functions. These functions appear in all cases where the phase is stationary; this is the case at the level for which φ becomes zero.

$$\phi(z) = w^{1/3}[AH_{1/3}^{(1)}(w) + BH_{1/3}^{(2)}(w)]$$

with $w = \frac{2}{3}(k/\alpha)\zeta^{3/2}$. For $z > z_0$, or $\zeta < 0$:

$$\phi(z) = [CH_{1/3}^{(1)}(w')]; \qquad w' = \frac{2}{3}\frac{k}{\alpha}(-\zeta)^{3/2}$$

The solution $H_{1/3}^{(2)}(w')$ must be discarded because of its behaviour when z tends to infinity.

The unknowns are B/A', C/A and $\mathfrak{R}$. They are found after adding two relations on $z = z_0$ (continuity of ϕ and $\partial\phi/\partial z$). It is found that

$$\mathfrak{R} = \frac{I_{-2/3} - I_{2/3} + \iota(I_{1/3} + I_{-1/3})}{I_{-2/3} - I_{2/3} - \iota(I_{1/3} + I_{-1/3})}$$

with $I_\nu(z)=$ modified Bessel function $= \exp(-\iota\nu\pi/2)J_\nu(ze^{\iota\pi/2})$. The argument of the I functions is w for $z = 0$. The phase of $\mathfrak{R}$ is $\gamma = (4k/3\alpha)\sin^3\varphi - \pi/2$, where the coefficient ι appears.

To summarize this section, it has been shown how to evaluate the complex reflection coefficients (along with the transmission coefficient) for sharp or soft boundaries between two semi-infinite media. It must be noticed that the method of phase integral is very general and the W.K.B. method is quite convenient for propagation in a slowly varying medium.

Reflection on an elastic thin plate

The elastic waveguide (plate, cylinder), immersed in a fluid (air, water) is studied with the same method used for the fluid waveguide, which is only a particular case of the elastic waveguide. The propagation is described using discrete modes and the expressions for the cut-off frequencies, phase and group velocities are given. Because it is possible to excite transverse waves, a mode is a combination of longitudinal and transverse waves.

If this thin elastic waveguide is the boundary of a fluid waveguide, the interesting case is when this boundary is reflecting. Generally speaking, for audio frequencies, the thicknesses of the boundaries of artificial waveguides are quite small (a few millimetres). Actually, these boundaries are made of plastic or metal and these thicknesses are large enough that the guide cannot be bent and is self-supporting.

Even at 10 kHz, the wavelength in the plastic or metal is greater than 25 cm and then only simple flexural waves must be taken into account (these frequencies are well below the cut-off frequency of the first transverse mode). This case is examined in chapter 8; the main result is that, for an infinite plate, the coincidence of the phase velocities in the fluid and in the plate leads to an almost total transparency. Then, to avoid this strong coupling effect between fluid and plate, one must try to move the coincidence frequency out of the frequency band of interest or, if necessary, to limit the propagation of the flexural wave by adding stiffeners or supports at particular locations. In the case of plates of finite dimensions (at least in one direction) perturbing flexural modes must be taken into account, especially on the parts of boundaries isolated by stiffeners or supports.

Finally, far from the resonance or coincidence frequencies, the transmission through the plate is given by the mass law and, even for thin plates, the reflection coefficient is close to 1. Let us go into more details.

The plate is characterized by a surface density M_s. Let p_i denote the incident pressure corresponding to a wave which impinges on the plate with an angle θ. Let p_r and p_t respectively denote the reflected pressure on the plate and the transmitted pressure through the plate. For symmetry reasons, the transmission angle is θ. The impedance of a plane wave in the fluid is ρc. The continuity conditions for the pressure and the normal velocity u lead to

$$p_i - p_r = p_t - \rho c \frac{u}{\cos\theta}; \qquad p_i + p_r - p_t \simeq \iota\omega M_s u$$

In the case of low rigidity, that is of a locally reacting plate:

$$\frac{p_r}{p_i} = -\frac{-\iota\omega M_S/\rho c}{(1 - \iota\omega M_S/\rho c)}$$

In the elastic plate, if ω is not too large, the velocity of flexural waves is approximated by[1]

$$c_f = \sqrt{1.8 c_L' f} \text{ with } c_L' = \frac{c_L}{1-\sigma^2}, \text{ and } c_L = \sqrt{\frac{E}{M_s}}$$

M_s is the density, σ the Poisson's ratio ($\sigma \simeq 0.3$), E the Young's modulus and h the thickness of the plate. c_L' is equal to several thousands of metres per second (5000 m s^{-1}, for example). It is easy to show that the mass impedance ($-\iota\omega M_S$) is replaced by the impedance:

$$-\iota\omega M_S\left[1 - \left(c_f \sin\frac{\theta}{c}\right)^4\right]$$

and then

$$\mathfrak{R} = \frac{(-\iota\omega M_S/\rho c)[1 - ((c_f/c)\sin\theta)^4 \cos\theta]}{1 - (-\iota\omega M_S/\rho c)[1 - (c_f/c)\sin\theta)^4 \cos\theta]}$$

[1] This is the reason why stiffeners and supports must be irregularly spaced in the direction of propagation.

The transmission is total for $c_f = c/\sin\theta$, this occurs at the critical frequency f_c:

$$f_c = \frac{c^2}{1.8hc_L' \sin^2\theta}$$

If $c_f < c$, the simplified formula of mass reactance is a correct approximation.

When the mass law applies, the losses by transmission are quite small, for example 10^{-3} of the incident energy. They must be compared with the classical losses in the propagation phenomenon.

7.2. The Problem of the Waveguide

7.2.1. General remarks

It has been seen in the previous section that, for particular conditions, natural or artificial boundaries can be considered as perfectly reflecting. This must be so in order that a wave can propagate a long distance, taking advantage of successive reflections. But it is also necessary that the waves reflected from either boundaries add in a constructive way. This means that all the components must have a common propagation term. In such conditions, a propagation mode appears. The simplest waveguide consists of a fluid domain bounded by two reflecting parallel planes. More precisely the planes are characterized by $|\mathcal{R}| = 1$, the phase change can be different on both boundaries.

7.2.2. The condition for mode propagation

Let φ still denote the angle $(\pi/2 - \theta)$. Let $\mathcal{R}_0$ and $\mathcal{R}_1$ be the complex reflection coefficients on boundaries $(z = 0)$ and $(z = H)$ respectively. An incident plane wave on $(z = 0)$ and the corresponding reflected plane wave on $(z = H)$ can be written

$$\phi_i(x, z) = \phi_0 e^{\iota k(x \cos\varphi + z \sin\varphi)} \quad \text{and}$$
$$\phi_{r,1}(x, z) = \phi_0 \mathcal{R}_1 e^{\iota k(x \cos\varphi - z \sin\varphi)} e^{2\iota kH \sin\varphi}$$

This corresponds to

$$\mathcal{R}_1 \simeq \frac{\phi_{r,1}}{\phi_i} \qquad \text{on } z = H$$

Similarly, when $\phi_{r,1}$ impinges on $(z = 0)$, the reflected wave must be identical to ϕ_i, and then

$$\mathcal{R}_0 \mathcal{R}_1 e^{2\iota kH \sin\varphi} = 1$$

This is the fundamental equation of the theory of modes in this waveguide. It is also written [2]:

$$\log \mathcal{R}_0 + \log \mathcal{R}_1 + 2\iota kH \sin\varphi = 2\pi n\iota \qquad \text{where } n \text{ is an integer}$$

The existence of modes is determined from the existence of solutions φ of this equation. In particular, real values of φ_n correspond to propagation modes. For $\mathcal{R}_0 = \mathcal{R}_1 = 1$, then $H \sin(\varphi)/\lambda = n/2$. For $\mathcal{R}_0 = -1$ and $\mathcal{R}_1 = 1$, then $H \sin(\varphi)/\lambda = (2n+1)/4$.

When $\mathcal{R}$ is expressed in the more general form with a phase integral, the possible waves (propagative, evanescent, inhomogeneous, ...) can be deduced from the study of the poles (real and complex), the branch integrals and the integrals on the imaginary axis. The next step is then to separate the waves which provide the main contribution. In the far-field the main contributions come from the propagative modes. This is why the simple problems (in air, for instance) can be solved by a straightforward method.

7.2.3. Solution of some simple problems

It is assumed in this section that $\mathcal{R} = \pm 1$, with possibly the condition that the angle of incidence θ is larger than θ_L (limit angle).

Infinite plane waveguide

The fluid medium is characterized by ρ and c and bounded by two perfectly reflecting, parallel planes. First, we solve the equation of propagation and then, by adding the boundary conditions, we find the eigenvalues which lead to the determination of the modes. The system of equations is

$$\begin{cases} \Delta\phi(y,z) + k^2\phi(y,z) = 0 \\ \dfrac{\partial\phi(y,z)}{\partial\vec{n}} = 0 \qquad \text{on } y = 0 \text{ and } y = H \end{cases} \tag{7.4}$$

The Neumann condition on $(y = 0)$ could be replaced by a Dirichlet condition $(\phi = 0)$ in the case of a layer of water. A more general form of boundary condition would be (mixed conditions)

$$\frac{\partial\phi}{\partial\vec{n}} + g\phi = 0$$

Particular solutions of (7.4) can be found by using the method of separation of variables:

$$\phi(y,z) = Y(y)Z(z)$$

This leads to

$$\frac{Y''}{Y}(y) + \frac{Z''}{Z}(z) + k^2 = 0; \qquad Y'(y) = 0 \text{ on } y = 0 \text{ and } y = H$$

and to

$$\frac{Y''}{Y}(y) = -\beta^2, \qquad \frac{Z''}{Z}(z) = -k^2 + \beta^2 = -K^2$$

β^2 can be complex. It is a separation constant still to be determined. Notice that writing $(+\beta^2)$ obviously leads to the same final result.

Y is then obtained as

$$Y(y) = A\ \cos\beta y + B\ \sin\beta y; \qquad Y'(y) = \beta(-A\ \sin\beta y + B\ \cos\beta y)$$

The boundary conditions lead to $B = 0$ and $\sin\beta H = 0$. The equation for the eigenvalues is $\beta_n = n\pi/H$ where n is an integer.

$$Z(z) = A_n e^{\iota k_n z} + B_n e^{-\iota k_n z}$$

$$\phi(y, z) = \sum_{n=0}^{\infty} (A_n e^{\iota k_n z} + B_n e^{-\iota k_n z}) \cos\frac{n\pi y}{H}$$

with $k_n^2 = (\omega/c)^2 - \beta_n^2$.

If k_n is imaginary, the mode is evanescent. The mode $(n = 0)$ is a plane wave with wavefronts perpendicular to the planes xy. If there is a reflection for some value of z, there are two plane waves travelling in opposite directions.

It must be noticed that the mode filtering which appears because of the definition of the cut-off frequencies implies that a reflection on any obstacle will provide, far from it, a plane wave (mode 0) in the opposite direction. Then the diffraction of the incident wave (mode 0) on an obstacle will be expressed by a modal decomposition in which all the modes except one are attenuated.

When solving a problem with real sources, A_n and B_n are still to be evaluated. They can be obtained, for example, by equating the normal velocities on the surface of the source and in the general expression of the sound field. The phase velocity $c_{\varphi,n}$ of mode n is such that

$$k_n = \frac{\omega}{c_{\varphi,n}} = \sqrt{k^2 - \frac{n^2\pi^2}{H^2}}$$

and then

$$c_{\varphi,n} = \frac{c}{\sqrt{1 - n^2\lambda^2/4H^2}}$$

It becomes infinite for the cut-off frequency $f_{c,n} = nc/2H$. For $c = 345\ \mathrm{m\,s^{-1}}$, and $H = 0.4$, $f_{c,1} = 431$ Hz and $f_{c,2} = 862$ Hz. The phase velocity is always greater than c and tends to c when f tends to infinity. For a fixed ω, there is at least one possible velocity (c for mode 0), and several in general. Figure 7.2 shows these properties. Figure 7.3 presents the transverse distribution of the pressure.

In general, the coefficients β_n, $c_{\varphi,n}$ and k_n have a small imaginary part, simply because of classical losses. This means that the propagative modes are 'slightly' attenuated and the evanescent modes are 'slightly' propagative.

Not all the modes are necessarily excited at given angular frequency ω_0. This depends on the spatial distribution of the source. Formally, the spatial repartition of the source can be decomposed on the $\cos(n\pi y/H)$ functions. For a complicated real source, it is easy to understand that information about the form of the source is

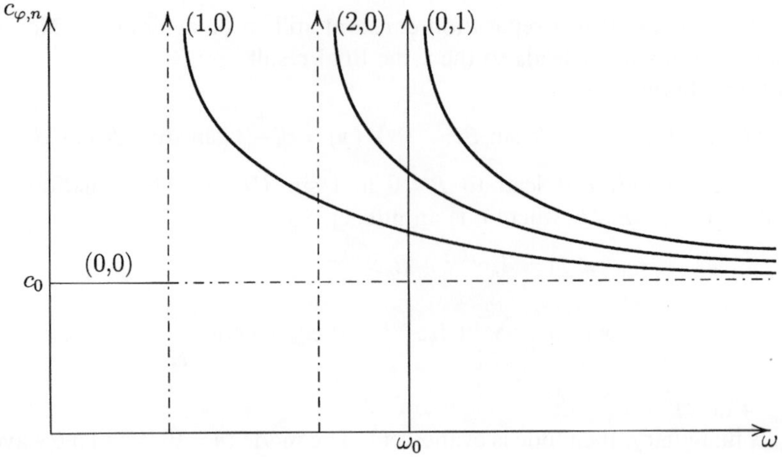

Fig. 7.2. Phase velocity.

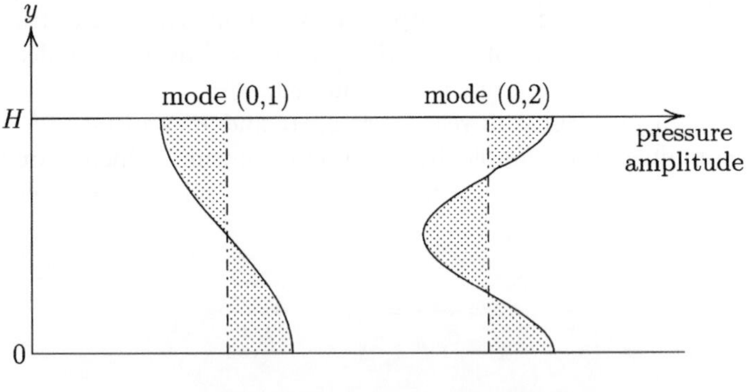

Fig. 7.3. Acoustic pressure in the waveguide.

partly lost because of the spatial filtering of the guiding phenomenon. For example, when a plane wave is observed in a wave guide, it is not possible to know whether the source is a point source or a plane section which vibrates like a piston.

The geometrical aspects of the guiding phenomenon can be understood with the help of Fig. 7.4. The wavefronts are represented by two plane waves symmetrical with z. At any intersection point, it is possible to draw planes parallel to Oz on which $\Re = \pm 1$. The following relations hold:

$$H \cos \theta_n = \frac{n\lambda}{2}; \qquad \lambda_{\varphi,n} = \frac{\lambda}{\sin \theta_n}$$

$$c_{\varphi,n} = \frac{c}{\sin \theta_n}; \qquad \theta_n = \arccos \frac{n\pi c}{\omega H} = \arccos \frac{n\lambda}{2H}$$

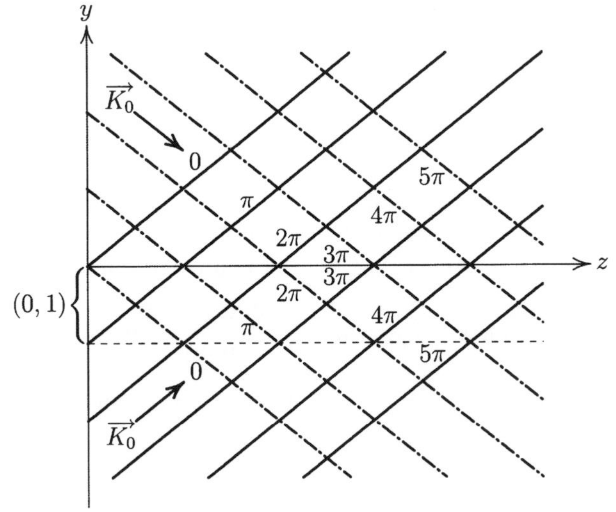

Fig. 7.4. Geometrical interpretation.

At a cut-off frequency, $c_{\varphi,n}$ is infinite and then all the points located on a parallel to Oz have the same phase. This is generally not observed, because of classical losses for instance. The discontinuities of θ_n are attenuated.

In the case of propagation of a pulse or a narrow-band signal ($\pm\Delta\omega$), the group velocity c_g corresponds to the velocity of energy circulation. By definition:

$$c_{g,n} = \frac{d\omega}{dk_n} \qquad \text{or} \qquad \frac{1}{c_{g,n}} = \frac{d}{df}\left(\frac{f}{c_{\varphi,n}}\right)$$

If $c_{\varphi,n}$ is replaced by its expression

$$c_{g,n} = c \sin\theta_n \qquad \text{and then} \qquad c_{g,n}\, c_{\varphi,n} = c^2$$

The notion of group velocity is interesting in the case of narrow-band signals or when the phase is stationary.

- Mode 0 is of great importance. From a historical point of view, it was the first studied. It is the only useful propagation mode in acoustical instruments or in artificial guides at low frequencies. For this mode, it is easy to deal with problems of perturbation of the propagation with z without necessarily considering the whole modal theory [5].
- It is interesting to evaluate the trajectory of the particle of fluid in the guide. Apart from a coefficient ($\iota\omega$) the velocity of the particle is given by

$$\frac{\partial\phi_n(y,z)}{\partial z} = \iota k_n (A_n e^{\iota k_n z} + B_n e^{-\iota k_n z}) \cos\frac{n\pi y}{H}$$

$$\frac{\partial\phi_n(y,z)}{\partial y} = -\frac{n\pi}{H}(A_n e^{\iota k_n z} + B_n e^{-\iota k_n z}) \sin\frac{n\pi y}{H}$$

If β_n is real, these two components are in quadrature: the path is elliptic. If β_n is imaginary, the two components have the same phase: the path is a part of a straight line.

Figure 7.5 shows this path for the mode $n = 1$, propagative or evanescent.

Rectangular duct with reflecting walls

Let $a \times b$ denote the section of the duct. The solution of the 3D Helmholtz equation is expressed as $\phi(x, y, z) = X(x)Y(y)Z(z)$. The boundary conditions are

$$\frac{\partial\phi(x, y, z)}{\partial x} = 0 \text{ on } x = 0, x = a; \qquad \frac{\partial\phi(x, y, z)}{\partial y} = 0 \text{ on } y = 0, y = b$$

With the same method as above, we find two separation constants α^2 and β^2:

$$\frac{X''}{X}(x) = -\alpha^2; \qquad \frac{Y''}{Y}(y) = -\beta^2; \qquad \alpha = \frac{m\pi}{a}; \qquad \beta = \frac{n\pi}{b} \ (m, n \text{ integers} \geqslant 0)$$

$$X_m(x)Y_n(y) = \cos\frac{m\pi x}{a} \cos\frac{n\pi y}{b} = \psi_{mn}(x, y)$$

$$\phi(x, y, z) = \sum_{m,n} \psi_{mn}(x, y)(A_{mn}e^{\iota k_{mn}z} + B_{mn}e^{-\iota k_{mn}z})$$

with

$$k_{mn}^2 = k^2 - \left(\frac{m^2\pi^2}{a^2} + \frac{n^2\pi^2}{b^2}\right) = \frac{\omega^2}{c_{\varphi,mn}^2}$$

$c_{\varphi,mn}$ is the phase velocity along Oz for the mode (m, n). It is given by

$$c_{\varphi,mn} = \frac{c}{\sqrt{1 - (\pi^2/k^2)(m^2/a^2 + n^2/b^2)}} = \frac{c}{\sqrt{1 - (\lambda^2/4)(m^2/a^2 + n^2/b^2)}}$$
$$= \frac{c}{\sqrt{1 - f_{c,mn}^2/f^2}}$$

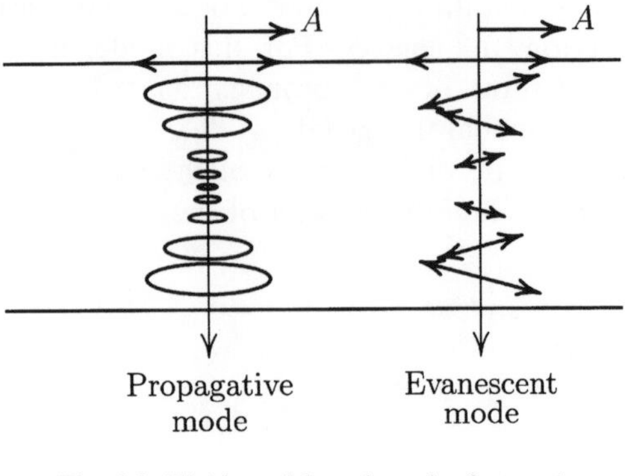

Fig. 7.5. Fluid particle trajectories for $n = 1$.

if $f_{c,mn}$ is the cut-off frequency for the mode (m, n),

$$f_{c,mn} = \frac{c}{2}\sqrt{\frac{m^2}{a^2} + \frac{n^2}{b^2}}$$

For example, with $c = 345\ \mathrm{m\,s^{-1}}$, $a = \sqrt{2}$, $b = 1$, $m = 3$, $n = 2$:

for $f = 1000$ Hz, $f_{c,32} = 572$ Hz and $c_{\varphi,32} = 420\ \mathrm{m\,s^{-1}}$;
for $f = 1144$ Hz, $c_{\varphi,32} = 398\ \mathrm{m\,s^{-1}}$.

From a geometrical point of view, there would be four incident plane waves with angles θ_m and θ_n such that $\theta_m = \arcsin(m\pi c/\omega a)$, $\theta_n = \arcsin(n\pi c/\omega b)$.

Figure 7.6 shows the stationary regime in the cross section.

If two perfectly reflecting conditions are added at $z = 0$ and $z = d$, they are used to evaluate the coefficients A_{mn} and B_{mn}. We find

$$\phi_{mnq}(x, y, z) = \cos\frac{m\pi x}{a}\cos\frac{n\pi y}{b}\cos\frac{q\pi z}{d}, \qquad m, n, q, \text{ integers} \geqslant 0$$

$$k_{mnq}^2 = k^2 - \left(\frac{m^2\pi^2}{a^2} + \frac{n^2\pi^2}{b^2} + \frac{q^2\pi^2}{d^2}\right)$$

These relations provide a description of the modes in a rectangular parallelepipedic enclosure. If the sound excitation inside the enclosure is represented by a plane wave and begins at time t_0, then a mode is established at $t_0 + \Delta t$ when the wavefront coincides again after several reflections with the incident wavefront. The eigenfrequencies are given by

$$f_{mnq} = \frac{c}{2}\sqrt{\frac{m^2}{a^2} + \frac{n^2}{b^2} + \frac{n^2}{b^2}}$$

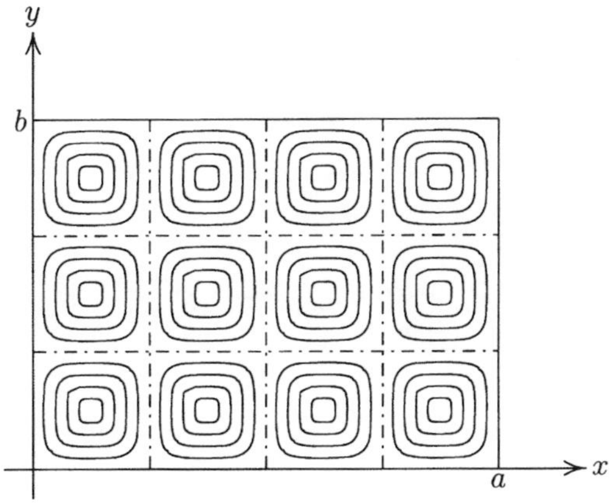

Fig. 7.6. Pressure distribution for the mode $m = 3$, $n = 2$.

The circular duct

Let us call Oz the axis of the cylinder and a the radius of the cross section. In cylindrical coordinates (r, θ, z), the system of equations for the field is

$$\begin{cases} \left(\dfrac{\partial^2}{\partial r^2}+\dfrac{1}{r}\dfrac{\partial}{\partial r}+\dfrac{\partial^2}{\partial \theta^2}+\dfrac{\partial^2}{\partial z^2}+k^2\right)\phi(r,\theta,z)=0 \\ \dfrac{\partial\phi(r,\theta,z)}{\partial r}=0 \text{ on } r=a \\ \phi \text{ bounded everywhere} \end{cases}$$

ϕ is written as $R(r)\Theta(\theta)Z(z)$. This leads to

$$\frac{1}{R(r)}(R''(r)+\frac{1}{r}R'(r))+\frac{1}{\Theta(\theta)}\frac{1}{r^2}\Theta''(\theta)+k^2=-\frac{1}{Z}\left(\frac{\partial^2 Z}{\partial z^2}\right)(z)$$

The right-hand side term is a function of z only. Both sides must be equal to a (separation) constant which is denoted $(-k_z^2)$. The equation for Z is a one-dimensional Helmholtz equation. Let us also remark that the solution must be 2π periodic with θ. This leads to:$\Theta''/\Theta=-m^2$ or $\Theta=(\sin m\theta)$ or $(\cos m\theta)$ with m an integer. The equation for R is then

$$\frac{R''}{R}(r)+\frac{1}{r}\frac{R'}{R}(r)+\left(\alpha^2-\frac{m^2}{r^2}\right)=0$$

where $\alpha^2=k^2-k_z^2$.

This is a Bessel equation. The general solution can be expressed as

$$R(r)=A_m J_m(\alpha r)+B_m Y_m(\alpha r)$$

The solution Y_m must be discarded because it tends to infinity when $r=0$ for all m. Let us notice that it must be kept if the propagation domain does not include the axis of the cylinder (co-axial duct). The boundary conditions for $r=a$ lead to the equation for the eigenvalues α_{mn}:

$$R'(r)=A_m\alpha J_m'(\alpha r)=0 \qquad \text{for } r=a$$

If α_{mn}' are the roots of J_m', then $\alpha_{mn}=\alpha_{mn}'/a$.

Finally, the general solution is written as

$$\phi(r,\theta,z)=\sum_m\sum_n(A_m e^{\iota k_{mn}z}+B_m e^{-\iota k_{mn}z})(C_1 e^{\iota n\theta}+C_2 e^{-\iota n\theta})J_m(\alpha_{mn}r)$$

with $k_{mn}^2=k^2-\alpha_{mn}^2$.

If a sound source emits a signal of increasing frequency, the first cut-off frequencies will be deduced successively from $\alpha_{10}'=1.84$, $\alpha_{20}'=3.05$, $\alpha_{01}'=3.83,\ldots$, $\alpha_{mn}'=\pi(n+m/2-3/4)$, as far as the source is able to excite the successive modes.

These cut-off frequencies are such that the phase velocity of $c_{\varphi,mn}$ is infinite, that is $k_{mn}^2 = \omega^2/c_{mn}^2$ is zero or

$$k^2 = \alpha_{mn}^2, \qquad f_{mn} = \alpha'_{mn}c/2\pi a, \qquad \lambda_{mn} = 2\pi a/\alpha'_{mn}$$

For $a = 5$ cm, $c = 345$ m s^{-1}, $f_{10} \simeq 2020$ Hz, $\lambda_{10} \simeq 17$ cm, the diameter is slightly larger than half the wavelength.

As for rectangular ducts, this approximation provides good estimates of the cut-off frequencies with very simple computations.

7.3. Radiation of Sources in Ducts with 'Sharp' Interfaces

7.3.1. General remarks

The general solutions obtained previously for several types of ducts in terms of modes provide a mathematical base and then they can be used to describe the radiation of a source inside a duct.

It has been mentioned previously (*bis repetita placent*) that far from the source the sound field can be expressed as the sum of some propagative modes (if they exist!; in a duct with free-boundary, there is no plane wave). The radiation of the source is filtered and replaced by a more or less accurate description depending on how many modes are taken into account. To determine the field close to the source, the attenuated modes must be taken into account (in practice, a 'sufficiently large' number of them).

Let us represent an omnidirectional point source located at M_0 by $\underset{M_0}{\delta}(M)$. From this elementary source, it is possible to represent any kind of more complicated sources. Let us recall that in the three-dimensional infinite space $\mathbb{R}^3$:

$$\underset{M_0}{\delta}(M) = \delta(x - x_0)\delta(y - y_0)\delta(z - z_0) \qquad \text{in rectangular coordinates}$$

$$\underset{M_0}{\delta}(M) = \frac{1}{2\pi r}\delta(r - r_0)\delta(\theta - \theta_0)\delta(z - z_0) \qquad \text{in cylindrical coordinates}$$

$$\underset{M_0}{\delta}(M) = \frac{1}{r^2 \sin\theta}\delta(r - r_0)\delta(\theta - \theta_0)\delta(\varphi - \varphi_0) \qquad \text{in spherical coordinates}$$

with classical notations.

The denominators are the Jacobians when changing the coordinate system. They actually appear because of the conservation of elementary surfaces and volumes. These expressions are quite suited to the separation method because the integrations on each δ are easy to evaluate.

7.3.2. Point source and Green's function in the duct

Let $a \times b$ denote the cross section of a rectangular duct. The source is located at point M_0. The sound field $\phi(x, y, z)$ is the solution of

$$\Delta\phi(x, y, z) + k^2\phi(x, y, z) = -\delta(x - x_0)\delta(y - y_0)\delta(z - z_0) \tag{7.5}$$

ϕ is called the Green's function of the problem. Let us write it as

$$\phi(x, y, z) = \sum_{m,n=0}^{\infty} Z(z)\psi_{mn}(x, y) \qquad \text{with } \psi_{mn}(x, y) = \cos\frac{m\pi x}{a}\cos\frac{n\pi y}{b}$$

Equation (7.5) becomes

$$\sum_{m,n=0}^{\infty} [Z''(z)\psi_{mn}(x, y) + k_{mn}^2 Z(z)\psi_{mn}(x, y)] = -\delta(x - x_0)\delta(y - y_0)\delta(z - z_0) \quad (7.6)$$

with

$$k_{mn}^2 = k^2 - \left(\frac{m^2\pi^2}{a^2} + \frac{n^2\pi^2}{b^2}\right)$$

Because the modes in (x, y) are orthogonal, the next step is to multiply both sides of (7.6) by $\psi_{mn}(x_0, y_0)$ and to integrate on the cross section. This leads to

$$Z''(z) + k_{mn}^2 Z(z) = -\frac{\psi_{mn}(x_0, y_0)}{\Lambda_{mn}}\delta(z - z_0)$$

where $\Lambda_{mn} = [ab(1 + \delta_0^n)(1 + \delta_0^m)/4]$ is the normalization coefficient with the Kronecker symbol, not to be confused with the Dirac function.

This expression of Λ_{mn} is easy to find from the equality

$$\int_0^a \cos^2(m\pi x/a)\,dx = \int_0^a \frac{1 + \cos(2m\pi x/a)}{2}\,dx = \frac{a}{2} + \frac{a}{2}\frac{[\sin(2m\pi x/a)]_0^a}{2m\pi/a}$$

The solution varies as $\exp(\pm\iota k_{mn}(z - z_0))$. When the wave is attenuated (for k_{mn} imaginary) the amplitude must be decreasing for both $z > z_0$ and $z < z_0$; the solution is then written as

$$Z(z) = A\ \exp(\iota k_{mn} \mid z - z_0 \mid)$$

After integration on the z-axis, ϕ is found as

$$\phi_G(x, y, z) = \frac{\iota}{2}\sum_{m,n=0}^{\infty} \frac{\psi_{mn}(x_0, y_0)\psi_{mn}(x, y)}{k_{mn}\Lambda_{mn}} \exp(\iota k_{mn} \mid z - z_0 \mid) \quad (7.7)$$

For the infinite waveguide, ϕ_G is symmetrical with M and M_0.

Remark: The presence of the term $1/2$ in the integration is not *a priori* obvious. It must be noted that, with this method, the singularity of the source in the three-dimensional space is replaced by a singularity on the z-axis only. Then the integration is carried out as if the source is an infinitely thin layer, located between two cross sections and which is infinite for $z > z_0$ and $z < z_0$. The velocity $v_z = -\nabla_z(\phi_G)$ is then discontinuous as can be seen by deriving ϕ_G.

When adding two perfectly reflecting boundaries at $z = 0$ and $z = d$, the Green's function can be found by the same method. The variables x, y, z play the same role.

The Green's function for a parallelepipedic enclosure with perfectly reflecting walls is given by

$$\phi_{GP}(x, y, z) = \frac{\iota}{2} \sum_{m,n=0}^{\infty} \frac{\psi_{mnq}(x_0, y_0)\psi_{mnq}(x, y)}{\Lambda_{mn}\sqrt{k_{mnq}^2 - (\omega/c)^2}}$$

$$\text{with } \psi_{mnq}(x, y, z) = \cos\frac{m\pi x}{a}\cos\frac{n\pi y}{b}\cos\frac{q\pi z}{d}$$

Similar results can be obtained for locally reacting boundaries, since the orthogonality of the modes is still true (see chapter 2).

7.3.3. Radiation impedance of the source in the duct

The radiation impedance of a real source gives information on the effect of the near field on the source. From a practical point of view, a source is a system with a vibrating surface, driven by a generator which has a finite internal resistance. The effect of the near field is to change this internal impedance in an electrical (or mechanical) circuit equivalent to the complete radiating system. This is the only way to model the radiation of a source in a closed domain, at least partially. Indeed, let us consider the example of a closed volume with reflecting walls. If the variation of the field had no effect on the motion of the source, that is if this were a forced motion with constant displacement, whatever the variation, the pressure and the temperature would increase. A non-linear regime would then appear and the results presented here would not apply. This assumption of forced motion with constant velocity must be used cautiously, except, in a steady regime, if all the losses by radiation or viscosity dissipation compensate the power delivered by the source at each time and if the system remains quasi-linear. In other words, for a steady regime such a balance is likely to happen, while for a transient regime the power given by the source varies while the regime is establishing. When a source begins to radiate in a closed volume, its behaviour is similar to that in infinite space until the emitted wave impinges on a boundary which then radiates a part of the incident energy towards the source.

To evaluate the radiation impedance of a point source, the source is described as a small pulsating sphere of radius r_0 which tends to zero, as done in free space. To determine the total field on the vibrating surface of the source, all the evanescent modes (an infinite number) and (some) propagative modes must be taken into account.

The comparison of the results for an infinite space and a closed volume is quite interesting. The very detailed computation is of no interest here. It consists in comparing the expansion of the singular term $(1/4\pi r)$ with the singular term of ϕ_G in the duct. It is found that the real part of the impedance for the first mode (plane wave) is $R_{duct} = \rho c(8\pi^2 r_0^4)/S$, where S is the area of the cross section.

This relation is still valid for r_0 small compared with the smallest transverse dimension of the duct. It is equal to a constant while the radiation resistance of a

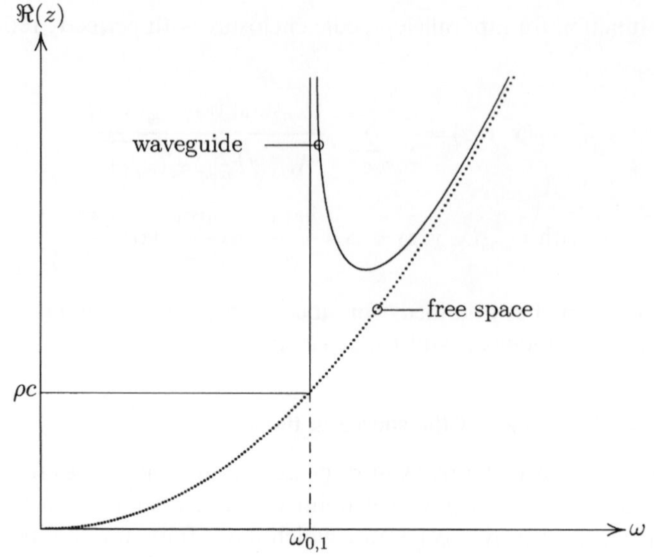

Fig. 7.7. Radiation resistance of a spherical source.

point source in free space is $R_L = \rho c k^2 r_0^2/(1 + k^2 r_0^2)$ when ω tends to zero. This means that R_L tends to zero with k.

It can also be shown that the imaginary part of the impedance is only slightly modified compared with the expression obtained in free space.

For higher modes, the impedance Z_{mn} is given by

$$Z_{mn} = \frac{-\iota\omega\rho\phi_{mn}}{-\nabla\phi_{mn}} = \rho c_{\varphi,\, mn}, \qquad \text{with } c_{\varphi,\, mn} = \omega/k_{mn}$$

The value of the impedance gives information on the efficiency of the radiation when ω varies. The variations of $c_{\varphi,\, mn}$ are known. Figure 7.7 presents the resistance of radiation of a small spherical source in a one-dimensional duct for the first modes and in free space. At very low frequencies, the source is still efficient in the duct.

7.3.4. Extended sources – Pistons

When the Green's function ϕ_G is known, it is possible, at least theoretically, to evaluate the sound field radiated by an extended source. Extended sources are mainly encountered in artificial guides but also sometimes in natural guides.

Piston at one end of the duct

Let us consider a plane rigid piston, with surface S_1, part of the cross section S at $z = 0$. $w(\omega)$ denotes its normal velocity. On S_1, $w(\omega)$ must be equal to the normal velocity in the fluid. The remaining part $(S - S_1)$ is assumed to be perfectly reflecting (see Fig. 7.8). One has

$$\phi(x, y, z > 0) = \sum_{m,n=0}^{\infty} A_{mn}\psi_{mn}(x, y)e^{\iota k_{mn}z} \equiv \phi_{mn}(x, y, z)$$

which leads to

$$V_z = -\nabla_z\phi(x, y, 0) = \sum_{m,n=0}^{\infty} (-\iota k_{mn})\phi_{mn}$$

$$V_z = w(\omega) \text{ on } S_1,$$
$$= 0 \text{ on } (S - S_1)$$

Multiplying V_z by ψ_{mn} and integrating on S leads to

$$A_{mn} = \iint_{S_1} \frac{w(\omega)\psi_{mn}(x, y)}{\iota k_{mn}\Lambda_{mn}} dS \equiv \frac{N}{D}$$

$\Lambda_{mn} = || \psi_{mn} ||^2$ is the norm of ψ_{mn}. With the notations defined in Fig. 7.8, for a rectangular duct:

$$N \equiv \int_e^{e+a_1} \int_d^{d+b_1} \cos\frac{m\pi x}{a} \cos\frac{n\pi y}{b}\, dx\, dy$$

$$= \frac{4ab}{mn\pi^2} \cos\left(\frac{m\pi}{a}\frac{2e+a_1}{2}\right) \sin\frac{n\pi b_1}{2b} \cos\left(\frac{n\pi}{b}\frac{2d+b_1}{2}\right) \sin\frac{m\pi a_1}{2a}$$

$$\Lambda_{mn} = \frac{ab}{4}(1+\delta_0^m)(1+\delta_0^m)$$

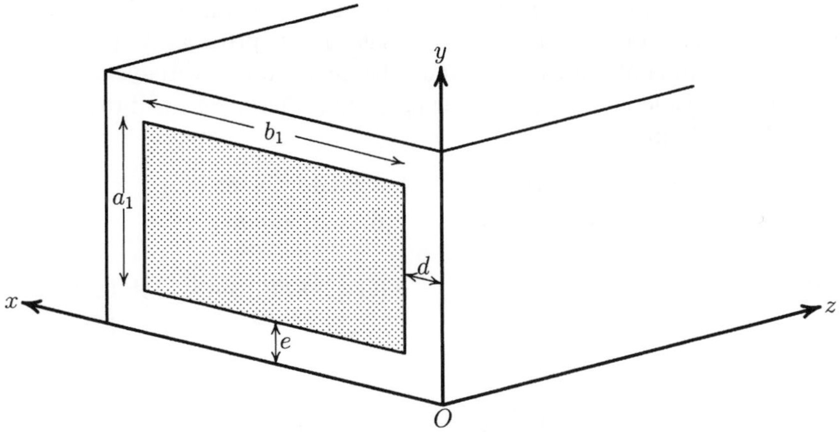

Fig. 7.8. Piston at the end of a waveguide.

Let Z_R denote the radiation impedance. For any ω, its real part is equal to ρc if $a_1 = a$ and $b_1 = b$. It is always less than ρc if the piston is smaller than the cross section, roughly in the proportion $a_1 b_1/ab$ for the plane mode. As mentioned before, to describe the diffraction on the (small) piston and evaluate Z_R, all the attenuated modes and some propagative modes must be taken into account. The global effect of the field on the source is

$$\frac{F}{w(\omega)} = \frac{1}{S_1} \int_{S_1} \frac{p(x, y, 0)}{w(\omega)} dS$$

The results for Z_R are not the same for this geometry and for a piston located in an infinite plane, especially when ω tends to zero. Indeed, because of the reflections on the walls of the duct, the model for the piston in an infinite plane is a plane with an infinite number of pistons with the same phase. This describes the diffraction pattern due to the images of the piston. This result must obviously be related to the result obtained for the point source in a duct.

Piston on a wall of the duct

Only the particular case of a rectangular duct with perfectly reflecting walls is studied here. The case of a piston clamped in the wall is especially interesting when the fluid is in motion. The presence of the source does not generate perturbation on the flow. As said above, only propagative modes must be taken into account in the far field. The position and the shape of the source have no significant effects on the mode (0, 0). This remark still holds, at least roughly, for sensors. However when the sensors are located in a wall they are inside the boundary layer and then a more detailed study of the problem is necessary.

The expression of ϕ_G has been given before (equation (7.7)). Let us consider a piston of length ℓ ($0 < z_0 \leqslant \ell$) and width b, the same width as the duct (see Fig. 7.9).

In the plane ($z = z_0$), the velocity has a discontinuity $D = -2\nabla\phi$. Since each point of the piston radiates in both directions $z > z_0$ and $z < z_0$, there is necessarily an interference between two elementary sources corresponding to different z. This leads to suppression of the 'absolute value' signs in the propagative terms, but the velocity must be discontinuous. Let P and V be some reduced variables of ϕ and $\nabla\phi$ for a mode (m, n) and let us consider the propagative term with z. If the indices are omitted, P and V are expressed as

$$P = Ae^{\iota kz} + Be^{-\iota kz}; \qquad V = \iota k(Ae^{\iota kz} - Be^{-\iota kz})$$

$$P(0) = A + B; \qquad V(0) = \iota k(A - B)$$

$$P(\ell) = P(0)\cos k\ell + \frac{V(0)}{k}\sin k\ell; \qquad V(\ell) = -kP(0)\sin k\ell + V(0)\cos k\ell$$

This can be written as

$$\begin{pmatrix} P \\ V \end{pmatrix}_{z=\ell} = T \begin{pmatrix} P \\ V \end{pmatrix}_{z=0}$$

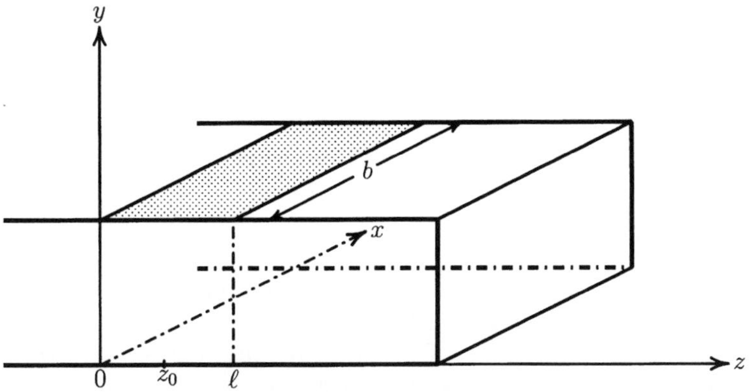

Fig. 7.9. Piston along the side of a waveguide.

if T is the matrix

$$T = \begin{pmatrix} \cos k\ell & (\sin k\ell)/k \\ -k \sin k\ell & \cos k\ell \end{pmatrix}$$

This relation is easily extended to the abscissas $(z+\ell)$ and z. Since there is a discontinuity at $(z = z_0)$, the relation must also be written between $(z_0 - \varepsilon)$ and $(z_0 + \varepsilon)$, with $\varepsilon \to 0$. If $D = -2\nabla_z \phi(z = z_0)$, we get

$$\begin{pmatrix} P \\ V \end{pmatrix}_{z=\ell} = T^{-1} \begin{pmatrix} P \\ V + D \end{pmatrix}_{z = z_0 + \varepsilon} + T^{-1} \begin{pmatrix} P \\ V \end{pmatrix}_{z=\ell}$$

The last step is to integrate over ℓ. It is necessary to take the coefficients of P and V into account to describe the evolution of the pressure and the velocity along the duct as well as the impedance $Z = \iota\omega\rho P/V$.

These expressions are quite convenient for numerical computations. For electrical lines, similar results have long been used.

7.3.5. Interference pattern in a duct

Let us add another piston, similar to the first one and located in the cross section $z = z_0$ (see Section 7.3.4), in order that they are placed symmetrically with $x = a/2$.

On the surfaces S_1 and S_2 of the pistons, their velocities $w_1(\omega)$ and $w_2(\omega)$ must be equal to the normal velocity $\nabla\phi_z(z = z_0)$ in the fluid.

With the method used in the previous sections, we find

$$A_{mn} = \frac{w_1 - (-1)^{m+n} w_2}{\iota k_{mn}(1 + \delta_0^m)(1 + \delta_0^n)} \frac{a_1 b_1}{mn\pi^2}$$

When $w_1 = w_2$, if the pistons have the same phase, the real part of the impedance is $\Re(Z(0,0)) \simeq 2a_1 b_1/ab$ but its imaginary part is not zero. If their phases are opposite, $\Re(Z(0,0))$ is zero, and there is no radiation below the cut-off frequency of the first mode, whatever the dimensions a_1 and b_1. The diffraction around the pistons is only described by attenuated modes.

It is possible to observe experimentally the decreasing resistance of radiation and the increasing reactance by examining the complex currents in the electrical circuit of the transducers when changing the phase of one of the pistons.

7.3.6. Impulse response in a duct

The impulse response plays a similar part in the time domain to that of the Green's function in space. With these two functions, it is possible, at least from a theoretical point of view, to solve any kind of problem. Because the phase velocity of the guided waves depends on frequency, the propagation of a transient signal depends on the modes excited. If $P(z, \omega)$ denotes the transfer function for the pressure in a duct, and $P_0(\omega)$ is the Fourier transform of the excitation signal $p_0(t)$, the time dependent pressure $p(z, t)$ is written

$$p(z, t) = \frac{1}{2\pi} \int_{-\infty}^{+\infty} P_0(\omega) P(z, \omega) e^{-\iota\omega t} \, d\omega$$

For a mode n (n can stand for one or two indices):

$$P_n(z, t) = \psi_n(x, y) \frac{1}{2\pi} \int_{-\infty}^{+\infty} P_0(\omega) e^{-\iota z\sqrt{\omega^2 - (ck_n)^2} - \iota\omega t} \, d\omega \tag{7.8}$$

If $p_0(t) = \delta(t)$, $P_0(\omega) = 1$ and

$$P_n(z, t) = \psi_n(x, y) \left[\delta\left(t - \frac{z}{c}\right) - Y\left(t - \frac{z}{c}\right) \right] k_n z \frac{J_1(k_n\sqrt{c^2t^2 - z^2})}{\sqrt{c^2t^2 - z^2}} \tag{7.9}$$

where Y is the Heaviside function.

If it were possible to describe the propagation of a Dirac function with one mode only, the signal observed far from the source would be similar to the curves presented in Fig. 7.10. This kind of theoretical response can be observed by using two loudspeakers (to simulate the pistons of Section 7.3.5) and sending them an impulse. It is then possible, for a sufficiently low frequency, to excite the mode (0, 1). Functions δ and Y are replaced by smoother functions. In the general case, several modes must be taken into account. The final result is that, at the observation point, there arrive successively several impulses and the higher the mode, the shorter the duration.

Remarks

(1) The experiment discussed in Sections 7.3.5 and 7.3.6 is easier to carry out if the width b of the duct is small. This corresponds to a duct with only one transverse dimension a ($b \ll a$).

(2) Adding the contributions of the modes can be cumbersome. It is easier to add the contributions of the images of the source through the perfectly reflecting walls.

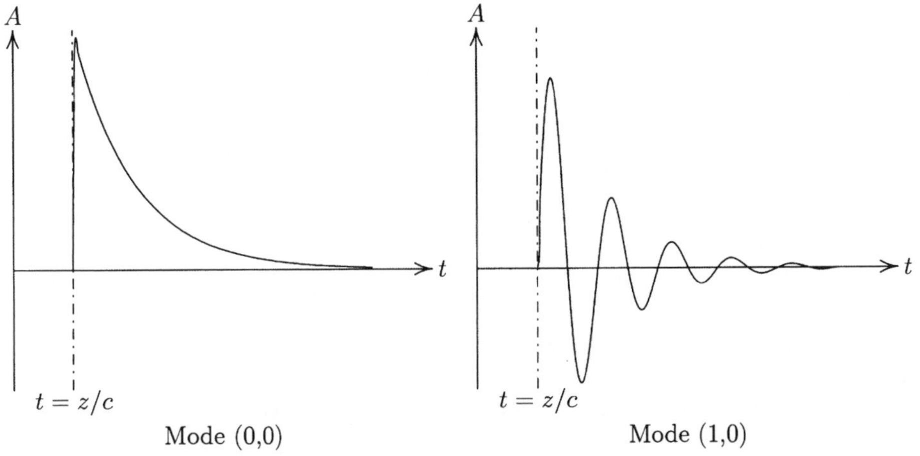

Fig. 7.10. Impulse responses for the modes (0,0) and (1,0).

If b is quite small, the source and its image are close to line sources. Their radiation is of the form $H_0^{(1)}(kz \cos \theta)$. The previous integration (7.8) of P_n is then easier [2]:

$$P = f(x, t) \frac{\cos(k\sqrt{c^2t^2 - z^2})}{\sqrt{c^2t^2 - z^2}}$$

The curve is then quite similar to the curve previously obtained since $J(\alpha)/\alpha$ is similar to $(\cos \alpha)/\alpha$.

(3) One way to explain the presence of the Dirac function in the exact solution (7.9) is the following: for high order modes, the group velocity is close to c. Adding all these contributions leads to a Dirac function. The curve corresponding to the attenuated oscillating part (with the Y term) depends on the curve $c_\varphi(\omega)$. For ducts with 'sharp' interfaces, c_g is a monotone function which increases from zero to c. Then as the observation time $(t + \Delta t)$ increases, c_φ decreases and ω tends to zero. Indeed at time $(t + \Delta t)$, the angular frequencies ω observed correspond to the group velocity $c_g = z/(t + \Delta t)$. This result no longer holds for an interface between two fluids (shallow water case).

7.4. Shallow Water Guide

7.4.1. Properties of the shallow water guide

Underwater measurements show that the velocity of the acoustic waves depends on the depth. More precisely, it depends on the salinity, the temperature and other local parameters. Let us call $c_j(z)$ this velocity in a medium j.

A very simple case is that of a shallow layer of water with depth H (H must be compared with $\lambda(\omega)$) where c_j and the density ρ_j are assumed to be constant. The bottom of the layer is supposed to be the interface with another fluid: this fluid can be, for example, a medium with water-saturated sand and with a thickness large compared with the wavelength. Generally speaking, the sound speed in this fluid is greater than the sound speed in the layer of water.

Some authors use the term 'shallow water' only for layers with thickness around $\lambda/4$, which is often the case at low frequency. Phenomena are then more difficult to analyse because of the different types of waves which can propagate around the first mode.

In the following, the parameters ρ_j and c_j are real. This assumption can be somewhat restrictive for providing a satisfactory interpretation of the phenomena observed. Actually, it leads to some academic aspects which are not useful for applications.

The case of the shallow water guide has been studied extensively because of the many applications in underwater propagation, seismology or E.M. propagation in the ionosphere. This last aspect was first developed during World War II.

7.4.2. The Pekeris model

The Pekeris model is the simple model presented in the previous section. The parameters are ρ_1 and c_1 for the layer of water (medium 1) and ρ_2 and c_2 for the fluid (medium 2), H is the thickness of medium (1) and z_0 is the z-coordinate of the point source in the layer. ρ_j and c_j are assumed to be constant. The domain $z<0$ is air. The boundary ($z=0$) with the layer (1) is then characterized by a homogeneous Dirichlet condition.

Let us note that the behaviour of the sound field is assumed to be, as before, ($e^{-\iota\omega t}$) as in [2], while the term ($e^{+\iota\omega t}$) is assumed in [1]. With this last choice, the fucntion $H_0^{(1)}$ is replaced by $H_0^{(2)}$ and the integration paths in the complex K plane are symmetrical with the axis $\Re(K)$, if K^2 is the separation constant (introduced below).

The unknowns of the problem are the scalar potentials ϕ_1 and ϕ_2. The coordinate system is (r, z). Particular solutions of the form $\phi_j(r, z) = R(r)\psi_j(z)$ can be found with the separation method. This method is not the one used by Pekeris but it leads to a much simpler analysis that is sufficient here.

From a theoretical point of view, it should be proved that this solution is quite general. Indeed, it would be more rigorous to write two different solutions in (1) and (2) and use continuity conditions at the boundary. It might be feared that the lateral wave which appears on the boundary $z=H$ would not appear in the analysis. This wave appears when studying the radiation of a source in the presence of a (sharp) interface. It is a consequence of the impedance change due to the presence of the interface; it is experimentally observed in seismology and underwater acoustics. Its amplitude decreases more rapidly than the surface wave amplitude which appears when the source is in the elastic medium. Its contribution becomes quite essential when there is no propagative wave ($H<\lambda/4$).

The system of equations is

$$\begin{cases} \left(\frac{1}{r}\frac{\partial}{\partial r}\left(r\frac{\partial}{\partial r}\right)+\frac{\partial^2}{\partial z^2}+k_j^2\right)\phi_j(r,z)=0 \\ \phi_1(r,z)=0 \qquad \text{on } z=0 \\ \rho_1\phi_1(r,z)=\rho_2\phi_2(r,z) \qquad \text{on } z=H \\ \frac{\partial\phi_1(r,z)}{\partial z}=\frac{\partial\phi_2(r,z)}{\partial z} \qquad \text{on } z=H \\ \text{Sommerfeld conditions} \end{cases}$$

with $k_j=\omega/c_j$. The condition of continuity of the normal velocities corresponds to non-viscous media. The separation method leads to

$$\psi_j''(z)+\left[\left(\frac{\omega}{c_j}\right)^2-K^2\right]\psi_j(z)=0$$

$$\frac{1}{r}\frac{d}{dr}(rR'(r))+K^2R(r)=0$$

where K^2 is a separation constant to be determined.

The general solutions are $H_0^{(1,2)}(Kr)$. For large Kr, these solutions take the simple asymptotic form $\sqrt{2/\pi KR}\exp(\pm\iota(KR-\pi/2))$: for fixed Kr, this implies that r is large enough ($r \gg \lambda$) and, thus, $r \gg H$.

7.4.3. Solutions $\psi_j(z)$

Let β_j^2 be defined by $\beta_j^2=(\omega/c_j)^2-K^2$.

If β_1 is real positive, $A_1 \sin\beta_1 z$ is the solution in (1).

If β_2^2 is negative, β_2 can be written $(\pm\iota\alpha)$. Only the solution $e^{-\alpha z}$ is kept since the other one, $e^{\alpha z}$, does not satisfy the conditions at infinity. It has previously been shown for the sharp interface that it corresponds to a total reflection. Then for all possible values for β_1, there exists a propagative mode in fluid (1).

If β_2^2 is positive, ψ_2 can be written $\psi_2=A_2\sin\beta_2 z$ for $z>H$. We keep this solution, which remains finite when z tends to $(-\infty)$.

7.4.4. Eigenvalues

First it is assumed that β_2^2 is negative. The eigenvalues are deduced from the boundary conditions written for the solutions $\psi_1(z)=A_1\sin\beta_1 z$ and $\psi_2(z)=A_2e^{-\iota\beta_2 z}$, ($\beta_2$ imaginary and negative). The continuity of the pressure leads to

$$\psi_2(z)=A_1\frac{\rho_1}{\rho_2}\sin(\beta_1 H)\,e^{-\iota\beta_2 H}$$

and the continuity of the normal velocity gives

$$A_1\beta_1 \cos(\beta_1 H) = -\iota\beta_2 A_2 = -\iota\beta_2 A_1 \frac{\rho_1}{\rho_2} \sin(\beta_1 H)$$

This leads to the eigenvalue equation:

$$\tan(\beta_1 H) = \iota \frac{\beta_1\rho_2}{\beta_2\rho_1}$$

The right-hand member is positive.

If β_1^2 is negative, there are no eigenvalues K^2, and then there is no guided wave in the layer (1).

Let us now assume that β_2^2 is positive. In this case ψ_2 can be written as $\psi_2(z) = A_2 \sin(\beta_2 z + B_2)$. The boundary conditions then lead to

$$\rho_1 A_1 \sin(\beta_1 H) = A_2\rho_2 \sin(\beta_2 H + B_2)$$
$$A_1\beta_1 \cos(\beta_1 H) = A_2\beta_2 \cos(\beta_2 H + B_2)$$

and the eigenvalue equation is

$$\tan(\beta_1 H) = \frac{\beta_1\rho_2}{\beta_2\rho_1}[\tan(\beta_2 H + B_2)]$$

This equation always has solutions and there are continuous modes. The eigenvalues K^2 are located on a part of the real axis and on the imaginary axis, as shown in Fig. 7.11.

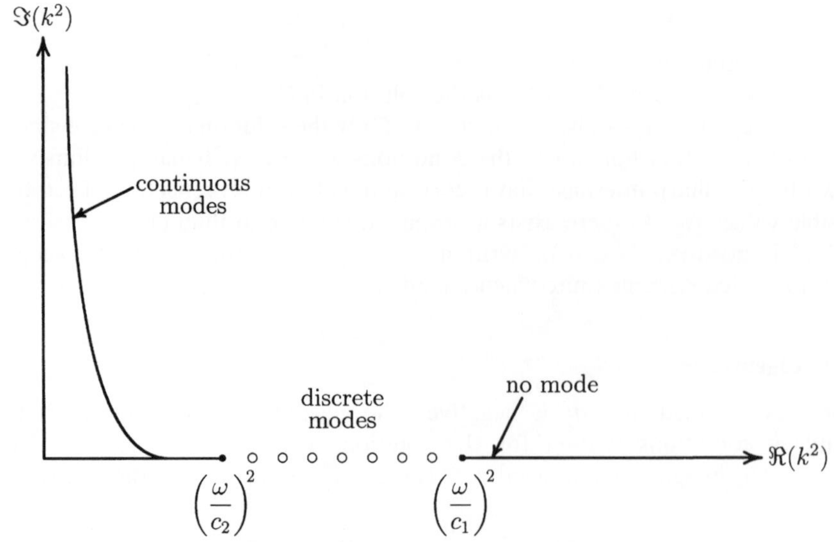

Fig. 7.11. Propagation in shallow water: localization of the eigenvalues.

7.4.5. Eigenmodes

If ρ_1 and ρ_2 are not constant functions, ψ_1 and ψ_2 associated with these eigenvalues K^2 are not orthonormal except if $\rho_1 = \rho_2$, which is generally not true. But it can be shown that functions $\sqrt{\rho_j}\psi_j$ provide a basis of functions useful to develop particular solutions [6]. These functions are normalized with

$$\int_0^{\infty} \rho_j\, N_{j,mn}^2\, \psi_{j,mn}^2\, dz = 1$$

7.4.6. Solutions *R*(*r*)

When K is known, the equation for R is quite classical:

$$\frac{1}{r}\frac{d}{dr}(rR'(r)) + K^2 R(r) = 0; \qquad K^2 = \left(\frac{\omega}{c_1}\right)^2 - \beta_1^2$$

To any real value of β_1^2 there corresponds a real or imaginary K and $R(r) = AH_0^{(1)}(Kr)$. The amplitude decreases as $1/\sqrt{r}$ when r increases.

7.4.7. Essential remark

In the complex K plane, there is no difficulty with the integration path. The method of Pekeris leads to more complicated paths with branches parallel to the negative imaginary axis $[0, -\iota\infty[$ for $K = \omega/c_1$ and ω/c_2. The integrals on these branches give the lateral wave. Obviously, the paths must be equivalent.

It remains to explain the presence of continuous modes on the contour shown in Fig. 7.11. They correspond to attenuated modes.

7.4.8. Radiation of a harmonic point source

The method consists in developing the singularity on the set of the orthonormal functions previously obtained. The point source M_0 is located at $z_0 < H$ (this is the usual case, but M_0 can be anywhere on the axis, as long as the presence of a fluid for $z < 0$ is taken into account).

$\phi_1(r, z)$ is the solution of a Helmholtz equation:

$$\frac{1}{r}\frac{\partial}{\partial r}\left(r\frac{\partial \phi_1(r,z)}{\partial r}\right) + \frac{\partial^2 \phi_1(r,z)}{\partial z^2} + \left(\frac{\omega}{c}\right)^2 \phi_1(r,z) = -2\frac{\delta(r)}{r}\delta(z - z_0) \qquad (7.10)$$

Let us write

$$\phi_1(r,z) = \sum_n R_n(r)\psi_n(z) + \int_\Gamma R(r)\psi(z)\, dK$$

where Γ is the path already discussed and presented in Fig. 7.11.

By replacing ϕ_1 with this expression in (7.10), multiplying by $\sqrt{\rho_j}\psi_n$ and integrating with z, it is found that

$$R_n(r) = \iota\pi\rho(M_0)\,\frac{\psi_n(z_0)}{\Lambda_{mn}}\,H_0^{(1)}(K_n r) \quad \text{(discrete modes)}$$

$$R(r) = 2\iota\,\frac{\rho(M_0)}{\rho_2}\,\frac{K}{\beta_2}\,A_1\psi(z_0)\,\frac{A_1^2}{A_2^2}\,H_0^{(1)}(Kr) \quad \text{(continuous modes)}$$

A_1/A_2 is known and A_1 is chosen to be equal to 1. Finally

$$\phi_1(r, z) = 2\iota\pi \sum_n \frac{\beta_{1,n}\ \sin(\beta_{1,n}z_0)\ \sin(\beta_{1,n}z)H_0^{(1)}(K_n r)}{\beta_{1,n}H\ \sin(\beta_{1,n}H)\ \cos(\beta_{1,n}H) - \rho_1^2\ \sin^2(\beta_{1,n}H)\ \tan(\beta_{1,n}H)/\rho_2^2}$$
$$- 2\iota\,\frac{\rho_1}{\rho_2}\int_\Gamma \frac{\beta_2\ \sin(\beta_1 z_0)\ \sin(\beta_1 z)}{\beta_1^2\ \cos^2(\beta_1 H) + \beta_1^2\rho_1^2\ \sin^2(\beta_1 H)/\rho_2^2}$$

ϕ_1 is again symmetrical with z and z_0.

7.4.9. Propagation of a pulse

To describe the propagation of a signal resulting from an explosion, it is necessary to know the group velocity. The eigenvalue equation for K^2 is implicitly an equation for $c_{\varphi,n}$ (through $\beta_{1,n}^2$) which can be solved by successive approximations. The first iteration can be obtained by introducing some values of $\beta_1 H$ corresponding to a reflecting boundary for $z = H$. Figure 7.12 presents some

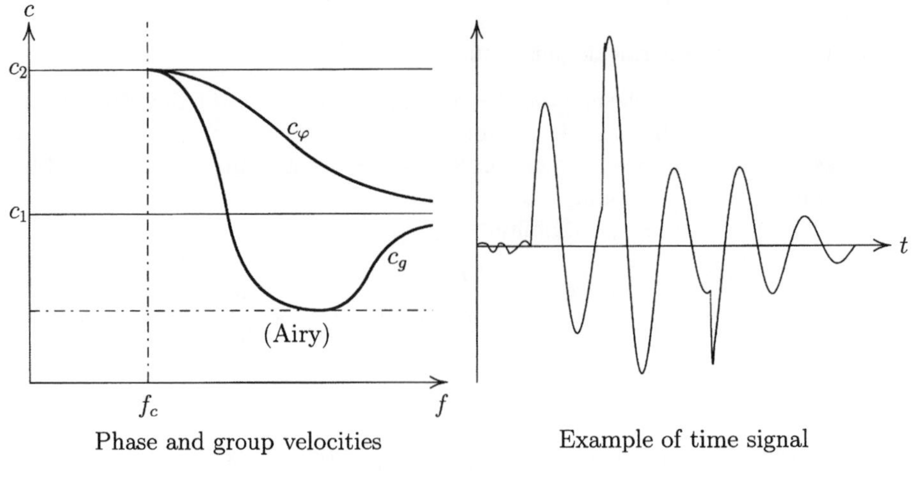

Fig. 7.12. Qualitative aspects of phase velocity c_φ, group velocity c_g and time signal.

classical results for the phase velocity $c_{\varphi,n}$ such that $K_n = \omega/c_{\varphi,n}$. $c_{\varphi,n}$ is equal to c_2 at the cut-off frequency of mode n. For example, for the first mode, this frequency is

$$f_c = \frac{c_2}{4H\sqrt{1 - c_1^2/c_2^2}}$$

which can also be written as a function of (H/λ_1). This simple result is obtained by considering the critical angle of total reflection for a plane wave (arcsin c_1/c_2). This may be done because in the far field the source can be approximated by a sum of plane waves.

The group velocity c_g is equal to

$$c_g = \frac{d\omega}{dK} = \frac{d}{dK}(Kc_\varphi) = c_\varphi + K\frac{dc_\varphi}{dK}$$

c_g has a minimum which is less than c_1. For c_g less than c_1, there are two solutions for ω. In the neighbourhood of the minimum, the phase is stationary. The contributions of the corresponding waves tend to add. These waves are called Airy waves and can clearly be observed in the signal $s(t)$ at any point.

The contributions which arrive first at the observation point have the highest velocity c_2. They correspond to the cut-off frequency $(f_c + \varepsilon)$ and to the lowest possible frequency for the mode: it is the ground wave. With a delay Δt $(c_2/z < \Delta t < c_1/z)$, low frequency waves arrive along with high frequency waves. The Airy wave arrives later.

These results are outlined in Fig. 7.12. They correspond to experimental results obtained in shallow water. An Airy wave is associated with each mode. These Airy waves come out of the signal enough to be used for multichannel transmission of signals (one each per mode which can be excited).

A classical application corresponds to

$$f(t) = \begin{cases} 0 & \text{if } t < 0 \\ \exp(-st) & \text{if not} \end{cases}$$

It describes a damped discharge of a transducer. The Laplace transform of $f(t)$ is well known and

$$f(t) = \frac{1}{2\pi}\int_0^\infty \frac{e^{-\iota\omega t}}{s - \iota\omega}\, d\omega$$

$$p(r, z, t) = \int_0^\infty \frac{S_n(\omega)}{s - \iota\omega}\, e^{-\iota\omega t + \iota K_n r - \pi/4}\, d\omega$$

$S_n(\omega)$ is the radiation of the source for mode n.

7.4.10. Extension to a stratified medium

In general, c_j is not constant in the whole layer of fluid. If c_j depends only on z, it is still possible to find a solution. The propagation medium ($z < 0$) is divided into sub-layers in which c_j can be assumed to be constant. This leads to

$$\psi_j''(z) + \left[\left(\frac{\omega}{c_j}\right)^2 - K^2\right]\psi_j(z) = 0 \tag{7.11}$$

Let us define as previously $\beta_j^2 = (\omega/c_j)^2 - K^2$. Then

$$\psi_j''(z) + (\beta_j^2 - V_j)\psi_j(z) = 0$$

with $V_j = (\omega/c_j)^2 - (\omega/c)^2$. This is a Schrödinger equation (with potentials V_j). The equation is solved in each layer. For the first and the last layers, the method is similar to the one used in the previous section.

Equation (7.11) written as above is interesting because solutions are known when V_j is a constant and also when V_j is of the form

$$V_j = az + b \qquad \text{or} \qquad V_j = az^2 + bz + c$$

with a, b, c equal to constants.

Because of this, it is possible to reduce the number of sub-layers to approximate correctly the sound speed profile as a function of z.

Nowadays, the power of computers is such that it could be simpler to approximate $c(z)$ by a piecewise constant or a piecewise linear function. This is an interesting problem, from a numerical point of view, which must take into account the errors introduced by each method, including artificial backscattering errors caused by the discontinuities of the approximations. Software based on such approximations has been developed and used for a long time.

7.5. Duct with Absorbing Walls

Generally speaking, the internal boundaries of a duct are not perfectly rigid. They have a finite impedance. In particular, the viscosity effects which cannot be neglected close to the wall can be taken into account by adding a correcting term in the normal admittance.

Also, sometimes, in order to attenuate the wave which propagates, some parts of the walls (or all the walls) are covered with absorbing materials or with resonators. Such a situation also appears in natural waveguides but the modelling is not so simple.

Let us assume that the acoustic properties of the wall are fully described by $Z(\omega)$, the normal impedance. It is often a correct approximation of more complicated cases, except perhaps for particular frequency bands for which there is a strong coupling between the vibrations in the fluid and inside the wall.

7.5.1. Duct with plane, absorbing walls

The fluid, in the duct, is characterized by ρ_0 and c_0. The sound field $\phi(y, z)$ is the solution of

$$\begin{cases} (\Delta + k_0^2)\phi(y, z) = 0 \\ Z_0 \dfrac{\partial \phi(y, z)}{\partial \vec{n}} + \iota\omega\rho_0\phi(y, z) = 0 \qquad \text{on } y = 0 \text{ and } y = b \\ \phi \text{ bounded} \end{cases}$$

where $k_0 = \omega / c_0$ and $\vec{n}$ is the normal vector exterior to the duct. By separating the variables, we find for $Y(y)$

$$Y''(y) = -\beta^2 Y(y); \qquad Z_0 Y'(0) = -\iota\omega\rho_0 Y(0); \qquad Z_b Y'(b) = +\iota\omega\rho_0 Y(b);$$

where $\beta^2 = k_0^2 - k_z^2$, k_z^2 is the separation constant.

The general form of the solution is $Y(y) = \cos(\beta y + \gamma)$. The boundary conditions are

for $y = 0$, $Z_0\beta \sin \gamma = \iota\omega\rho_0 \cos \gamma$ or $\tan \gamma = \iota\omega\rho_0 / (\beta Z_0)$;
for $y = b$, $-Z_b\beta \sin(\beta b + \gamma) = \iota\omega\rho_0 \cos(\beta b + \gamma)$.

This leads to

$$-Z_b\beta[\tan(\beta b) + \tan \gamma] = \iota\omega\rho_0[1 - \tan(\beta b) \tan \gamma]$$

In the simple case $Z_0 = Z_b$:

$$\tan(\beta b)\left(Z_0\beta + \frac{\omega^2\rho_0^2}{Z_0\beta} \right) = -2\iota\omega\rho_0$$

or

$$\tan(\beta b)\left(\frac{Z_0}{\rho_0 c_0} \frac{\beta}{k_0} + \frac{k_0}{\beta} \frac{\rho_0 c_0}{Z_0} \right) = -2\iota$$

The reduced impedance $Z_0/\rho_0 c_0$ of an absorbing layer on a perfectly reflecting surface can be evaluated from Kundt's tube measurements, for example. The solutions of the equation for the eigenvalues β_n are complex numbers in the general case. γ_n is deduced from β_n by using the boundary condition at $y = 0$.

If the surface $y = 0$ is rigid, then $\tan \gamma = 0$ and the eigenvalue equation becomes

$$\beta \tan(\beta b) = -\frac{\iota\omega\rho_0}{Z_b}$$

and the propagative terms which depend on z are expressed with $k_n = \sqrt{k_0^2 - \beta_n^2}$.

In the case of a low admittance and for only one absorbing wall $y=b$, the eigenvalue equation becomes

$$\beta \tan(\beta b) = -\iota k_0 \frac{\rho_0 c_0}{Z_b}$$

$$\text{with } \tan(\beta b) \simeq \beta b, \qquad \beta^2 = -\frac{\iota k_0}{b}\frac{\rho_0 c_0}{Z_b} \qquad \text{and} \qquad k_z^2 = k_0^2 + \frac{\iota k_0}{b}\frac{\rho_0 c_0}{Z_b}$$

For example, if $Z_0/\rho_0 c_0 = 0.2 + 10\iota$, for $f = 100$ Hz, $b = 0.1$ and $c_0 = 345$ m s^{-1}, then $k_z \simeq 1.82(1 + \iota 5 \times 10^{-3})$.

It is easy to extend these results to three dimensions and, for a duct with circular cross section with radius a, the eigenvalue equation is

$$\beta a J_m'(\beta a) = -\left[\iota k_0 a \Big/ \left(\frac{Z(a)}{\rho_0 c_0}\right)\right] J_m(\beta a) \tag{7.12}$$

If the wall $r=a$ is rigid ($Z_0(a)=\infty$) and $m=0$, (7.12) is the equation which is obtained for the circular waveguide (7.2.3). The eigenvalues are the roots of $J_m'(\beta a)$.

If only a part of the wall is covered with an absorbing coating (let us say, for $0 < \theta < \theta_0$) the study is much more complicated. The discontinuous function of θ would have first to be decomposed on the basis $(\cos(m\theta), \sin(m\theta))$ before solving for the eigenvalues. An interesting application of this problem is to obtain an absorbing effect with a minimum of surface coating. It can be shown that the absorbing effect rapidly increases when θ_0 increases above a small value θ_s and becomes 'not so interesting' when θ_0 tends to 2π. In other words, it is not necessary to cover the whole internal surface of the duct.

7.5.2. Orthogonality of the modes

The method and computations previously shown are quite efficient if the basis functions associated with the eigenvalues k_n are orthogonal. If they are not, an orthogonalization procedure can be used but the physical meaning of the orthogonal projection should be considered. For the locally reacting boundary conditions it can be shown that the functions are orthogonal. Each ψ_n is the solution of

$$\begin{cases} (\Delta + \beta_n^2)\psi_n(y) = 0 \\ \dfrac{\partial \psi_n(y)}{\partial \vec{n}} \dfrac{Z}{\rho_0 c_0} = \iota k_0 \psi_n(y) \qquad \text{on } y=0 \text{ and } y=b \end{cases}$$

Z may have different values on $y=0$ and $y=b$.

Let us consider two functions ψ_n and ψ_p, with $n \neq p$. By multiplying their corresponding Helmholtz equations by ψ_p and ψ_n respectively, integrating on $[0, b]$ and subtracting, the following expression is obtained:

$$\int_0^b (\psi_p \Delta \psi_n - \psi_n \Delta \psi_p)\, dy = (\beta_n^2 - \beta_p^2) \int_0^b \psi_n \psi_m \, dy$$

Applying the Green's formula, the right-hand side becomes

$$\left[\psi_n \frac{\partial \psi_p}{\partial \vec{n}} - \psi_p \frac{\partial \psi_n}{\partial \vec{n}}\right]_0^b$$

At $y = b$, this term is equal to

$$\rho_0 c_0 \left[\psi_n(b) \frac{\iota k_0}{Z_b} \psi_p(b) - \psi_p(b) \frac{\iota k_0}{Z_b} \psi_n(b)\right] = 0$$

A similar expression is obtained for $y = 0$. It is obviously equal to zero also. This proves the orthogonality.

Then in the general case where β_n is complex

$$\int_0^b \psi_n \psi_p^* \, dy = 0, \qquad n \neq p$$

Let us now evaluate the norm Λ_{pp} of ψ in the case of an absorbing surface ($y = b$) and a rigid surface ($y = 0$). In this case, $\gamma = 0$

$$\int_0^b \psi_n^2 \, dy \equiv \int_0^b \cos^2(\beta_n y) \, dy = \frac{1}{2}\left[y + \frac{\sin(2\beta_n y)}{2\beta_n}\right]_0^b$$

7.5.3. Losses on the walls of the duct

In an impedance tube (Kundt's tube), it is easy to observe that the minima of sound pressure become smoother further from the reflecting end. This is the effect of the walls. Close to the walls, the gradient of the tangential velocity leads to losses by viscosity while energy is transmitted from the incident wave to the wall by thermal conductivity.

Theory [1] shows that these effects can be characterized by equivalent layer boundaries of thickness d_{visc} and d_{th}. If it is assumed that the normal of the gradient of the tangential velocity and the amplitude of the thermal mode (which represents thermal losses) are exponentially decreasing functions, the thickness of the boundary layers corresponds to a decrease of (1/e), that is 1 neper or 8.6 dB: $(1 - 1/(2.71)^2) = 0.86$ part of the energy is dissipated in the layer. In classical situations,

$$d_{visc} = \frac{0.25}{\sqrt{f}} \qquad \text{and} \qquad d_{th} = \frac{0.21}{\sqrt{f}}$$

expressed in centimetres.

In a rigid tube with smooth boundaries, $d_{visc} = 0.1$ mm at 625 Hz. This is about the size of the irregularities of the surface. In a natural guide, the most interesting phenomena often correspond to the lowest frequencies. For a wave emitted by a thermonuclear explosion at 10^{-4} Hz and propagating to the antipodes, $d_{th} \simeq 20$ cm, which is again the size of the 'small' irregularities of the surfaces. The thickness of these layers is very small compared with the wavelength (about 10^{-2} to $10^{-3}\lambda$). It is then necessary to modify the impedance of the wall to take into account this kind of loss.

First, let us remark that the viscosity effect depends on the tangential particle velocity (not to be confused with flow velocity). In a one-dimensional duct, it has been shown previously that the angle between the propagative wave vector $\vec{k}_n$ and the normal is such that $\cos\theta_n \simeq n\lambda/2H$. For a high propagative frequency, θ_n tends to $\pi/2$. On the contrary, for mode (0,0) the particle velocity is purely tangential so that, if this mode is present and n is small, the viscosity losses can be taken into account with the mode (0,0) only (this is, of course, an approximation).

The losses by thermal conductivity depend only on the pressure close to the wall. The pressure is constant in the boundary layer because its thickness d is small compared with λ. Then the losses do not depend on the angle of incidence θ.

A formula [1] for the corrected admittance of the wall has been obtained:

$$\frac{1}{Z_c} = \beta_{corr} = \beta + (1 - \iota)\left[\frac{\omega d_{visc}}{2c_0}\sin^2\theta + (\gamma - 1)\frac{\omega d_{th}}{2c_0}\right]$$

θ is the angle of incidence (θ_n for mode n). $\theta = \pi/2$ for the plane wave. γ is the ratio of specific heats for the fluid ($\gamma = 1.4$ in air).

For example, for $Z = 10 + \iota 0.1$, $f = 625$ Hz and $c_0 = 345$ m s^{-1}, $\beta_{corr} \simeq 0.16 - 0.06\iota$ and $Z_c \simeq 6.25 + 2.34\iota$.

Then, because of the losses, an impedance $|Z| > 100$ for the wall is not realistic for the audio-frequency band in an artificial duct.

7.6. Ducts with Varying Cross Section

7.6.1. Linear duct with discontinuous cross section

Let us consider a duct made of two ducts of parallel axes, of cross sections S_I and S_{II}. The interface is the cross section on $(z = 0)$ with a reflecting condition on $(S_{II} - S_I)$. In each duct, it is assumed that the separation method applies. This leads to

$$\phi^I(u_1, v_1, z) = \sum_m \sum_n (A_{mn} e^{\iota k_{mn} z} + B_{mn} e^{-\iota k_{mn} z})\psi^I_{mn}(u_1, v_1)$$

$$\phi^{II}(u_2, v_2, z) = \sum_p \sum_q (C_{pq} e^{\iota k_{pq} z} + D_{pq} e^{-\iota k_{pq} z})\psi^{II}_{pq}(u_2, v_2)$$

with unambiguous notation. The mode decomposition is different for $z < 0$ and $z > 0$.

At $(z=0)$, a point of the cross section can be written (u_1, v_1) or (u_2, v_2) and relations are available to go from one coordinate system to the other. It is then possible to write all the functions in the same system. Let us write

$$u_1 = U_1(u_2, v_2); \qquad v_1 = V_1(u_2, v_2)$$
$$u_2 = U_2(u_1, v_1); \qquad v_2 = V_2(u_1, v_1)$$

or globally $(u_2, v_2) = T^{21}(u_1, v_1)$. T^{21} and T^{12} can be easily evaluated for simple geometries.

On the cross section $z=0$, the pressures and normal velocities can be written

$$p^I = \sum_{m,n} (-\iota\omega\rho_I)(A_{mn} + B_{mn})\psi^I_{mn}; \qquad v^I = \sum_{m,n} (-\iota k^I_{mn})(A_{mn} - B_{mn})\psi^I_{mn}$$

$$p^{II} = \sum_{p,q} (-\iota\omega\rho_{II})(C_{pq} + D_{pq})\psi^{II}_{pq}; \qquad v^{II} = \sum_{p,q} (-\iota k^{II}_{pq})(C_{pq} - D_{pq})\psi^{II}_{pq}$$

The condition of continuity for the pressure leads to

$$\rho_1 \sum_m \sum_n (A_{mn} + B_{mn})\psi^I_{mn}(u_1, v_1) = \rho_2 \sum_p \sum_q (C_{pq} + D_{pq})\psi^{II}_{pq}(T^{21}(u_1, v_1))$$

By multiplying on both sides by $\psi^{I^*}_{rs}(u_1, v_1)$ and integrating on S_I, we find

$$A_{rs} + B_{rs} = \frac{\rho_2}{\rho_1} \int_{S_I} \sum_p \sum_q \frac{\psi^I_{pq}(T^{21}(u_1, v_1))\psi^{I^*}_{rs}(u_1, v_1)(C_{pq} + D_{pq})}{\Lambda^I_{rs}} dS_I \tag{7.13}$$

$$\text{with } \Lambda^I_{rs} = \| \psi^I_{rs} \|^2 = \textstyle\int_{S_I} \psi^I_{rs}\psi^{I^*}_{rs}\, dS_I$$

The same kind of formula is obtained for the velocity. But in order to make the relations symmetrical with ψ^I and ψ^{II}, use is made of the orthogonality property of ψ^{II}_{pq}:

$$\sum_m \sum_n \iota k_{mn}(A_{mn} - B_{mn}) \int_{S_{II}} \frac{\psi^I_{mn}(T^{12}(u_2, v_2))\psi^{II^*}_{pq}(u_2, v_2)T^{21}}{\Lambda^{II}_{pq}} dS_{II}$$
$$= \iota k^{II}_{pq}(C_{pq} - D_{pq}) \tag{7.14}$$

Except for the simplest cases, the integrals must be evaluated numerically. The expressions for the pressure and the velocity are quite adapted to a transfer matrix formulation, as done in a previous section to take into account the presence of a source. Then numerical computations can be developed to take into account geometrical discontinuities and sources and also tube ends described by an impedance *Z(M)*.

The case of an end radiating in free space has not been studied here. It is a complex diffraction problem which can be solved for simple (cylindrical) ducts for mode (0,0). For higher modes, semi-infinite elements should be used to match with free space. Also, approximations are available by considering that the free end is clamped in a reflecting infinite plane.

A particular case can be correctly solved: a tube of small diameter connected to the reflecting end of a duct. First, it must be checked that, for all the modes to be taken into account, the phase varies slowly in the opening. It is then assumed that the normal velocity is constant, that is it is assumed that the opening can be modelled as if there were a small piston with no mass. This problem has been solved. To obtain better accuracy, its solution can be used as an initial guess in an iterative procedure.

7.6.2. Rectangular and circular cross sections

As an example, let us consider two ducts with rectangular cross sections $a_1 \times b_1$ and $a_2 \times b_2$ respectively. The walls are parallel. Then T^{21} is given by

$$x_2 = x_1 + e \qquad \text{and} \qquad y_2 = y_1 + d$$

and the integral on S_I in (7.13) can be written

$$I^I = 4 \int_0^{a_1} \int_0^{b_1} \cos \frac{p\pi(x_1 + e)}{a_2} \cos \frac{q\pi(y_1 + d)}{b_2} \cos \frac{m\pi x_1}{a_1} \cos \frac{n\pi y_1}{b_1}$$

$$\times \frac{1}{(1 + \delta_0^m)} \frac{1}{(1 + \delta_0^n)} \frac{a_1}{b_1} \, dx_1 \, dy_1$$

The same kind of result can be obtained for I^{II} in (7.14) and it is shown that

$$a_2 b_2 (1 + \delta_0^p)(1 + \delta_0^q) I^I = a_1 b_1 (1 + \delta_0^n)(1 + \delta_0^m) I^{II}$$

In the case of two ducts with circular cross sections of radius r_1 and r_2, Fig. 7.13 shows that

$$\frac{\sin \theta_2}{r_1} = -\frac{\sin \theta_1}{r_2}$$

With $\varphi = (\overrightarrow{O_1 O_2}, \overrightarrow{Ox})$, it is then possible to evaluate the integrals on S_I and S_{II}.

7.6.3. Connection between an absorbing duct and a lossless duct

This is an application of the results previously shown. It corresponds to the absorption of sound in a duct of fluid. The wall is partially coated.

To describe the general procedure, let us consider a two-dimensional problem (Oy,Oz) and two semi-infinite ducts connected at $z = 0$. The incident field which comes from the ($z < 0$) duct can be decomposed on a basis associated with the variable y ($0 < y < b$). To take advantage of the orthogonality of the eigenfunctions, it is assumed that the absorbing surface can be described by a local reaction condition. For simplicity, it is also assumed that there are no waves propagating backwards in the ($z > 0$) duct. This is quite realistic if the coating is efficient. If the coating is not used on an (almost) infinite length, a more general form of the sound

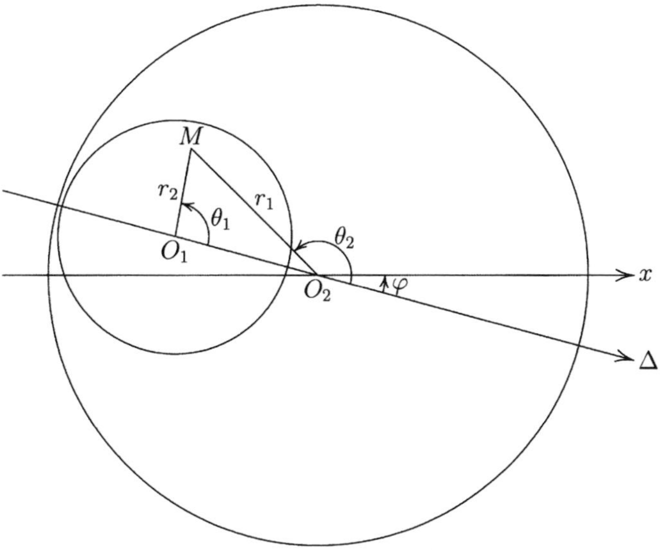

Fig. 7.13. Junction between two cylindrical waveguides.

field must be introduced. It would be easily seen that the mode decomposition is significantly modified in the case of a finite length coating part.

The continuity of the field at $z = 0$ leads to

$$\sum_{n=0}^{\infty} (A_n + B_n) \cos \frac{n\pi y}{b} = \sum_{p=0}^{\infty} C_p \cos(\beta_p y)$$

$$\sum_{n=0}^{\infty} (A_n - B_n) \cos \frac{n\pi y}{b} k^I_{n,z} = \sum_{p=0}^{\infty} C_p \cos(\beta_p y) k^{II}_{p,z}$$

$$k_0 = \frac{\omega}{c_0}; \qquad k^I_{n,z} = \sqrt{k_0^2 - \frac{n^2\pi^2}{b^2}}; \qquad k^{II}_{p,z} = \sqrt{k_0^2 - \beta_p^2}$$

In the ($z < 0$) duct, the A_n are known (they depend on the source). The β_n and C_p, for n and p finite, are then deduced. The convergence can be checked by computing again with $n' > n$ and $p' > p$. The computation takes advantage of the orthogonality of the cosine functions. The norm is

$$\Lambda_{pp} = \int_0^b \cos^2(\beta_p y)\, dy = \frac{b}{2}\left[1 + \frac{\sin(2\beta_p b)}{2\beta_p}\right]$$

$\sin(2\beta_p b)$ is then expressed as a function of $\tan(\beta_p b)$ which is known from the eigenvalue equation (see Section 7.5.1). If the wall ($z = 0$) is rigid and the wall

($z = b$) characterized by a normal impedance Z, it has been seen previously that

$$\tan(\beta b) = -\frac{\iota k_0}{Z_b \beta / \rho_0 c_0} = -\frac{\iota \omega \rho_0}{Z_b \beta}$$

7.6.4. Changes in direction

A natural or artificial waveguide with an axis moderately bent (radius of curvature $\gg \lambda$) in terms of a wavelength can be often modelled as a straight duct. If long distances are considered, the geometry must be taken into account. This is the case of propagation in the ionosphere (for electromagnetic waves) and in the troposphere (for infra-sound).

Also, artificial guides (such as ducts of fluid or smoke) can present sharp variations of section and direction.

For some simple cases (there are quite a number), satisfactory approximate solutions can be obtained by determining a connecting volume in which the variables can be separated. Figure 7.14 shows this procedure. When this connecting volume is not so simple, the equations can be solved by a finite element method in the connecting zone and matched with mode decomposition at the boundaries. The problem is quite a good example to validate an algorithm, at least in cases I and IV of Fig. 7.14. In cases II and III, the connecting zone is not correctly taken into account by the calculus.

The analysis becomes quite complicated if a fluid flow must be taken into account in the duct, which is a classical case in industrial applications. There are then two possibilities. In the first one, the flow is correctly introduced with a realistic velocity profile and only the plane wave case is considered. In the second, the flow is assumed to be laminar with a uniform velocity U in the cross section; it is then easy to find the corrections to extend the computation for a fluid at rest, for all the modes. Indeed, the flow velocities in ducts are generally kept low to avoid pressure drops. The turbulence, if it exists, can be neglected except perhaps around the discontinuities. The boundary layer is quite small compared with the transverse dimensions. For these reasons, the uniform flow velocity U is assumed to be much

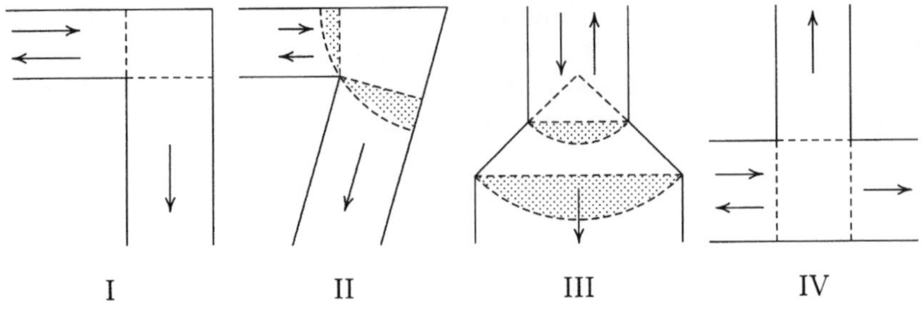

Fig. 7.14. Examples of waveguides junctions.

smaller than c, ($U \simeq 15\ \text{m s}^{-1}$, for instance); this corresponds to a Mach number $M = U/c$ no greater than 0.05. Even if this assumption does not clearly appear in the calculus, the results cannot be extended to higher M, since the assumption of uniform flow would not be correct for large M. Let us consider a duct with rectangular cross section $a \times b$. The flow, of velocity U, goes in the $(z > 0)$ direction. Use is made of a description of the phenomenon related to the observer, along with the changes of variables:

$$z = z' + Ut'; \qquad t' \mapsto t; \qquad (x, y) \mapsto (x, y)$$

Only the time derivatives are modified:

$$\frac{d\phi}{dt'} = \frac{\partial \phi}{\partial t}\frac{\partial t}{\partial t'} + \frac{\partial \phi}{\partial z}\frac{\partial z}{\partial t'} = \frac{\partial \phi}{\partial t} + U\frac{\partial \phi}{\partial z}$$

and the d'Alembert equation becomes

$$\frac{1}{c^2}\left(\frac{\partial}{\partial t} + U\frac{\partial}{\partial z}\right)^2 \phi = \Delta\phi \qquad \text{or} \qquad \frac{\partial^2 \phi}{\partial z^2}(1 - M^2) - \frac{2M}{c}\frac{\partial^2 \phi}{\partial t \partial z} - \frac{1}{c^2}\frac{\partial^2 \phi}{\partial t^2} = \Delta\phi$$

In a rectangular duct, separating the transverse variables leads to

$$\frac{\partial^2 \phi}{\partial z^2} + 2\iota k_0 \frac{M}{1 - M^2}\frac{\partial \phi}{\partial z} + \frac{k_0^2}{1 - M^2}\phi = \left(\frac{m^2\pi^2}{a^2} + \frac{n^2\pi^2}{b^2}\right)\phi; \qquad k_0 = \omega/c_0$$

$$k_{mn} = \frac{\omega}{c_{mn}} = \frac{k_0}{1 - M^2}\left\{-M \pm \iota\left[\frac{1 - M^2}{k_0^2}\left(\frac{m^2\pi^2}{a^2} + \frac{n^2\pi^2}{b^2}\right) - 1\right]\right\}^{1/2}$$

For $M = 0$, the formula is the one previously obtained. The expression of k_{mn} shows that there is always a propagative term, even for evanescent waves. The cut-off frequency $f_{c,mn}$ of mode (m, n) is changed. If the propagative term of k_{mn} is set to zero, then

$$\sqrt{1 - M^2}\sqrt{\frac{m^2\pi^2}{a^2} + \frac{n^2\pi^2}{b^2}} = k_0' = \frac{2\pi f'_{c,mn}}{c_0} = \frac{2\pi}{\lambda'_{mn}}$$

If M is small, $\sqrt{1 - M^2} \simeq 1 - M^2/2$, then $f'_{c,mn} = (1 - M^2/2) f_{c,mn}$. For $M = 0.06$ and $f = 1000$ Hz, $f'_{c,mn} = 1000(1 - 18 \times 10^{-4})$. The shift is equal to 1.8 Hz only.

The changes in the representation of the transient regime are more complex. This representation can be obtained in a straightforward way or from the Fourier transform of the solution $\phi(\omega)$ (obtained as in Section 7.3.6).

The time excitation for a mode (m, n) can be expressed as

$$A\psi_{mn}(x, y)\delta(t) = A \cos\frac{m\pi x}{a} \cos\frac{n\pi y}{b}\,\delta(t)$$

Because the solution in z does not depend on x and y, it must have the general form: $Z(z)A\psi_{mn}(x, y)$. It must be a solution of the d'Alembert equation (written above) and initial conditions, for example $\phi = 0$ and $\partial\phi/\partial t = 0$ at $t = 0$. The solution is obtained through a Laplace transform. It is similar to the expression obtained for a fluid at rest, with the following changes:

- the arrival times $(t - x/c)$ and $(t + x/c)$ are replaced by

$$\left(t - \frac{x}{c(1+M)}\right) \quad \text{and} \quad \left(t + \frac{x}{c(1+M)}\right);$$

- the eigenvalues are multiplied by $(1 - M^2)$.

The formula is not quite so simple to interpret. Let us notice, however, that these changes are quite small if M is quite small and cannot be observed experimentally.

Finally, let us point out that the rigorous study of a flow when the acoustic mode (0, 0) is present is a problem of fluid mechanics.

7.7. Conclusion

Many problems have been studied in this chapter. Many more problems have not even been mentioned. For example, only deterministic cases have been considered. What can be said for a non-periodic signal? for a random signal? How do we deal with the irregularities of a surface when they are not too small and not periodically spaced, when they are randomly spaced? What must be added to a model if the fluid is highly turbulent?

Clearly, the models presented in this chapter are more or less convenient. The methods proposed previously cannot be used to study random propagation of a random signal in a duct with such a geometry that variables cannot be separated. Global energy methods must be developed to obtain qualitative information on propagation. This was not the purpose of this chapter.

Bibliography

[1] MORSE, P. and INGARD, U., 1968. *Theoretical acoustics.* McGraw-Hill, New York.

[2] BUDDEN, K.G., 1961. *The wave guide mode theory of wave propagation.* Academic Press, New York.

[3] BREKHOVSKIKH, L.M., 1980. *Waves in layered media.* Academic Press, New York.

[4] ABRAMOWITZ, M. and STEGUN, I., 1972. *Handbook of mathematical functions.* Dover Publications, New York.

[5] LEVINE, H., 1981. On a problem of sound radiation in a duct. Contract NASA JIAA TR42, August 1981, Stanford University (USA).

[6] ROURE A., 1976. Propagation guidée; étude des discontinuités. Thèse de 3ème cycle, Université Aix-Marseille 1.

CHAPTER 8

Transmission and Radiation of Sound by Thin Plates

Paul J.T. Filippi

Introduction

In many practical situations, a sound wave is generated by the vibrations of an elastic structure: let us mention musical instruments like a guitar, a violin, a piano, a drum, etc.; industrial machines like an electric transformer, a tool machine, all sorts of motors, etc.; in a building, noise is transmitted through the structures (windows, floors, walls) because, for some reason, they are set into vibrations; there are also a lot of domestic noise and sound sources, like a door bell, a loudspeaker, a washing machine, a coffee grinder, etc., the vibrations of which generate noise. This is a very short list of noise generation by vibrating structures.

A reciprocal aspect of this phenomenon is the vibration of a structure generated by an incident acoustic wave. A classical example is commonly reported. A flautist or violinist is playing in a party while people are drinking champagne in crystal glasses; suddenly he plays a note exactly tuned on a resonance frequency of one of the glasses which starts vibrating with such a large amplitude that it breaks apart and the champagne is for the rug. Another very old and excessively common example is the mechanics of the ear: the air, excited by an acoustic wave, makes the eardrum move and, in its turn, the eardrum generates a motion of the ear bones which excite the auditory nerve. Finally, let us mention that the fundamental part of a microphone is a membrane which is set into vibrations by sound fields.

These two aspects of the phenomenon can occur at the same time: a sound field excites a structure which then creates a second sound field. This is why it is possible to hear external noises inside a room.

All these phenomena are, in fact, various aspects of a single one: the interaction between a compressible fluid and a vibrating structure. This part of mechanics is often called *vibro-acoustics*.

This chapter starts with a very simple one-dimensional example: a wave guide separated into two parts by a mass/spring system which can move in the axial direction. Sound can be generated by this structure or transmitted through it. The interest of this academic case is twofold: the equations can be solved analytically; and the behaviour of this system provides the fundamental laws which govern vibro-acoustic phenomena. The second example is that of an infinite plate, embedded in a fluid; it is excited either by an incident plane wave or by a point force. Then the example of a plate of finite dimension extended by an infinite perfectly rigid plane (baffled plate) is examined. In a first step, the solution of the problem is expanded as series of either the eigenmodes or the resonance modes. When the fluid is a gas, a light fluid approximation is presented which accounts for the fact that the presence of the fluid has a very small influence on the vibration of the plate, and can, thus, be considered as a small perturbation. Finally, it is shown how the problem is reduced to a set of boundary integral equations which can be solved by various numerical methods similar to those described for acoustics problems: as an example, the approximation of the unknown functions by truncated series of orthogonal polynomials is presented and discussed.

8.1. A Simple One-dimensional Example

The system is an infinite waveguide with constant cross section of unit area. The abscissa of a cross section is denoted by x. The guide is separated into two parts by an infinitely thin but perfectly rigid panel located at $x = 0$. This wall is supported by a system of springs which exerts a force on it when it is moved back and forth in the x-direction (see Fig. 8.1).

It is assumed that the two half-spaces $x < 0$ and $x > 0$ are filled with a fluid characterized by a density ρ and a sound speed c. The panel has a mass m and the springs system has a rigidity r. It is assumed that the panel can move in the x direction only, which implies that it can generate only plane acoustic waves.

8.1.1. Governing equations

Let $\tilde{F}(t)$ be the total force acting on the wall in the x-direction. The wall displacement $\tilde{u}(t)$ obeys the following differential equation:

$$m \frac{d^2\tilde{u}(t)}{dt^2} + r\tilde{u}(t) = \tilde{F}(t) \tag{8.1}$$

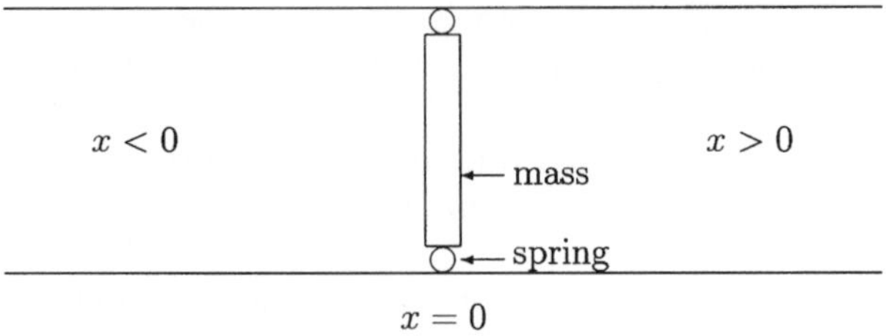

Fig. 8.1. Scheme of the one-dimensional vibro-acoustic system.

In general, the force $\tilde{F}(t)$ is the sum of two terms: an external force $\tilde{F}_e(t)$ (which can be zero); the difference $\tilde{P}(t)$ between the acoustic pressure exerted by the fluid contained in the half-guide $x < 0$ and that exerted by the fluid contained in the $x > 0$ half-guide.

Let $\tilde{S}^-(x, t)$ and $\tilde{S}^+(x, t)$ be acoustic sources located in $x < 0$ and $x > 0$ respectively. The acoustic pressures in the two half-guides $\tilde{p}^-(x, t)$ and $\tilde{p}^+(x, t)$ satisfy wave equations:

$$\begin{aligned} \frac{\partial^2 \tilde{p}^-(x,t)}{\partial x^2} - \frac{1}{c^2}\frac{\partial^2 \tilde{p}^-(x,t)}{\partial t^2} &= \tilde{S}^-(x,t) \qquad \text{in } x<0 \\ \frac{\partial^2 \tilde{p}^+(x,t)}{\partial x^2} - \frac{1}{c^2}\frac{\partial^2 \tilde{p}^+(x,t)}{\partial t^2} &= \tilde{S}^+(x,t) \qquad \text{in } x>0 \end{aligned} \tag{8.2}$$

We thus have

$$\tilde{P}(t) = \tilde{p}^-(0, t) - \tilde{p}^+(0, t) \tag{8.3}$$

A continuity relationship at $x = 0$ is required to express the fact that the fluid particles remain in contact with the elastically supported piston. This is achieved by letting the particle acceleration be equal to the piston acceleration, that is

$$\frac{d^2\tilde{u}(t)}{dt^2} = \tilde{\gamma}^-(0, t) = \tilde{\gamma}^+(0, t) \tag{8.4}$$

where $\tilde{\gamma}^-(x, t)$ (resp. $\tilde{\gamma}^+(x, t)$) is the fluid particle acceleration in the half-guide $x < 0$ (resp. $x > 0$). Then, it is necessary to express the fact that the energy conservation principle is satisfied. This is done by requiring that the acoustic waves radiated and reflected by the panel are purely outgoing waves in the two half-guides. Finally, initial conditions must be given. It will be assumed that, for $t < 0$, all source terms are zero and the system is at rest.

With these conditions, the solution of equations (8.1, 8.2) exists and is unique. An easy method for obtaining the solution is first to solve the time Fourier-transformed equations (this is done in the following subsection) and then to go back to the time dependent solution by an inverse transform (last subsection).

8.1.2. Fourier-transformed Equations and Response of the System to a Harmonic Excitation

The Fourier transform classically adopted in acoustics is defined by

$$f(\omega) = \int_{-\infty}^{+\infty} \tilde{f}(t) e^{\iota\omega t}\, dt$$

$$\tilde{f}(t) = \frac{1}{2\pi}\int_{-\infty}^{+\infty} f(\omega) e^{-\iota\omega t}\, d\omega$$

Let $F_e(\omega)$, $S^-(x,\omega)$ and $S^+(x,\omega)$ be the Fourier transforms of the source terms. The Fourier transform $(u(\omega), p^-(x,\omega), p^+(x,\omega))$ of the solution satisfies the following system of equations:

$$(-m\omega^2 + r)u(\omega) = F_e(\omega) + p^-(0,\omega) - p^+(0,\omega) \tag{8.5}$$

$$\begin{cases} \dfrac{d^2p^-(x,\omega)}{dx^2} + k^2p^-(x,\omega) = S^-(x,\omega), & x \in]-\infty, 0[\\ \dfrac{d^2p^+(x,\omega)}{dx^2} + k^2p^+(x,\omega) = S^+(x,\omega), & x \in]0, +\infty[\end{cases} \tag{8.6}$$

with $k = \omega/c$. Introducing the momentum equation into (8.4), these relationships become:

$$\omega^2 \rho u(\omega) = \frac{dp^-(0,\omega)}{dx} = \frac{dp^+(0,\omega)}{dx} \tag{8.7}$$

Remark. The solution $(u(\omega), p^-(x,\omega), p^+(x,\omega))$ is the response of the system to the harmonic excitation $(F_e(\omega)e^{-\iota\omega t}, S^-(x,\omega)e^{-\iota\omega t}, S^+(x,\omega)e^{-\iota\omega t})$.

Equation (8.5) points out a particular angular frequency

$$\omega_0 = \sqrt{r/m} \tag{8.8}$$

which is the *in vacuo* angular resonance frequency of the system composed of the mass and the spring system. It will be seen that it plays an important role in the vibro-acoustic behaviour of this simple system.

The analysis of the phenomenon is simplified by distinguishing two cases. First, it is assumed that there are no acoustic sources ($S^- \equiv S^+ \equiv 0$) and the energy is due to the force $F_e(\omega)$ applied to the panel: this aspect is referred to as *acoustic radiation by a vibrating structure*. Then it is assumed that there is no external force acting on the panel and that the acoustic source in the half-guide $x > 0$ is zero: this aspect of the phenomenon is referred to as *sound transmission through an elastic structure* and *acoustic excitation of an elastic structure.*

Acoustic radiation of an elastically supported piston in a waveguide

In this section, it is assumed that the only energy source is the external force $F_e(\omega)$ acting on the panel. The energy conservation principle implies that the acoustic waves must travel away from the piston. Thus, they have the form

$$p^-(x,\omega) = A^-e^{-\iota kx}, \qquad p^+(x,\omega) = A^+e^{\iota kx}$$

Equations (8.5) governing the mass displacement and the continuity conditions (8.7) are written

$$\begin{aligned} A^+ - A^- - m(\omega^2 - \omega_0^2)u &= F_e(\omega) \\ \iota kA^+ - \rho\omega^2 u &= 0 \\ -\iota kA^- - \rho\omega^2 u &= 0 \end{aligned} \tag{8.9}$$

The determinant of this system is

$$D = \left(\frac{\iota\omega}{c}\right)^2 \left[2\iota\omega\frac{\rho c}{m} + \omega^2 - \omega_0^2\right] m$$

It is different from zero for any real value of the angular frequency ω, that is for any physical angular frequency of the excitation. Thus, the solution of (8.9) exists and is unique, and is given by:

$$\begin{aligned} u &= -F_e(\omega)m^{-1}\left[2\iota\omega\frac{\rho c}{m} + \omega^2 - \omega_0^2\right]^{-1} \\ A^- &= -F_e(\omega)\,\frac{\iota\omega\rho c}{m}\left[2\iota\omega\frac{\rho c}{m} + \omega^2 - \omega_0^2\right]^{-1} \\ A^+ &= F_e(\omega)\,\frac{\iota\omega\rho c}{m}\left[2\iota\omega\frac{\rho c}{m} + \omega^2 - \omega_0^2\right]^{-1} \end{aligned} \tag{8.10}$$

It can first be noted that the pressures $p^-(x)$ and $p^+(x)$ have the same modulus, which was *a priori* obvious. A second remark is that the pressures $p^-(-x)$ and $p^+(x)$ have phases with opposite signs, which could also be found *a priori*: indeed, when the piston creates a positive pressure in the $x > 0$ half-guide, it automatically creates a negative pressure in the other half-guide.

It is useful to look at the mean value, over any integer number of periods, of the various powers involved. The instantaneous power provided by the excitation force to the system is

$$\Re[F_e(\omega)e^{-\iota\omega t}]\,\frac{d}{dt}\,\Re[ue^{-\iota\omega t}]$$

which is the product of the instantaneous external force by the instantaneous panel velocity. The mean value over one period is

$$\mathcal{P}^0 = \frac{1}{T}\int_0^T \Re[F_e(\omega)e^{-\iota\omega t}]\Re[-\iota\omega ue^{-\iota\omega t}]\,dt$$

The instantaneous power flow which crosses over the guide cross section with abscissa x in the $x > 0$ direction is

$$\Re[p^+(x)e^{-\iota\omega t}]\Re[v^+(x)e^{-\iota\omega t}]$$

where $v^+(x)$ is the complex amplitude of the particle velocity. The momentum equation

$$v^+(x) = \frac{1}{\iota\omega\rho}\,\frac{dp^+(x)}{dx}$$

leads to the expression of the mean power flow:

$$\mathcal{P}^{+} = \frac{1}{T}\int_0^T \Re[p^{+}(x)\mathrm{e}^{-\iota\omega t}]\Re\left[\frac{1}{\iota\omega\rho}\frac{dp^{+}(x)}{dx}\mathrm{e}^{-\iota\omega t}\right] dt$$

The mean power flow $\mathcal{P}^{-}$ across any cross section of the half-guide in the $x<0$ direction is defined in a similar way. The result is:

$$\mathcal{P}^{0} = \frac{|F_e(\omega)|^2\rho c}{m^2[4\rho^2c^2/m^2 + \omega^2(1-\omega_0^2/\omega^2)]}, \qquad \mathcal{P}^{-} = \mathcal{P}^{+} = \tfrac{1}{2}\mathcal{P}^{0} \tag{8.11}$$

Expressions (8.10) and (8.11) point out that the *in vacuo* resonance angular frequency ω_0 of the system panel/spring has a particular role in the behaviour of the coupled system. Indeed, the quantities $|u|$, $|A^{-}|$, $|A^{+}|$ and $\mathcal{P}^{0}$ have a maximum at $\omega = \omega_0$.

If the fluid is a gas, and the panel a solid, the parameter $\varepsilon = \rho c/m$ is small compared to unity. It is, thus, tempting to expand the expressions of the system response into a Taylor series of this parameter and to keep the first order term only: this is commonly called the light fluid approximation of order one. The result is

$$\begin{aligned} u &= -\frac{F_e(\omega)}{m(\omega^2-\omega_0^2)}\left[1 - \frac{2\iota\omega\varepsilon}{\omega^2-\omega_0^2} + \mathcal{O}(\varepsilon^2)\right] \\ A^{+} = -A^{-} &= \frac{F_e(\omega)}{(\omega^2-\omega_0^2)}\,\iota\omega\varepsilon[1+\mathcal{O}(\varepsilon^2)] \end{aligned} \tag{8.12}$$

It is clear from this result that the condition $\varepsilon \ll 1$ is not sufficient to ensure the convergence of the Taylor series expansion. The parameter which is involved is $2\iota\omega\varepsilon/(\omega^2-\omega_0^2)$. Thus the condition for the Taylor series to converge is

$$\varepsilon < \frac{\omega}{2}\left|1 - \frac{\omega_0^2}{\omega^2}\right|$$

This shows that the light fluid approximation is meaningless for $\omega = \omega_0$.

This approximation can equally be introduced in a more intuitive way. If the fluid density is small compared to the panel mass, the panel displacement is, as a first approximation, the same as in the absence of the fluid, that is

$$u \simeq u_0 = -\frac{F_e(\omega)}{m(\omega^2-\omega_0^2)}$$

Such a displacement induces pressure fields with amplitudes

$$A^{+} = -A^{-} \simeq A_1^{+} = -A_1^{-} = -\iota\omega\rho c u_0 = \frac{\iota\omega\varepsilon F_e(\omega)}{\omega^2-\omega_0^2}$$

Then, from the corresponding acoustic fields, a first order correcting term for the panel displacement is obtained

$$u_1 = -u_0 \frac{2\iota\omega\varepsilon F_e(\omega)}{\omega^2 - \omega_0^2}$$

The iterative calculation can be continued to get higher order correcting terms. It is to be noticed that the Taylor series expansion and the iterative process provide identical results. For this particular one-dimensional example, correction terms of any order are easily evaluated. But it seems that, for two- or three-dimensional structures, only the first order correcting term can be used; the higher order ones are generally difficult to evaluate numerically.

Acoustic transmission through an elastically supported piston in a waveguide

All sources are assumed to be identically zero. The energy is provided by an incident plane wave of unit amplitude in the half-guide $x < 0$; thus, the acoustic pressure is written

$$p^-(x, \omega) = \mathrm{e}^{\iota kx} + p_r^-(x, \omega)$$

where $p_r^-(x, \omega)$ is the wave reflected by the panel. This function and the transmitted acoustic pressure $p^+(x, \omega)$ satisfy a homogeneous Helmholtz equation. The energy conservation principle implies that these acoustic fields have the following forms:

$$p_r^-(x, \omega) = R(\omega)\mathrm{e}^{-\iota kx}, \qquad p^+(x, \omega) = T(\omega)\mathrm{e}^{\iota kx} \tag{8.13}$$

With such a choice, the reflected wave travels in the direction $x < 0$, while the transmitted one travels in the $x > 0$ direction. The panel displacement $u(\omega)$, the reflection coefficient $R(\omega)$ and the transmission coefficient $T(\omega)$ are the solutions of the following system of linear equations:

$$\begin{aligned} R(\omega) - T(\omega) + m(\omega^2 - \omega_0^2)u(\omega) &= -1 \\ \iota k R(\omega) + \omega^2 \rho u(\omega) &= \iota k \\ -\iota k T(\omega) + \omega^2 \rho u(\omega) &= 0 \end{aligned} \tag{8.14}$$

One gets

$$\begin{aligned} u(\omega) &= -\frac{2}{m}\left[2\iota\omega\frac{\rho c}{m} + \omega^2 - \omega_0^2\right]^{-1} \\ R(\omega) &= \left(\omega^2 - \omega_0^2\right)\left[2\iota\omega\frac{\rho c}{m} + \omega^2 - \omega_0^2\right]^{-1} \\ T(\omega) &= 2\iota\omega\frac{\rho c}{m}\left[2\iota\omega\frac{\rho c}{m} + \omega^2 - \omega_0^2\right]^{-1} \end{aligned} \tag{8.15}$$

Here again, the *in vacuo* resonance angular frequency of the mass/spring system plays an important role. Indeed for $\omega = \omega_0$, one has

$$u(\omega_0) = \frac{1}{\iota\omega_0 \rho c}, \qquad R(\omega_0) = 0, \qquad T(\omega_0) = 1$$

The system behaves as if there were no separating wall: the mass displacement is identical to the fluid particle displacement induced by the incident wave only.

Let us now introduce basic notions of architectural acoustics:

- **Definition 1:** The ratio $\tau(\omega)$ of the transmitted mean power flow to the incident mean power flow is called the *energy transmission rate* of the wall.
- **Definition 2:** The difference $\mathcal{I}(\omega)$ between the incident mean power level and the transmitted mean power level is called the *insertion loss index* of the wall.

These concepts are rigorously defined for a one-dimensional system. For walls in a three-dimensional space, their definitions cannot be as rigorous and require some approximations; nevertheless, they are very helpful, and, probably, necessary to qualify the insulation properties of walls and floors. From formulae (8.15), the following result is established:

$$\tau(\omega) = \frac{4\omega^2(\rho c/m)^2}{4\omega^2(\rho c/m)^2 + (\omega^2 - \omega_0^2)^2}$$
$$\mathcal{I}(\omega) = 10 \log (1/\tau(\omega)) \tag{8.16}$$

It is interesting to introduce two reduced parameters $\mu = \rho c/m\omega_0$ and $\Omega = \omega/\omega_0$. Then the energy transmission rate is written

$$\tau = \frac{4\mu^2}{4\mu^2 + \Omega^2(1 - \Omega^2)}$$

Figure 8.2 shows the insertion loss index as a function of Ω for $\mu = 10^{-3}$.

It is interesting to examine the asymptotic case $\varepsilon = \rho c/m \ll 1$, for which the fluid is a gas and the wall is an elastic structure. The lowest order term of the Taylor series expansion of the transmission coefficient leads to the approximation

$$\tau(\omega) \simeq \frac{4\rho^2 c^2}{m^2\omega^2(1 - \omega_0^2/\omega^2)^2}$$

Under the hypothesis $\omega_0^2/\omega^2 \ll 1$, that is for frequencies much higher than the *in vacuo* resonance frequency of the wall, one gets

$$\tau(\omega) \simeq \frac{4\rho^2 c^2}{m^2\omega^2}$$

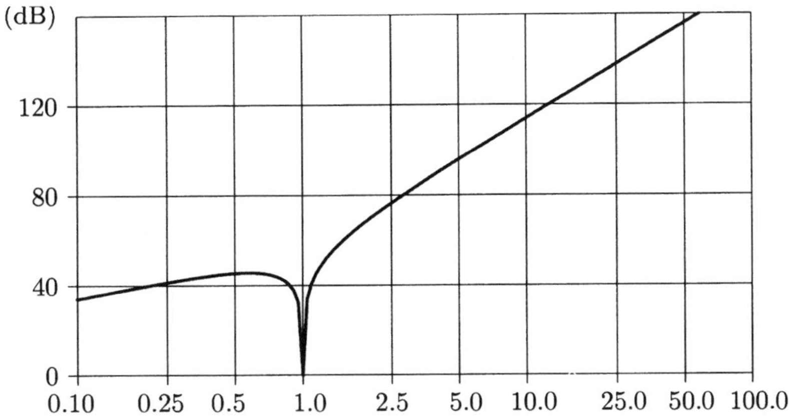

Fig. 8.2. Insertion loss index of an elastically supported piston in a waveguide as a function of the reduced frequency parameter Ω.

This approximation shows that, asymptotically, the amplitude of the transmitted acoustic wave is inversely proportional to the wall mass. This is called the *mass law*, which can be defined as a good approximation of the high frequency behaviour of building walls. In the example shown, the mass law provides an almost perfect prediction for $\Omega > 2.5$.

8.1.3. Transient response of the system to a force acting on the panel

Let us now find the solution of equations (8.1) to (8.4) with $\tilde{S}^+(x, t) \equiv \tilde{S}^-(x, t) \equiv 0$, and $\tilde{F}_e(t)$ non-zero on a bounded time interval $]0, T[$. The panel displacement is given by the Fourier integral

$$\tilde{u}(t) = -\frac{1}{2\pi}\int_{-\infty}^{+\infty} \frac{F_e(\omega)}{m[2\iota\omega\rho c/m + \omega^2 - \omega_0^2]}\, \mathrm{e}^{-\iota\omega t}\, d\omega \tag{8.17}$$

The integrand being a meromorphic function, the integration can be performed by using the residue theorem. *A priori*, two integration contours are necessary: for $t < 0$, the real axis is closed by a half-circle with radius $R \to \infty$ in the half-plane $\Im(\omega) > 0$; for $t > 0$, it is closed by a half-circle in the other half-plane. The denominator of the integrand is a polynomial of order 2 in ω with a discriminant equal to $(\omega_0^2 - \rho^2 c^2/m^2)$ that we assume to be positive (the other case is left as an exercise). The roots of the denominator are

$$\Omega^+ = -\iota\frac{\rho c}{m} + \sqrt{\omega_0^2 - \frac{\rho^2 c^2}{m^2}} = \dot{\Omega} - \iota\hat{\Omega}, \qquad \omega^- = -\iota\frac{\rho c}{m} - \sqrt{\omega_0^2 - \frac{\rho^2 c^2}{m^2}} = -\dot{\Omega} - \iota\hat{\Omega}$$

These two roots lie in the complex half-plane $\Im(\omega) < 0$, thus, $\tilde{u}(t)$ is zero for $t < 0$. For $t > 0$, one has

$$\tilde{u}(t) = -\frac{1}{m}\Im\left[\frac{F_e(\Omega^+)}{\Omega^+ + \iota\rho c/m}\,\mathrm{e}^{-\iota\dot{\Omega}t}\right]\mathrm{e}^{-\hat{\Omega}t}, \qquad t > 0 \tag{8.18}$$

(the proof is left to the reader). This is a damped oscillation.

The acoustic pressure $\tilde{p}^+(x, t)$ is given by the integral

$$\tilde{p}^+(x, t) = -\frac{1}{2\pi}\int_{-\infty}^{+\infty}\frac{\iota\omega\rho c F_e(\omega)}{m[2\iota\omega\rho c/m + \omega^2 - \omega_0^2]}\,\mathrm{e}^{-\iota\omega(t - x/c)}\,d\omega \tag{8.19}$$

Owing to the term $\exp(-\iota\omega(t - x/c))$, this signal is zero for $t < x/c$: this time corresponds to the delay which is necessary for the wavefront to reach the point of abscissa x. For $t > x/c$, the acoustic pressure is

$$\tilde{p}^+(x, t) = -\frac{\rho c}{m}\Re\left[\frac{\Omega^+ F_e(\Omega^+)}{\Omega^+ + \iota\rho c/m}\,\mathrm{e}^{-\iota\dot{\Omega}(t - x/c)}\right]\mathrm{e}^{-\hat{\Omega}(t - x/c)}, \qquad t > x/c \tag{8.20}$$

(the proof is again left to the reader).

These results require a few comments. The poles Ω^+ and Ω^- of the integrands in (8.17) and (8.19) have a physical interpretation. Let us look for free oscillations of the system, that is for solutions of equations (8.1) to (8.4) in the absence of any excitation. It is easily proved that the only non-zero solutions have a time dependence of the form $\exp(-\iota\Omega^+ t)$ or $\exp(-\iota\Omega^- t)$. For that reason, these angular frequencies are called *resonance angular frequencies*. The corresponding responses of the system are

$$U_\pm(t) = \mathrm{e}^{-\iota\Omega^\pm t}, \qquad P_\pm(x, t) = \mathrm{sgn}(x)\rho c\Omega^\pm \mathrm{e}^{-\iota\Omega^\pm(t - |x|/c)}$$

They are called the *resonance modes* of the system composed of the elastically supported piston coupled to the fluid.

8.2. Equation Governing the Normal Displacement of a Thin Elastic Plate

An elastic solid is called a *thin plate* if one of its three dimensions – called the thickness of the plate – is small compared to the other two and if all physical quantities (displacement vector, stress, ...) vary very slowly through the thickness. Geometrically, a plate is a cylinder with base a surface Σ, bounded by a contour $\partial\Sigma$ – called the *mean surface* – and with a height – called the *thickness* – described by a positive function $h(x_1, x_2)$, the domain of variation of the variable x_3 being $]-h(x_1, x_2)/2, +h(x_1, x_2)/2[$. The thickness is equal to a few per cent of the dimensions of Σ and, here, it will be assumed to be constant. The elastic solid is characterized by a density ρ, a Young's modulus E and a Poisson ratio ν.

The equations of elasticity are obtained through the Hamilton principle by taking the variation of the total energy of the solid. To get an approximation of these equations under the thin plate hypothesis, it is necessary to have approximations of the strain and stress tensors.

Let (u_1, u_2, u_3) be the components of the displacement vector of the plate, d_{ij} the strain tensor components and σ_{ij} the stress tensor components. Using the classical notation $f,_i$ for the derivative $\partial f/\partial x_i$, one has

$$d_{ij} = \frac{1}{2}(u_{i,j} + u_{j,i})$$

$$\sigma_{11} = \frac{E}{(1+\nu)(1-2\nu)}[(1-\nu)d_{11} + \nu(d_{22} + d_{33})]$$

$$\sigma_{22} = \frac{E}{(1+\nu)(1-2\nu)}[(1-\nu)d_{22} + \nu(d_{33} + d_{11})]$$

$$\sigma_{33} = \frac{E}{(1+\nu)(1-2\nu)}[(1-\nu)d_{33} + \nu(d_{11} + d_{22})]$$

$$\sigma_{12} = \frac{E}{1+\nu} d_{12} = \sigma_{21}, \qquad \sigma_{13} = \frac{E}{1+\nu} d_{13} = \sigma_{31}, \qquad \sigma_{23} = \frac{E}{1+\nu} d_{23} = \sigma_{32}$$

The first hypothesis is that both boundaries $x_3 = -h/2$ and $x_3 = +h/2$ are free, which leads to $\sigma_{i3} = 0$ along these surfaces. The second hypothesis which is used is that, h being small, all quantities can be expanded into Taylor series of the variable x_3 around zero and terms of order 1 are sufficient to describe the potential energy. From this assumption, it follows that the displacement components u_1 and u_2 can be approximated by expressions which involve the lowest order approximation $\tilde{w}$ of the component u_3 only (which, of course, does not depend on x_3):

$$u_1 \simeq -x_3 \tilde{w},_1, \qquad u_2 \simeq -x_3 \tilde{w},_2$$

This leads to the following approximation of the strain tensor:

$$\begin{gathered} d_{11} \simeq -x_3 \tilde{w},_{11}, \qquad d_{22} \simeq -x_3 \tilde{w},_{22} \\ d_{12} \simeq -x_3 \tilde{w},_{12}, \qquad d_{13} \simeq 0, \qquad d_{23} \simeq 0 \\ d_{33} \simeq \frac{x_3 \nu}{1-\nu}(\tilde{w},_{11} + \tilde{w},_{22}) \end{gathered} \tag{8.21}$$

The potential energy density e is thus approximated by

$$e = \sum_{i,j} \sigma_{ij} d_{ij} \simeq x_3^2 \frac{E}{24(1-\nu^2)} \{[\tilde{w},_{11} + \tilde{w},_{22}]^2 + 2(1-\nu)[\tilde{w},_{12}^2 - \tilde{w},_{11}\tilde{w},_{22}]\}$$

and the total potential and kinetic energies of the plate are approximated by

$$\mathscr{E} = \int_{-h/2}^{+h/2} dx_3 \int_\Sigma e \, d\Sigma$$

$$\simeq \frac{Eh^3}{24(1-\nu^2)} \int_\Sigma \{[\tilde{w},_{11} + \tilde{w},_{22}]^2 + 2(1-\nu)[\tilde{w},_{12}^2 - \tilde{w},_{11}\tilde{w},_{22}]\} \, d\Sigma \qquad (8.22)$$

$$\mathscr{C} = \frac{1}{2} \int_{-h/2}^{+h/2} dx_3 \int_\Sigma \rho(\dot{u}_1^2 + \dot{u}_2^2 + \dot{u}_3^2) \, d\Sigma \simeq \frac{\rho h}{2} \int_\Sigma \dot{\tilde{w}}^2 \, d\Sigma$$

where $\dot{f}$ is the derivative of f with respect to the time variable. Let $\tilde{F}$ be a force density applied to the plate in the normal direction; it is also assumed that no effort is applied along the plate boundary. The Hamilton principle gives

$$\delta\mathscr{E} + \delta\mathscr{C} = \int_\Sigma \tilde{F} \, \delta\tilde{w} \, d\Sigma$$

where δf stands for the variation of the quantity f. Using expressions (8.22), one obtains

$$\int_\Sigma \left\{ \frac{Eh^3}{12(1-\nu^2)} [(\tilde{w},_{11} + \tilde{w},_{22})(\delta\tilde{w},_{11} + \delta\tilde{w},_{22}) + (1-\nu)(2\tilde{w},_{12}\,\delta\tilde{w},_{12} - \tilde{w},_{11}\,\delta\tilde{w},_{22} - \tilde{w},_{22}\,\delta\tilde{w},_{11})] + \rho h \ddot{\tilde{w}} \, \delta\tilde{w} \right\} \delta\Sigma = \int_\Sigma \tilde{F} \, \delta\tilde{w} \, d\Sigma \qquad (8.23)$$

Assume that the boundary $\partial\Sigma$ of the plate is piecewise differentiable (for example, it is composed of arcs of analytical curves). Thus, a unit normal vector $\vec{n}$ and a unit tangent vector $\vec{s}$ can be defined almost everywhere. Performing integrations by parts in the first term of equation (8.23), one gets

$$\int_\Sigma \left\{ \frac{Eh^3}{12(1-\nu^2)} \Delta^2\tilde{w} + \rho h \ddot{\tilde{w}} \right\} \delta\tilde{w} \, d\Sigma + \frac{Eh^3}{12(1-\nu^2)} \int_{\partial\Sigma} [\ell_1(\tilde{w}) \operatorname{Tr} \partial_n \delta\tilde{w} - \operatorname{Tr} \partial_n \Delta\tilde{w} \operatorname{Tr} \delta\tilde{w} + \ell_2(\tilde{w}) \, \partial_s \operatorname{Tr} \delta\tilde{w}] \, ds = \int_\Sigma \tilde{F} \, \delta\tilde{w} \, d\Sigma \qquad (8.24)$$

In this equation, the various symbols are defined as follows:

$$\Delta^2 = \frac{\partial^4}{\partial x_1^4} + 2 \frac{\partial^4}{\partial x_1^2 \, \partial x_2^2} + \frac{\partial^4}{\partial x_2^4}$$

$$\operatorname{Tr} \tilde{w}(M) = \lim_{P \in \Sigma \to M \in \partial\Sigma} \tilde{w}(P), \qquad \operatorname{Tr} \partial_n \tilde{w} = \lim_{P \in \Sigma \to M \in \partial\Sigma} \vec{n}(M) \cdot \nabla_P \tilde{w}(P)$$

$$\text{Tr } \partial_s \tilde{w} = \lim_{P \in \Sigma \to M \in \partial\Sigma} \vec{s}(M) \cdot \nabla_P \tilde{w}(P)$$

$$\text{Tr } \partial_{s^2} \tilde{w} = \lim_{P \in \Sigma \to M \in \partial\Sigma} \vec{s}(M) \cdot \nabla_P [\vec{s}(M) \cdot \nabla_P \tilde{w}(P)]$$

$$\text{Tr } \partial_n \, \partial_s \tilde{w} = \lim_{P \in \Sigma \to M \in \partial\Sigma} \vec{n}(M) \cdot \nabla_P [\vec{s}(M) \cdot \nabla_P \tilde{w}(P)]$$

$$\ell_1(\tilde{w}) = \text{Tr } \Delta\tilde{w} - (1-\nu) \text{ Tr } \partial_{s^2} \tilde{w}, \qquad \ell_2(\tilde{w}) = (1-\nu) \text{ Tr } \partial_n \, \partial_s \tilde{w}$$

These results are very easy to obtain for rectangular or circular plates; they are established for any boundary in classical textbooks (see the short bibliography at the end of the chapter). The terms which occur in the boundary integral represent different densities of work; their components have an interpretation in terms of boundary efforts:

- $Eh^3/12(1-\nu^2)$ Tr $\partial_n \Delta\tilde{w}$, being the factor of $\delta\tilde{w}$, represents the density of shearing forces that the plate boundary exerts on its support.
- $-Eh^3/12(1-\nu^2)[\text{Tr } \Delta\tilde{w} - (1-\nu) \text{ Tr } \partial_{s^2}\tilde{w}]$, being the factor of Tr $\partial_n \delta\tilde{w}$, represents the density of bending moments (rotation around the tangential direction).
- $-(1-\nu)Eh^3/12(1-\nu^2)$ Tr $\partial_n \, \partial_s \tilde{w}$, being the factor of ∂_s Tr $\delta\tilde{w}$, represents the density of twisting moments (rotation around the normal direction).

If the boundary $\partial\Sigma$ has no angular point, the tangent unit vector $\vec{s}$ is defined everywhere and the last integral in the left-hand side of (8.24) can be integrated by parts, leading to

$$\int_\Sigma \left\{ \frac{Eh^3}{12(1-\nu^2)} \Delta^2 \tilde{w} + \rho h \ddot{\tilde{w}} \right\} \delta\tilde{w} \, d\Sigma$$
$$+ \frac{Eh^3}{12(1-\nu^2)} \int_{\partial\Sigma} \{ [\text{Tr } \Delta\tilde{w} - (1-\nu) \text{Tr} \partial_{s^2} \tilde{w}] \text{Tr } \partial_n \delta\tilde{w}$$
$$- [(1-\nu) \, \partial_s \text{Tr } \partial_n \partial_s \tilde{w} + \text{Tr } \partial_n \Delta\tilde{w}] \text{ Tr } \delta\tilde{w} \} \, ds = \int_\Sigma \tilde{F} \, \delta\tilde{w} \, d\Sigma \quad (8.25)$$

This equality must be satisfied for any displacement variation $\delta\tilde{w}$, which implies that the sum of integrals over Σ and the sum of integrals over $\partial\Sigma$ must be zero independently of each other. This first leads to the well-known time dependent plate equation and harmonic plate equation

$$\begin{cases} \left(D\Delta^2 + \mu \dfrac{\partial^2}{\partial t^2} \right) \tilde{w} = \tilde{F} \\ \text{with} \quad D = \dfrac{Eh^3}{12(1-\nu^2)}, \qquad \mu = \rho h \end{cases} \quad (8.26)$$

$$\left\{\begin{array}{l} (\Delta^2 - \lambda^4)w = \dfrac{F}{D} \\ \text{with} \quad w = \displaystyle\int_{-\infty}^{+\infty} \tilde{w} e^{\iota\omega t} dt, \qquad F = \int_{-\infty}^{+\infty} \tilde{F} e^{\iota\omega t} dt, \qquad \lambda^4 = \dfrac{\mu\omega^2}{D} \end{array}\right. \tag{8.27}$$

By analogy with the Helmholtz equation, the parameter λ is called the plate wavenumber and a pseudo-velocity of flexural waves (generally simply called 'flexural wave velocity') is defined by $c_p = \omega/\lambda$: it is frequency dependent and it does not correspond to a genuine wave velocity, that is to the speed of a wavefront. The cancellation of the boundary integrals is obtained by assuming that the displacement satisfies two independent boundary conditions. The most classical ones are:

- **Clamped boundary:**

$$\text{Tr } w = 0, \qquad \text{Tr } \partial_n w = 0$$

- **Free boundary:**

$$\text{Tr } \Delta w - (1-\nu) \text{ Tr } \partial_{s^2} w = 0, \qquad \text{Tr } \partial_n \Delta w + (1-\nu)\, \partial_s \text{ Tr } \partial_n \partial_s = 0$$

- **Simply supported boundary:**

$$\text{Tr } w = 0, \qquad \text{Tr } \Delta w - (1-\nu) \text{ Tr } \partial_{s^2} w = 0$$

8.3. Infinite Fluid-loaded Thin Plate

The second example of a fluid-loaded structure is that of an infinite thin plate embedded in a fluid extending to infinity. In a first subsection, attention is paid to free waves which can propagate along the plate. A second subsection is devoted to the transmission of a plane acoustic wave through the plate. And finally, the radiation of the fluid-loaded plate excited by a point harmonic force is studied.

It will be seen that the behaviour of such a structure is very similar to that of the elastically supported piston in the waveguide. In particular, for a given incident plane wave, there exists a frequency (which depends on the angle of incidence) for which the reflection coefficient is zero while the transmission coefficient is equal to one: this frequency is analogous to the *in vacuo* resonance frequency of the elastically supported piston. If the fluid is a gas, it is possible to use a light fluid approximation which leads, at high frequencies, to a mass law.

Let us consider an infinite thin plate, characterized by a rigidity D and a surface mass μ, located in the plane $\Sigma \equiv (z = 0)$. The two half-spaces $\Omega^+ \equiv (z > 0)$ and $\Omega^- \equiv (z < 0)$ are occupied by a fluid characterized by a density μ_0 and a sound speed c_0. The time-dependent displacement of the plate is $\tilde{w}(x, y; t)$; the corresponding pressure fields are denoted by $\tilde{p}^+(x, y, z;\ t)$ in Ω^+ and $\tilde{p}^-(x, y, z;\ t)$ in Ω^-. The harmonic regimes are denoted by $w(x, y)$, $p^+(x, y, z)$ and $p^-(x, y, z)$. The source terms are $\tilde{F}(x, y;\ t)$ (or $F(x, y)$) on the plate, $\tilde{S}^+(x, y, z;\ t)$ and $\tilde{S}^-(x, y, z; t)$ (or $S^+(x, y, z)$ and $S^-(x, y, z)$) in the fluid. In what follows, we sometimes adopt the notation $f(Q)$ for $f(x, y, z)$ and $f(M)$ for $f(x, y)$.

The governing equations are

$$\left(D\Delta^2 + \mu \frac{\partial^2}{\partial t^2}\right)\tilde{w}(M;\ t) + \tilde{p}^+(M;\ t) - \tilde{p}^-(M;\ t) = \tilde{F}(M;\ t), \qquad M \in \Sigma$$

$$\left(\Delta - \frac{1}{c_0}\frac{\partial^2}{\partial t^2}\right)\tilde{p}^\pm(Q;\ t) = \tilde{S}^\pm(Q;\ t), \qquad Q \in \Omega^\pm \tag{8.28}$$

$$\left.\frac{\partial \tilde{p}^\pm(x, y, z;\ t)}{\partial z}\right|_{z=0} = -\mu_0 \frac{\partial^2 \tilde{w}(x, y;\ t)}{\partial t^2}, \qquad \text{on } \Sigma$$

$$D(\Delta^2 - \lambda^4)w(M) + p^+(M) - p^-(M) = F(M), \qquad M \in \Sigma$$

$$(\Delta - k^2)p^\pm(Q) = S^\pm(Q), \qquad Q \in \Omega^\pm \tag{8.29}$$

$$\left.\frac{\partial p^\pm(x, y, z)}{\partial z}\right|_{z=0} = \mu_0\omega^2 w(x, y), \qquad \text{on } \Sigma$$

The plate equations express the fact that the plate excitation is partly due to the difference between the acoustic pressures existing in the two domains Ω^- and Ω^+. The continuity equations express the fact that the acoustic pressures satisfy a non-homogeneous Neumann boundary condition on Σ, the excitation term being proportional to the plate acceleration. To ensure the uniqueness of the solution, it is necessary to add an 'outgoing wave condition' for both the acoustic fields and the displacement field; for harmonic regimes, this is expressed by a Sommerfeld condition or, equivalently, by the limit absorption principle. Finally, for transient regimes, initial conditions must be given.

8.3.1. Free plane waves in the plate

Let us look for free plane waves propagating within the fluid-loaded plate. They must be solutions of equations (8.29) with no sources (homogeneous equations). The plate displacement is sought in the following form:

$$w = e^{-\iota\Lambda x} \tag{8.30}$$

Then, the sound pressure fields satisfy the following boundary conditions:

$$\partial_z p^-(x, y, 0) = \partial_z p^+(x, y, 0) = \omega^2\mu_0 e^{-\iota\Lambda x} \tag{8.31}$$

The functions $p^-(x, y, z)$ and $p^+(x, y, z)$ are solutions of a homogeneous Helmholtz equation which, in fact, are independent of y, and one must have

$$p^-(x, y, z) = p^-(x, z) = -\omega^2 \mu_0 \frac{e^{\iota(\Lambda x - \alpha z)}}{\iota\alpha}$$

$$p^+(x, y, z) = p^+(x, z) = \omega^2 \mu_0 \frac{e^{\iota(\Lambda x + \alpha z)}}{\iota\alpha} \tag{8.32}$$

$$\alpha \text{ defined by} \qquad k^2 = \Lambda^2 + \alpha^2$$

Introducing these expressions of the plate displacement and of the acoustic pressures into the plate equation, one gets the *fluid-loaded plate dispersion equation*:

$$\mathcal{H}(\Lambda) = \iota\alpha D(\Lambda^4 - \lambda^4) + 2\omega^2\mu_0 = 0 \tag{8.33}$$

The roots of this equation give the possible wavenumbers Λ and α of the free plane waves. It can be remarked first that if Λ is a solution, $-\Lambda$ is also a solution: thus, only one set of solutions needs to be determined. Then, we notice that α^2 is the parameter which is known in terms of Λ, so that α can be arbitrarily chosen as $\sqrt{\alpha^2}$ or $-\sqrt{\alpha^2}$: thus equation (8.33) can be replaced by

$$(k^2 - \Lambda^2)(\Lambda^4 - \lambda^4)^2 - 4\frac{\omega^4\mu_0^2}{D^2} = 0 \tag{8.33$'$}$$

Roots of the dispersion equation

Considered as an equation in Λ^2, the dispersion equation has five roots. It is useful to determine the number of real roots and their respective positions on the real axis relative to k and λ. This is easily done by drawing the curve

$$\mathcal{D}(\Lambda) = \frac{(k^2 - \Lambda^2)(\Lambda^4 - \lambda^4)^2}{k^2\lambda^8}$$

It is obvious that two cases must be distinguished: $k < \lambda$ and $k > \lambda$. The value of the angular frequency for which the wavenumber in the fluid is equal to that of the flexural wave in the plate obviously has a particular role. The corresponding frequency f_c is called the *critical frequency* and is given by

$$f_c = \frac{c_0^2}{2\pi}\sqrt{\frac{\mu}{D}} \tag{8.34}$$

Both situations are shown in Fig. 8.3. The plate is made of compressed wood characterized by $h = 2$ cm, $\mu = 13$ kg m^{-2}, $E = 4.6\ 10^9$ Pa, $\nu = 0.3$. The fluid is air, with $\mu_0 = 1.29$ kg m^{-3}, $c_0 = 340$ m s^{-1}. The critical frequency is 1143 Hz. Equation (8.33$'$) is satisfied for positive values of the function $\mathcal{D}(\Lambda)$. This function has two zeros $\Lambda = k$ and $\Lambda = \lambda$, and, for Λ larger than both k and λ, it is positive

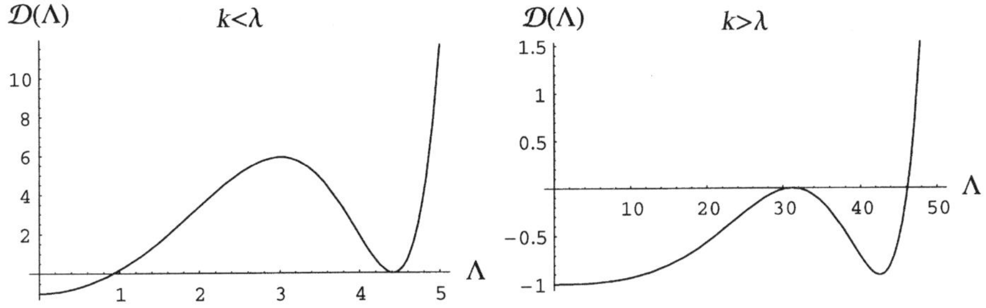

Fig. 8.3. Dispersion curves for $k < \lambda$ and $k > \lambda$.

and grows to infinity. Thus, the roots of equation (8.33′) are

- in any situation, there is a pair of real roots $\pm\Lambda_1^r$ with modulus larger than both k and λ;
- for $k < \lambda$, there are either two other pairs of real roots, $\pm\Lambda_2^r$ and $\pm\Lambda_3^r$ such that $k < \Lambda_2^r < \Lambda_3^r < \lambda$ and two pairs of complex roots $\pm\Lambda_4^c$ and $\pm\Lambda_4^{c*}$, or four pairs of complex roots $\pm\Lambda_2^c$, $\pm\Lambda_2^{c*}$, $\pm\Lambda_3^c$ and $\pm\Lambda_3^{c*}$;
- for $k > \lambda$, there are four pairs of complex roots $\pm\Lambda_2^c$, $\pm\Lambda_2^{c*}$, $\pm\Lambda_3^c$ and $\pm\Lambda_3^{c*}$.

Interpretation of free waves corresponding to the different roots

For the following discussion, it is sufficient to consider only the roots with a positive real part, that is those which correspond to waves propagating in the direction $x > 0$; the waves corresponding to the other roots propagate in the reverse direction.

The flexural wave which corresponds to the largest real root, $w_1 = \exp \iota\Lambda_1^r x$, does not lose any energy. It behaves as a purely propagating wave with a pseudo-velocity $c_1 = \omega/\Lambda_1^r$ which is smaller than c_0 and than the pseudo-velocity ω/λ of the *in vacuo* free waves which can propagate in the plate. The corresponding acoustic pressure in Ω is

$$p_1^{\pm}(x, z) = \omega^2 \mu_0 \frac{e^{\iota(\Lambda_1^r x + \alpha_1 z)}}{\iota\alpha_1}, \qquad \text{with} \qquad \alpha_1^2 = k^2 - (\Lambda_1^r)^2 < 0$$

If we choose $\alpha_1 = \iota\sqrt{(\Lambda_1^r 2 - k^2)}$, the pressure decreases exponentially with respect to the variable z: it is interpreted as an evanescent wave. For the other possible choice, $\alpha_1 = -\iota\sqrt{(\Lambda_1^r 2 - k^2)}$, the pressure is exponentially increasing with z, and there does not seem to be any interesting physical interpretation.

The other real roots, Λ_2^r and Λ_3^r, when they exist, correspond to flexural waves which propagate with a pseudo-velocity smaller than the sound speed but larger than the *in vacuo* structural free waves. As in the previous case, the acoustic fields can be either evanescent or exponentially increasing with the variable z.

A complex root, for example $\Lambda_4^c = \dot{\Lambda}_4^c + \iota\hat{\Lambda}_4^c$, induces a damped structural wave

$$w_4 = \exp(\iota\dot{\Lambda}_4^c x) \exp(-\hat{\Lambda}_4^c x)$$

which propagates in the $x > 0$ direction. Two solutions of the equation $\alpha^2 = k^2 - (\Lambda_4^c)^2$ are associated with this root:

$$\alpha' = \dot{\alpha} - \iota\hat{\alpha}, \qquad \alpha'' = -\dot{\alpha} + \iota\hat{\alpha}, \qquad \text{with } \dot{\alpha} > 0 \text{ and } \hat{\alpha} > 0$$

The corresponding acoustic pressures in Ω^+ are

$$p_4^{+\prime} = \omega^2\mu_0 \frac{e^{\iota\Lambda_4^c x} e^{+\iota\dot{\alpha}z} e^{+\hat{\alpha}z}}{\iota\alpha'}, \qquad p_4^{+\prime\prime} = -\omega^2\mu_0 \frac{e^{\iota\Lambda_4^c x} e^{-\iota\dot{\alpha}z} e^{-\hat{\alpha}z}}{\iota\alpha'}$$

Both are damped in the direction $x > 0$. The first one propagates towards the positive z and increases exponentially, while the second one propagates in the reverse direction and has an amplitude which decreases exponentially with z. It is interesting to calculate the mean energy density exchanged by the fluid and the plate during one period. It is defined by

$$\delta\mathscr{E} = \frac{1}{T}\int_0^T \Re\left[\mathrm{Tr}\, p^+ e^{-\iota\omega t}\right] \Re\left[-\iota\omega w e^{-\iota\omega t}\right] dt$$

For both possible pressure fields, one obtains

$$\delta\mathscr{E}' = \frac{\omega^3\mu_0}{2} e^{-2\hat{\Lambda}_4^c x} \frac{\dot{\alpha}}{|\alpha|^2} > 0$$

$$\delta\mathscr{E}'' = -\frac{\omega^3\mu_0}{2} e^{-2\hat{\Lambda}_4^c x} \frac{\dot{\alpha}}{|\alpha|^2} < 0$$

This result has the following interpretation: in the first case, the plate gives energy to the fluid; in the second case, the fluid provides energy to the plate.

The *light fluid* approximation

When the fluid is a gas, it is intuitive that it has, in most situations, a very small influence on the vibrations of the elastic solid. Let us introduce the parameter $\varepsilon = \mu_0/\mu$, the ratio between the fluid density and the surface mass of the plate, and assume that it is 'small'. The dispersion equation (8.33) can be rewritten as

$$\iota\alpha(\Lambda^4 - \mu\omega^2) = -2\lambda^4\varepsilon \qquad (8.33'')$$

The roots of this equation are obviously close to $\pm\lambda$, $\pm\iota\lambda$ and $\pm k$. A classical perturbation method for finding approximations of these roots is to look for a formal expansion in successive powers of ε^a where a is a constant to be determined. Here, we will calculate the lowest order correcting terms for $k < \lambda$ (the case $k > \lambda$ is left as an exercise).

Let us first examine the possible roots close to λ, that is roots of the form $\Lambda \simeq \lambda(1 + \gamma_1\varepsilon^a + \gamma_2\varepsilon^{2a} + \cdots)$. This expansion is introduced into equation (8.33″) and the successive powers of ε^a are set equal to zero. Recalling that α has two possible determinations, the first order equation is

$$\pm 4\sqrt{\lambda^2 - k^2}\, \lambda^4 \gamma_1 \varepsilon^a = -2\lambda^4\varepsilon$$

This leads to $a=1$ and the corresponding roots are

$$\Lambda_1^+ \simeq \lambda\left(1+\frac{\varepsilon}{2\sqrt{\lambda^2-k^2}}\right), \qquad \Lambda_1^- \simeq \lambda\left(1-\frac{\varepsilon}{2\sqrt{\lambda^2-k^2}}\right)$$

It appears clearly that, due to the term $\sqrt{\lambda^2-k^2}$ in the denominator, the light fluid approximation cannot be valid in the neighbourhood of the critical frequency. Similarly, we get

$$\Lambda_2^+ \simeq \iota\lambda\left(1+\frac{\iota\varepsilon}{2\sqrt{\lambda^2+k^2}}\right), \qquad \Lambda_2^- \simeq \iota\lambda\left(1-\frac{\iota\varepsilon}{2\sqrt{\lambda^2+k^2}}\right)$$

Let us now look for a root of the form $\Lambda \simeq k(1-\gamma_1\varepsilon^a+\gamma_2\varepsilon^{2a}+\cdots)$. The first order equation is:

$$\pm\iota k(k^2-\lambda^2)(k^2+\lambda^2)\sqrt{\gamma_1}\varepsilon^{a/2} = -2\lambda^4\varepsilon$$

which implies $a=2$ and leads to a unique root

$$\Lambda_3 = k\left[1+\left(\frac{\lambda^4}{k(\lambda^2-k^2)(\lambda^2+k^2)}\right)^2\frac{\varepsilon^2}{2}\right]$$

In accordance with the analysis made in the previous section, we have found two real roots between k and λ and a third real root larger than λ. There are also five other roots which are opposite in sign to those which have been defined above.

To conclude this subsection, let us say that the solutions of the dispersion equation have a fundamental importance in solving the problems of sound radiation and sound transmission by plates.

8.3.2. Reflection and transmission of a harmonic plane acoustic wave

It is again assumed that there is no source at finite distance but there exists an incident harmonic plane wave in the fluid domain Ω^+, with wavenumber k; thus, the acoustic pressure p^+ is

$$p^+(x,y,z) = e^{\iota k(x\sin\theta - z\cos\theta)} + p^r(x,y,z) \tag{8.35}$$

where θ is the angle of incidence of the incoming wave.

It is intuitive that, because the incident field is independent of the variable y, the reflected and transmitted pressures, together with the plate displacement, are independent of this variable also. A mathematical reason which can be used is that the equations and the data are unchanged after any translation parallel to the y axis. It is equally intuitive that the reflected and transmitted pressures are plane waves: this result can be obtained by looking for a representation of these functions as plane waves. Nevertheless, it seems better to establish this result directly. To this end, use can be made of the two-dimensional Fourier transform of the equations.

Let $\tilde{f}(\xi, \eta; z)$ denote the Fourier transform of a function $f(x, y, z)$ defined by

$$\tilde{f}(\xi, \eta; z) = \iint_{\mathbb{R}^2} f(x, y, z) e^{-2\iota\pi(x\xi + y\eta)} \, dx \, dy$$

The Fourier transform of the incident field is

$$\mathcal{F} e^{\iota k(x \sin\theta - z \cos\theta)} = \delta(\xi - k \sin\theta/2\pi) \otimes \delta(\eta) e^{-\iota kz \cos\theta}$$

where $\delta(u - a)$ stands for the one-dimensional Dirac measure at the point $u = a$, and $\otimes$ denotes the tensor product of two distributions. Then, equations (8.29) become

$$\left[\frac{d^2}{dz^2} + k^2 - 4\pi^2(\xi^2 + \eta^2)\right] \tilde{p}^r(\xi, \eta; z) = 0, \qquad z > 0$$

$$\left[\frac{d^2}{dz^2} + k^2 - 4\pi^2(\xi^2 + \eta^2)\right] \tilde{p}^-(\xi, \eta; z) = 0, \qquad z < 0$$

$$[16\pi^4(\xi^4 + \eta^4) - \lambda^4]\tilde{w}(\xi, \eta) + \tilde{p}^r(\xi, \eta; 0) - \tilde{p}^-(\xi, \eta; 0) = -\delta(\xi - k \sin\theta/2\pi) \otimes \delta(\eta) e^{-\iota kz \cos\theta} \tag{8.36}$$

$$\frac{d\tilde{p}^-(\xi, \eta; 0)}{dz} = -\iota k \cos\theta \, \delta(\xi - k \sin\theta/2\pi) \otimes \delta(\eta) + \frac{d\tilde{p}^+(\xi, \eta; 0)}{dz} = \omega^2 \mu_0 \tilde{w}(\xi, \eta)$$

The solution of this system of equations is obviously proportional to $\delta(\eta)$. Thus, its inverse Fourier transform is independent of the variable y. The solution is also proportional to $\delta(\xi - k \sin\theta/2\pi)$: this implies that its inverse Fourier transform is proportional to $e^{\iota kx \sin\theta}$. We can conclude that the solution of equations (8.29) has the following form:

$$\begin{aligned} p^r(x, y, z) &= p^r(x, z) = R e^{\iota(kx \sin\theta + \alpha^+ z)}, \qquad \alpha^+ > 0 \\ p^-(x, y, z) &= p^-(x, z) = T e^{\iota(kx \sin\theta - \alpha^- z)}, \qquad \alpha^- > 0 \\ w(x, y) &= w(x) = W e^{\iota kx \sin\theta} \end{aligned} \tag{8.37}$$

Using the plate equation and the continuity conditions, one obtains the following solution:

$$\begin{aligned} p^r(x, z) &= \frac{\iota k \cos\theta(k^4 \sin^4\theta - \lambda^4)}{\mathcal{H}(k \sin\theta)} e^{\iota k(x \sin\theta + z \cos\theta)} \\ p^-(x, z) &= \frac{2\omega^2 \mu_0}{\mathcal{H}(k \sin\theta)} e^{\iota k(x \sin\theta - z \cos\theta)} \\ w(x) &= -\frac{2\iota k \cos\theta}{\mathcal{H}(k \sin\theta)} e^{\iota kx \sin\theta} \end{aligned} \tag{8.38}$$

with $\quad \mathcal{H}(k \sin\theta) = 2\omega^2 \mu_0 + \iota kD \cos\theta(k^4 \sin^4\theta - \lambda^4)$

The function $\mathcal{H}(k \sin\theta)$ is the same as in the dispersion equation.

The energy transmission factor $|T|^2$ is the squared modulus of the ratio between the transmitted wave amplitude and the incident wave amplitude. The insertion loss index is defined as the level in dB of $1/|T|^2$, that is

$$\mathcal{I}(\omega, \theta) = -10 \log \left| \frac{2\omega^2 \mu_0}{2\omega^2 \mu_0 + \iota k D \, \cos \theta (k^4 \sin^4 \theta - \lambda^4)} \right|^2$$

It depends on the frequency and on the angle of incidence. Its behaviour is similar to that of the elastically supported piston that was studied at the beginning of this chapter. Indeed, it is equal to zero if

$$k^4 \sin^4 \theta - \lambda^4 = 0 \qquad \text{that is for} \qquad \omega = \Omega_c(\theta) = \frac{c_0^2}{\sin^2 \theta} \sqrt{\frac{\mu}{D}}$$

where $\Omega_c(\theta)$ is called the *coincidence angular frequency*. This frequency depends on the angle of incidence and does not exist for $\theta = 0$. At the coincidence frequency, the *in vacuo* pseudo-wavelength of the plate flexural waves is equal to the product of the wavelength in the fluid and $\sin \theta$. Figure 8.4 shows the function $\mathcal{I}(\omega, \theta)$ for three values of the angle of incidence: the mechanical data of the fluid and of the plate are the same as in Fig. 8.3 (compressed wood in air).

At high frequency, the insertion loss index takes the asymptotic form

$$\mathcal{I}(\omega, \theta) \simeq -20 \log \left| \frac{2\mu_0 c_0}{\mu \omega \, \cos \theta} \right|$$

This is the mass law which is identical to that which has been found for the elastically supported piston. Figure 8.4 shows that this asymptotic behaviour is

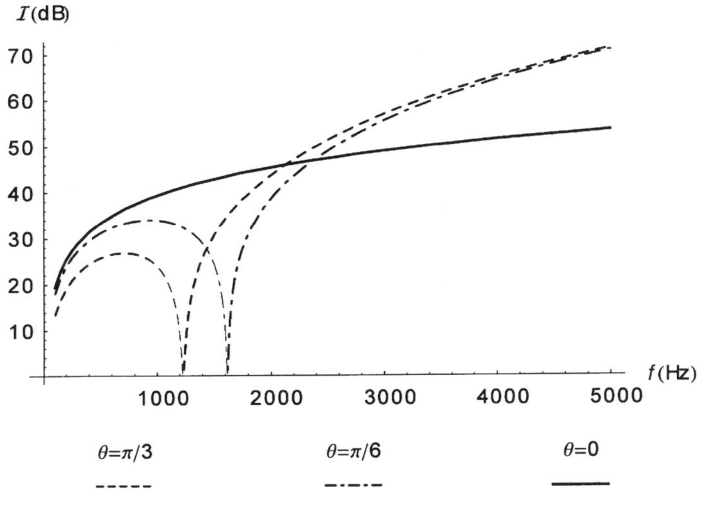

Fig. 8.4. Insertion loss index of an infinite plate for three angles of incidence.

reached around 2500 Hz for $\theta = 0$, while it appears for frequencies larger than 5000 Hz for the other two angles of incidence.

8.3.3. Infinite plate excited by a point harmonic force

We consider now the case of an infinite fluid-loaded plate excited by a point harmonic force of amplitude F and located at the coordinate origin. It is embedded in a fluid which extends to infinity and does not contain any acoustic source. Owing to the simple geometry, the Green's representation of the acoustic fields $p^+(Q)$ and $p^-(Q)$ in terms of the plate displacement is known:

$$p^{\pm}(Q) = \mp 2\omega^2 \mu_0 \int_\Sigma \frac{e^{\iota k r(Q, M')}}{4\pi r(Q, M')} w(M')\, d\Sigma(M') \tag{8.39}$$

Introducing this expression into the first equation (8.29), one gets an integro-differential equation for the plate displacement:

$$D(\Delta^2 + \lambda^4) w(M) - 4\omega^2 \mu_0 \int_\Sigma \frac{e^{\iota k r(M, M')}}{4\pi r(M, M')} w(M')\, d\Sigma(M') = F\delta, \qquad M \in \Sigma \tag{8.40}$$

Fourier transform of the solution

To solve this equation, use is made of the space Fourier transform. Because the governing equations and the data have a cylindrical symmetry, the solution (w, p^-, p^+) has the same symmetry. It is known that the two-dimensional Fourier transform $\hat{f}$ of a function f, depending on the radial variable ρ only, is a function of the dual radial variable ξ only; furthermore, the corresponding direct and inverse Fourier transforms are expressed as Bessel transforms:

$$\hat{f}(\xi) = 2\pi \int_0^\infty f(\rho) J_0(2\pi\xi\rho)\rho\, d\rho$$

$$f(\rho) = 2\pi \int_0^\infty \hat{f}(\xi) J_0(2\pi\xi\rho)\xi\, d\xi = \pi \int_{\infty e^{\iota\pi}}^\infty \hat{f}(\xi) H_0(2\pi\xi\rho)\xi\, d\xi$$

where $J_0(u)$ is the Bessel function of order 0 and $H_0(u)$ is the Hankel function of order 0 and of the first kind.

The Fourier transform of the squared Laplace operator is $16\pi^4\xi^4$. The Fourier transform of the integral operator is not so straightforward to get. Let us remark that the integral term can be written as

$$-\int_\Sigma \frac{e^{\iota k r(M, M')}}{4\pi r(M, M')} w(M')\, d\Sigma(M') = \lim_{Q \in \Omega^+ \to M \in \Sigma} \left[\left(-\frac{e^{\iota k r}}{4\pi r} \right) * w \otimes \delta_z \right](Q)$$

where $*$ stands for the space convolution product and $w \otimes \delta_z$ is the simple layer source supported by the plane Σ and with density w. Thus, the Fourier transform of this integral involves the product of the Fourier transforms of each term, that is

$$\mathcal{F}\left(-\frac{e^{\iota kr}}{4\pi r}\right) = \frac{e^{\iota K|z|}}{2\iota K}, \qquad \text{with } K^2 = k^2 - 4\pi^2\xi^2$$

$$\mathcal{F}w \otimes \delta_z = \hat{w}(\xi) \otimes \delta_z \tag{8.41}$$

$$\mathcal{F}\int_\Sigma \frac{e^{\iota kr(M,M')}}{4\pi r(M,M')} w(M')\, d\Sigma(M') = -\frac{1}{2\iota K}\hat{w}(\xi)$$

The determination of K is chosen by applying the limit absorption principle. Finally, the Fourier transformed equation is written

$$\left[D(16\pi^4\xi^4 - \lambda^4) + \frac{2\omega^2\mu_0}{\iota K}\right]\hat{w}(\xi) = F$$

which leads to the Fourier transform of the plate displacement:

$$\hat{w}(\xi) = \frac{F}{D}\frac{\iota KD}{\iota KD(16\pi^4\xi^4 - \lambda^4) + 2\omega^2\mu_0} \tag{8.42}$$

Expression of the plate displacement

The inverse Fourier transform of the last expression is written

$$w(r) = \frac{F}{D}\pi\int_{\infty e^{\iota\pi}}^{\infty} \frac{\iota KD}{\iota KD(16\pi^4\xi^4 - \lambda^4) + 2\omega^2\mu_0} H_0(2\pi\xi\rho)\xi\, d\xi \tag{8.43}$$

It can be evaluated by a contour integration method. First, the angular frequency is assumed to have a small positive imaginary part – that is to be of the form $\omega(1+\iota\varepsilon)$ – which is then decreased to zero (limit absorption principle). The integration contour is shown in Fig. 8.5: it is composed of the parts of the real axis $-R < \xi < -\varepsilon'$ and $\varepsilon' < \xi < R$, the half circle $|\xi| = R$ in the half-plane $\Im\xi > 0$ (where R grows up to infinity), and a branch contour which turns around the point $k = \omega(1+\iota\varepsilon)/c_0$. This last part of the contour defines the branch integral due to the term K which is multiply defined. It is obvious that there are residue terms which correspond to the poles of the integrand inside the contour, that is to the zeros of the dispersion equation which have a positive imaginary part. It is easily seen that, if $2\pi\xi$ is real and larger than both k and λ, then the term $\iota KD(16\pi^4\xi^4 - \lambda^4)$ is real and negative: this implies that the only real positive pole is larger than both $k/2\pi$ and $\lambda/2\pi$. It is also easily shown that if Ξ is a complex root of the dispersion equation, then $-\Xi^*$ is also a root. As a conclusion, three roots will give residues (the detailed proof is left as an exercise):

$$\Xi_1 > k/2\pi \quad \text{and} \quad \lambda/2\pi, \qquad \Xi_2 \text{ and } \Xi_3 = -\Xi_2^* \text{ with } \Im\Xi_2 > 0$$

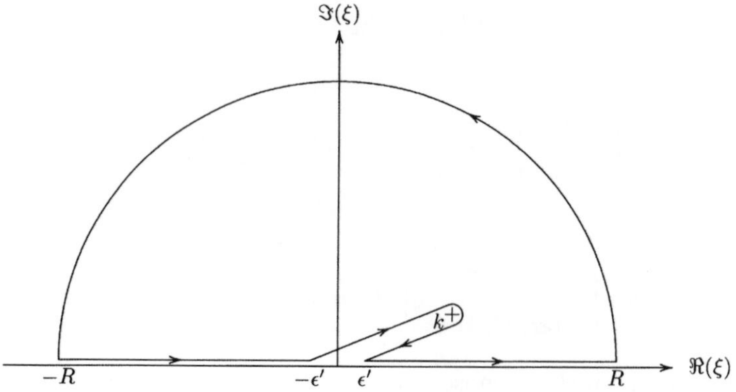

Fig. 8.5. Integration contour for the evaluation of expression (8.43).

The plate displacement can, thus, be written as follows

$$w(r) = 2\iota\pi^2 \frac{F}{D} \sum_{n=1}^{3} A_n H_0(2\pi\Xi_n r) + \text{branch integral}$$

$$A_n = \left[\frac{\iota\xi KD}{(\partial/\partial\xi)[\iota KD(16\pi^4\xi^4 - \lambda^4) + 2\omega^2\mu_0]} \right]_{\xi=\Xi_n} \tag{8.44}$$

$$\text{with } K > 0 \text{ if } K^2 > 0, \Im\, K > 0 \text{ else}$$

The branch integral cannot be calculated analytically. Different expressions, in terms of layer potentials for example, are found in the literature. But a numerical computation is always requested. It is outside the scope of this textbook to undertake such a difficult task. Let us just mention that the contribution of the poles becomes predominant a few wavelengths away from the source.

The far field acoustic pressure

It is, of course, sufficient to pay attention to one of the two pressure fields since they are simply opposite in sign. The pressure field $p^+(Q)$ cannot be expressed simply in terms of known special functions: various expressions, suitable for numerical computation or analytical approximations, can be found in the literature which all involve integrals which require numerical computation. But the large distance asymptotic expansion of this function can be analytically expressed in terms of the Fourier transform of the plate displacement.

Let (R, θ, φ) be the spherical coordinates of a point Q in a coordinate system centred at the origin of the initial Cartesian axes, the plane Σ being defined by $(\theta = \pi/2, 0 < \varphi \leqslant 2\pi)$. The asymptotic series for $R \to \infty$ of $p^+(Q)$ is obtained through the following classical steps (see chapter 5):

- The kernel $\exp[\iota kr(Q, M')]/4\pi r(Q, M')$, involved in expression (8.39), is expanded into a series of spherical functions $h_n(kR) j_n(kr') P_n^{|m|}(\cos\theta) P_n^{|m|}(0)$ $\exp[\iota m(\varphi - \varphi')]$ where $(R', \pi/2, \varphi')$ are the coordinates of M'

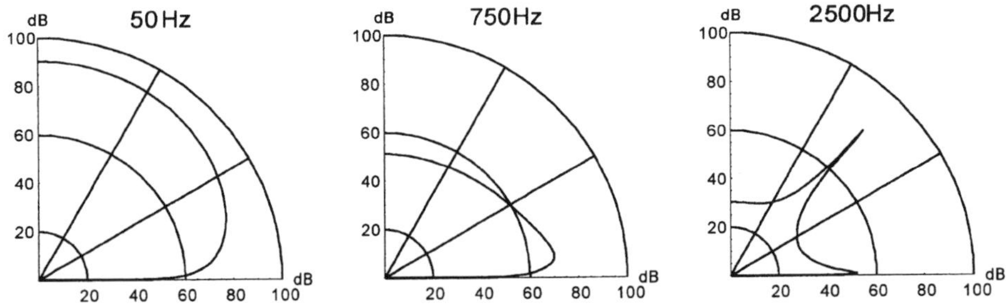

Fig. 8.6. Directivity pattern of an infinite plate excited by a point force for three different frequencies.

- The spherical Hankel functions $h_n(kR)$ are expressed as the product of $\exp(\iota kR)/kR$ and a polynomial of degree n in $1/kR$.
- The series is reordered as a series of the successive powers of $1/kR$ and then integrated term by term leading to an asymptotic series representation.

The first order term can be written very simply

$$p^+(Q) = -2\omega^2\mu_0 \frac{e^{ikR}}{4\pi R} \hat{w}(k \sin\theta/2\pi) + \mathbb{O}(R^{-2}) \tag{8.45}$$

The directivity pattern of the radiating plate is the coefficient of the term $-\exp(\iota kR)/4\pi R$. It is commonly represented in a dB scaling. Figure 8.6 presents the directivity pattern of an infinite plate excited by a point unit force at three different frequencies: the data are the same as in the former numerical examples. The first two diagrams are smooth while the third one presents a sharp peak for $\theta = \pi/4.23$. The presence of such a peak corresponds to a real minimum of the following expression

$$\iota k \cos\theta D(k^4 \sin\theta^4 - \lambda^4) + 2\omega^2\mu_0$$

which occurs in the denominator of $\hat{w}(k \sin\theta)$. This function is nothing but the function $\mathcal{H}(k \sin\theta)$ which appears in the dispersion equation (8.33). Such a minimum corresponds to a real zero of $(k^4 \sin\theta^4 - \lambda^4)$. It is obvious that real zeros occur for frequencies larger that the critical frequency only.

8.4. Finite-dimension Baffled Plate: Expansions of the Solution into a Series of Eigenmodes and Resonance Modes

Let us consider a thin elastic plate which occupies a region Σ of the plane $z = 0$. The plane complement $\complement\Sigma$ of Σ is perfectly rigid. The system is immersed in a gas extending to infinity. The boundary of the plate has almost everywhere a unit normal vector $\vec{n}$ pointing out to $\complement\Sigma$ and a unit tangent vector $\vec{s}$ which makes an angle $+\pi/2$ with the normal vector. The notation is the same as in the previous section.

It is assumed that the time dependence is harmonic and that the only source is $S^-(Q)$ in Ω^-. The governing equations are

$$\begin{aligned}
&(\Delta + k^2)p^+Q = 0, \qquad Q \in \Omega^+ \\
&(\Delta + k^2)p^-Q = S^-(Q), \qquad Q \in \Omega^- \\
&(D\Delta^2 - \mu\omega^2)w(M) + P(M) = 0, \qquad M \in \Sigma \\
&\text{Tr } \partial_z p^-(M) = \text{Tr } \partial_z p^+(M) = \omega^2 \mu_0 w(M), \qquad M \in \Sigma \\
&\qquad = 0, \qquad M \in \complement\Sigma \\
&\ell w(M) = \ell' w(M) = 0, \qquad M \in \partial\Sigma \\
&\text{Sommerfeld condition on } p^+ \text{ and } p^-
\end{aligned} \tag{8.46}$$

where $P(M) = \text{Tr } p^+(M) - \text{Tr } p^-(M)$, and ℓ and ℓ' are two boundary operators which express the boundary conditions for the plate.

The non-dimensional equations

Very often in the literature use is made of what are called *non-dimensional equations*. They are obtained by the following change of units:

- the time unit is the reciprocal of the critical frequency, that is $T = 2\pi c_0^{-2}\sqrt{D/\mu}$;
- the length unit is the wavelength in the fluid at the critical frequency, that is $L = 2\pi c_0^{-1}\sqrt{D/\mu}$.

Introducing these units into equations (8.46), one obtains a system of equations which has exactly the same form. Of course, the quantities involved have the same physical dimensions and the equations are not really dimensionless. The advantage of adopting such units is that, instead of dealing with a particular problem, we consider a particular class of problems. So, in what follows, the equations used can be considered as written with any system of units, in particular with the system which is defined on the basis of the critical frequency.

Let $\mathcal{G}_\omega(Q, Q')$ be the Green's function of the Neumann problem in Ω^+, that is

$$\mathcal{G}_\omega(Q, Q') = -\frac{e^{ikr(Q, Q')}}{4\pi r(Q, Q')} - \frac{e^{ikr(Q, Q'')}}{4\pi r(Q, Q'')}$$

where $r(Q, Q')$ stands for the distance between the two points Q and Q'; Q'' and Q' are symmetrical points with respect to the plane $z = 0$. The Green's representations of the acoustic fields are

$$\begin{aligned}
&p^+(Q) = \omega^2\mu_0 \int_\Sigma w(M')\mathcal{G}_\omega(Q, M')\, d\sigma(M'), \qquad Q \in \Omega^+ \\
&p^-(Q) = p_0^-(Q) - \omega^2\mu_0 \int_\Sigma w(M')\mathcal{G}_\omega(Q, M')\, d\sigma(M'), \qquad Q \in \Omega^- \\
&\text{with } p_0^-(Q) = \int_{\Omega^-} S^-(Q')\mathcal{G}_\omega(Q, Q')\, d\Omega(Q')
\end{aligned} \tag{8.47}$$

By introducing these expressions into the plate equation, we are left with an integro-differential equation for the plate displacement only:

$$(D\Delta^2 - \mu\omega^2)w(M) + 2\omega^2\mu_0 \int_\Sigma w(M')\mathcal{G}_\omega(M, M')\, d\sigma(M') = p_0^-(M), \qquad M \in \Sigma \tag{8.48}$$

Associated with the boundary conditions, this equation has a unique solution.

We intend to expand the displacement $w(M)$ into series of either the eigenmodes or the resonance modes. So, it is better to use the weak form of the equation, that is the integral form deduced from the principle of energy conservation. Let us define the following bilinear forms:

$$\langle w, v\rangle = \int_\Sigma w(M)v^*(M)\, d\sigma(M)$$

$$a(w, v) = D\int_\Sigma \left\{ \Delta w \Delta v^* + (1-\nu) \times \left[2\frac{\partial^2 w}{\partial x\partial y}\frac{\partial^2 v^*}{\partial x\partial y} - \frac{\partial^2 w}{\partial x^2}\frac{\partial^2 v^*}{\partial y^2} - \frac{\partial^2 w}{\partial y^2}\frac{\partial^2 v^*}{\partial x^2}\right]\right\} d\sigma(M) \tag{8.49}$$

$$\beta_\omega(w, v) = \int_\Sigma\int_\Sigma w(M)\mathcal{G}_\omega(M, M')v^*(M')\, d\sigma(M)\, d\sigma(M')$$

where v is any function in the functional space which w belongs to, that is the space of functions which are square integrable together with their derivatives up to order 2 and which satisfy the same boundary conditions as the plate displacement. The weak form of the fluid-loaded plate equation is

$$a(w, v) - \mu\omega^2\left[\langle w, v\rangle - 2\frac{\mu_0}{\mu}\beta_\omega(w, v)\right] = \langle p_0^-, v\rangle \tag{8.50}$$

For $v = w$, the first term represents the potential energy of the plate, the second one is the kinetic energy and the third one is the radiated energy.

8.4.1. Expansion of the solution in terms of the fluid-loaded plate eigenmodes

Let us introduce the parameter $\varepsilon = \mu_0/\mu$. The eigenvalues and the eigenmodes are defined by

$$a(U_n, v) = \Lambda_n[\langle U_n, v\rangle - 2\varepsilon\beta_\omega(U_n, v)] \tag{8.51}$$

The first remark is that these eigenvalues and eigenmodes depend on the angular frequency which appears in the coupling term $\beta_\omega(U_n, v)$.

For simplicity, it is assumed that to each eigenvalue there corresponds only one eigenmode. The eigenmodes satisfy an orthogonality relationship:

$$\langle U_n, U_m^* \rangle - 2\varepsilon\beta_\omega(U_n, U_m^*) = 0 \qquad \text{for} \qquad m \neq n$$
$$\text{or} \qquad a(U_n, U_m^*) = 0 \qquad \text{for} \qquad m \neq n$$

The quantity

$$a(U_n, U_n^*) = \Lambda_n[\langle U_n, U_n^* \rangle - 2\varepsilon\beta_\omega(U_n, U_n^*)]$$

plays the role of a norm. A pressure field is associated with each eigenmode by

$$p_n(Q) = \omega^2 \mu_0 \int_\Sigma U_n(M')\mathcal{G}_\omega(Q, M')\, d\sigma(M'), \qquad Q \in \Omega^\pm$$

Then the plate displacement and the pressure fields are expanded into a series of eigenmodes. The following result is easy to establish and the proof is left to the reader:

$$w(M) = \sum_{n=1}^{\infty} \frac{\Lambda_n}{\Lambda_n - \mu\omega^2} \frac{\langle p_0^-, U_n^* \rangle}{a(U_n, U_n^*)} U_n(M)$$
$$p^+(Q) = \sum_{n=1}^{\infty} \frac{\Lambda_n}{\Lambda_n - \mu\omega^2} \frac{\langle p_0^-, U_n^* \rangle}{a(U_n, U_n^*)} p_n(Q), \qquad Q \in \Omega^+ \tag{8.52}$$
$$p^-(Q) = p_0^-(Q) - \sum_{n=1}^{\infty} \frac{\Lambda_n}{\Lambda_n - \mu\omega^2} \frac{\langle p_0^-, U_n^* \rangle}{a(U_n, U_n^*)} p_n(Q), \qquad Q \in \Omega^-$$

The interest of this expansion is that the coefficients are expressed analytically in terms of the eigenvalues and the eigenmodes. But it has the disadvantage that the functions involved are frequency dependent. It is, thus, useful to look at an expansion in terms of the resonance modes, that is the free oscillations, which depend only on the geometrical and mechanical properties of the system.

8.4.2. Expansion of the solution as a series of the resonance modes

The resonance frequencies and resonance modes of the fluid-loaded plate, corresponding to free oscillations, are defined by

$$a(w_n, v) = \mu\omega_n^2[\langle w_n, v \rangle - 2\varepsilon\beta_{\omega_n}(w_n, v)] \tag{8.53}$$

Comparing this equation to equation (8.51), it is obvious that the resonance frequencies are solutions of the equations

$$\Lambda_n(\omega) = \mu\omega^2$$

It can be shown that each of these equations has two solutions

$$\omega_n = \tilde{\omega}_n - \iota\tau_n, \qquad \omega_{-n} = -\tilde{\omega}_n - \iota\tau_n$$

which are symmetrical with respect to the imaginary axis. Furthermore, because the free oscillations must be damped (energy loss into the fluid), their imaginary part must be negative ($\tau_n > 0$). The resonance modes are related to the eigenmodes by

$$w_n = U_n(\omega_n)$$

To each resonance mode corresponds a pressure field given by

$$\Psi_n(Q) = \mu_0 \omega_n^2 \int_\Sigma w_n(M') \mathcal{G}_{\omega_n}(Q, M')\, d\Sigma(M')$$

The following relationships are easily established:

$$w_{-n}(M) = w_n^*(M), \qquad \Psi_{-n}(Q) = \Psi_n^*(Q)$$

If we look for an expansion of the plate displacement in terms of the resonance modes, we are left with an infinite full system of linear equations because no orthogonality relationship, similar to that satisfied by the resonance modes, exists. This system can, of course, be approximately solved by a classical truncation method. There is another method which leads to an analytical result.

Let us first look for the response of the system to a transient real excitation $\tilde{S}^-(Q, t)$. It is obtained by taking the inverse Fourier transform, with respect to ω, of expression (8.52). Assuming that the source is zero for $t < 0$, and applying the residue theorem, one finds that the response of the system starts for $t > 0$ and is given by

$$\tilde{w}(M, t) = -\iota \sum_{n=1}^{\infty} \left\{ \frac{\mu\omega_n^2}{\Lambda_n'(\omega_n) - 2\mu\omega_n} \frac{\langle p_0^-(\omega_n), w_n^* \rangle}{a(w_n, w_n^*)} w_n(M) e^{-\iota\tilde{\omega}_n t - \tau_n t} \right.$$
$$\left. - \frac{\mu\omega_n^{*2}}{\Lambda_n'^*(\omega_n) - 2\mu\omega_n^*} \frac{\langle p_0^-(\omega_n), w_n^* \rangle^*}{a(w_n, w_n^*)^*} w_n^*(M) e^{\iota\tilde{\omega}_n t - \tau_n t} \right\} \tag{8.54}$$

$$\tilde{p}^+(Q, t) = -\iota \sum_{n=1}^{\infty} \left\{ \frac{\mu\omega_n^2}{\Lambda_n'(\omega_n) - 2\mu\omega_n} \frac{\langle p_0^-(\omega_n), w_n^* \rangle}{a(w_n, w_n^*)} \Psi_n(Q) e^{-\iota\tilde{\omega}_n t - \tau_n t} \right.$$
$$\left. - \frac{\mu\omega_n^{*2}}{\Lambda_n'^*(\omega_n) - 2\mu\omega_n^*} \frac{\langle p_0^-(\omega_n), w_n^* \rangle^*}{a(w_n, w_n^*)^*} \Psi_n^*(Q) e^{\iota\tilde{\omega}_n t - \tau_n t} \right\} \tag{8.54$'$}$$

$$\tilde{p}^-(Q, t) = \tilde{p}_0^-(Q, t) + \iota \sum_{n=1}^{\infty} \left\{ \frac{\mu\omega_n^2}{\Lambda_n'(\omega_n) - 2\mu\omega_n} \frac{\langle p_0^-(\omega_n), w_n^* \rangle}{a(w_n, w_n^*)} \Psi_n(Q) e^{-\iota\tilde{\omega}_n t - \tau_n t} \right.$$
$$\left. - \frac{\mu\omega_n^{*2}}{\Lambda_n'^*(\omega_n) - 2\mu\omega_n^*} \frac{\langle p_0^-(\omega_n), w_n^* \rangle^*}{a(w_n, w_n^*)^*} \Psi_n^*(Q) e^{\iota\tilde{\omega}_n t - \tau_n t} \right\} \tag{8.54$''$}$$

In these expressions, $\Lambda_n'(\omega_n)$ is the value, at ω_n, of the derivative of Λ_n with respect to ω. The notation $p_0^-(\omega_n)$ has been used to point out that, in the function p_0^-, the angular frequency is $\omega = \omega_n$. It must be noted that, if the source term is a real function (and this is always the case in physics), then the response of the system is real.

The expansion of the response of the system to a harmonic excitation in terms of the resonance modes is readily obtained by taking the Fourier transform of expression (8.54):

$$w(M) = \iota \sum_{n=1}^{\infty} \left\{ \frac{\mu\omega_n^2}{\Lambda_n'(\omega_n) - 2\mu\omega_n} \frac{\langle p_0^-(\omega_n), w_n^* \rangle}{a(w_n, w_n^*)} \frac{w_n(M)}{\iota(\omega - \tilde{\omega}_n) - \tau_n} - \frac{\mu\omega_n^{*2}}{\Lambda_n'^*(\omega_n) - 2\mu\omega_n^*} \frac{\langle p_0^-(\omega_n), w_n^* \rangle^*}{a(w_n, w_n^*)^*} \frac{w_n^*(M)}{\iota(\omega + \tilde{\omega}_n) - \tau_n} \right\} \tag{8.55}$$

$$p^+(Q) = \iota \sum_{n=1}^{\infty} \left\{ \frac{\mu\omega_n^2}{\Lambda_n'(\omega_n) - 2\mu\omega_n} \frac{\langle p_0^-(\omega_n), w_n^* \rangle}{a(w_n, w_n^*)} \frac{\Psi_n(Q)}{\iota(\omega - \tilde{\omega}_n) - \tau_n} - \frac{\mu\omega_n^{*2}}{\Lambda_n'^*(\omega_n) - 2\mu\omega_n^*} \frac{\langle p_0^-(\omega_n), w_n^* \rangle^*}{a(w_n, w_n^*)^*} \frac{\Psi_n^*(Q)}{\iota(\omega + \tilde{\omega}_n) - \tau_n} \right\} \tag{8.55'}$$

$$p^-(Q) = p_0^-(Q) - \iota \sum_{n=1}^{\infty} \left\{ \frac{\mu\omega_n^2}{\Lambda_n'(\omega_n) - 2\mu\omega_n} \frac{\langle p_0^-(\omega_n), w_n^* \rangle}{a(w_n, w_n^*)} \frac{\Psi_n(Q)}{\iota(\omega - \tilde{\omega}_n) - \tau_n} - \frac{\mu\omega_n^{*2}}{\Lambda_n'^*(\omega_n) - 2\mu\omega_n^*} \frac{\langle p_0^-(\omega_n), w_n^* \rangle^*}{a(w_n, w_n^*)^*} \frac{\Psi_n^*(Q)}{\iota(\omega + \tilde{\omega}_n) - \tau_n} \right\} \tag{8.55''}$$

This result is not as interesting as it seems. Indeed, it requires evaluation of the derivative of the eigenvalues with respect to the angular frequency. This can be achieved approximately by a numerical procedure which is, in general, rather time consuming. Nevertheless, in the case of a gas, a perturbation method can be developed and the series (8.55) are expressed analytically in terms of the *in vacuo* eigenfrequencies and eigenmodes of the plate.

8.4.3. The *light fluid* approximation

The difficulty in using the resonance modes series is the evaluation of the resonance frequencies and the resonance modes. Assume that the *in vacuo* resonance frequencies and resonance modes of the plate are known either analytically or through any numerical approximation. Are there situations for which the fluid-loaded resonance frequencies and modes can be deduced simply from the *in vacuo* ones?

As we did for the elementary example of the elastically supported piston, we can look for a simple approximation when the fluid density is small compared to the surface mass of the plate, that is when $\varepsilon = \mu_0/\mu \ll 1$.

Let Ω_n be an *in vacuo* angular resonance frequency of the plate and W_n the corresponding resonance mode. It must be noticed first that $-\Omega_n^*$ is a resonance frequency, too, which corresponds to the same resonance mode. The resonance frequencies and modes of the fluid-loaded plate are sought as formal Taylor series of the small parameter ε:

$$\omega_n^2 = \Omega_n^{(0)^2}(1 + \varepsilon\delta_n^{(1)} + \varepsilon^2\delta_n^{(2)} + \mathcal{O}(\varepsilon^3)), \qquad w_n = W_n^{(0)} + \varepsilon W_n^{(1)} + \varepsilon^2 W_n^{(2)} + \mathcal{O}(\varepsilon^3)$$

These formal expansions are introduced into the fluid-loaded resonance equation (8.53) and the successive powers of ε are set equal to zero. The first two equations are

$$\begin{aligned} a(W_n^{(0)}, v) - \mu\Omega_n^{(0)^2}\langle W_n^{(0)}, v\rangle &= 0 \\ a(W_n^{(1)}, v) - \mu\Omega_n^{(0)^2}[\langle W_n^{(1)}, v\rangle + \delta_n^{(1)}\langle W_n^{(0)}, v\rangle - 2\beta_{\Omega_n^{(0)}}(W_n^{(0)}, v)] &= 0 \end{aligned} \tag{8.56}$$

Obviously we have

$$\Omega_n^{(0)^2} = \Omega_n^2, \qquad W_n^{(0)} = W_n$$

that is the zero-order approximations of the fluid-loaded resonance frequency and resonance mode coincide with the *in vacuo* ones. We first determine the parameter $\delta_n^{(1)}$ by writing the second equation (8.56) for $v = W_n$: the first two terms cancel and one gets

$$\delta_n^{(1)} = \frac{2\beta_{\Omega_n}(W_n, W_n)}{\langle W_n, W_n\rangle} = \eta_n + \iota\zeta_n$$

with the corresponding first order approximation of the resonance frequency

$$\omega_n \simeq \Omega_n((1 + \varepsilon\eta_n) - \iota\varepsilon \mid \zeta_n \mid) \tag{8.57}$$

Then, the function $W_n^{(1)}$ is sought as a series of the *in vacuo* modes W_q; the coefficients α_n^q are given by:

$$\sum_{q=1}^{\infty} \alpha_n^q[a(W_q, v) - \mu\Omega_n^2\langle W_q, v\rangle] = 2\mu\Omega_n^2\left[\frac{\beta_{\Omega_n}(W_n, W_n)}{\langle W_n, W_n\rangle}\langle W_n, v\rangle - \beta_{\Omega_n}(W_n, v)\right]$$
$$= 0$$

Let us recall that the *in vacuo* resonance modes W_q form an orthogonal basis of the functional space to which any plate displacement belongs. The last equation is satisfied for any function v if it is satisfied for all the W_q. This leads to the result

$$\alpha_n^q = \frac{2\mu\Omega_n^2\beta_{\Omega_n}(W_n, W_q)}{(\Omega_n^2 - \Omega_q^2)\langle W_q, W_q\rangle} \qquad \text{for } q \neq n$$

For $q = n$ the equation has the form $0 \times \alpha_n^n = 0$. Thus, we can choose $\alpha_n^n = 0$: this only changes the norm of the approximation of the corresponding fluid-loaded mode which is, thus, given by

$$w_n \simeq W_n + \sum_{q \neq n} \frac{2\mu_0 \Omega_n^2}{(\Omega_n^2 - \Omega_q^2)} \frac{\beta_{\Omega_n}(W_n, W_q)}{\langle W_q, W_q \rangle} W_q \tag{8.58}$$

These approximations of the resonance frequencies and resonance modes can easily be used for any shape of plate and any boundary conditions for which an analytical exact or approximate method provides the *in vacuo* resonance frequencies and resonance modes. For example, for a rectangular clamped plate, Warburton's method provides an excellent approximation of the *in vacuo* resonance regimes. This solution can be corrected as described here to get an approximation of the fluid-loaded resonance regimes. Then introducing (8.57, 8.58) into the expressions (8.55), one gets a good approximation of the forced response of the fluid-loaded baffled plate. Figures 8.7 and 8.8 show a comparison between the light fluid approximation and experimental results. The plate is made of stainless steel and has the following geometrical and mechanical characteristics: dimensions = $1.54 \times 1.00 \times 0.0019$ m^3, $E = 2.21\ 10^{11}$ Pa, $\nu = 0.3$, $\mu = 14.8$ kg m^{-2}; its critical frequency is 6012 Hz; it is centred at the coordinates origin. The sound source is located at $x = 0.26$ m, $y = -0.17$ m, $z = -3.0$ m; a first microphone, which is located at $x = -0.26$ m, $y = -0.17$ m, $z = -0.25$ m, records the sum of the incident and reflected fields; a second microphone, which is located at $x = -0.26$ m, $y = -0.17$ m, $z = 0.25$ m, records the transmitted field. In Fig. 8.7, the difference between the pressure level in dB on the microphone in $z > 0$ and the pressure level on the microphone in $z < 0$ is computed and compared with the experimental data. In Fig. 8.8, a similar comparison is made between third octave mean levels. This second curve is much more significant from the point of view of architectural acoustics: indeed, in practice pure tones are almost never present; most of the noises encountered in real life have a large bandwidth spectrum

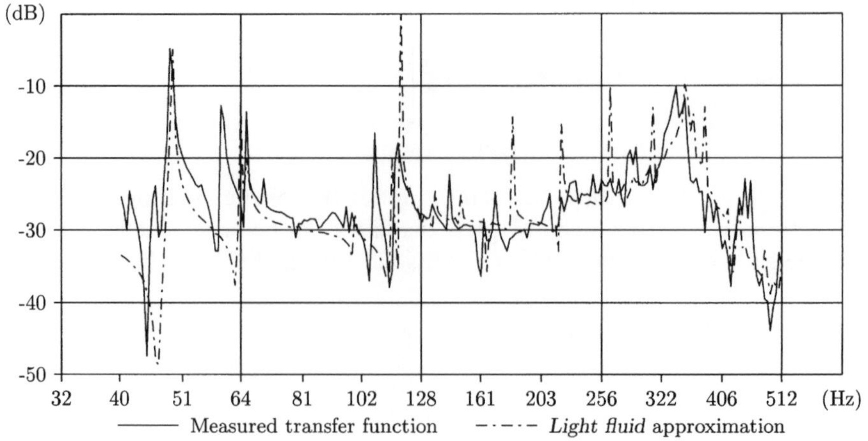

Fig. 8.7. Comparison between the measured transfer function and the light fluid approximation.

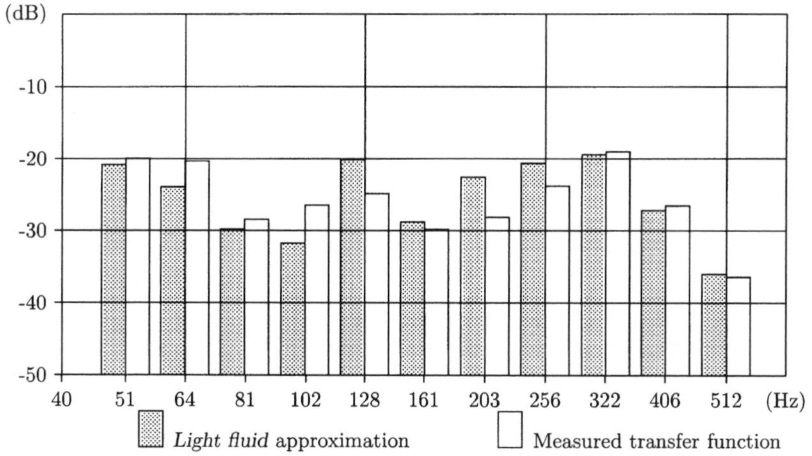

Fig. 8.8. Comparison between the measured mean (third octaves) transfer function and the light fluid approximation.

and third octave or octave analysis is more suitable for describing the nuisance produced by noises.

The light fluid approximation is, of course, a very powerful tool, but it has a restricted domain of validity. It has been mentioned that the parameter $\varepsilon = \mu_0/\mu$ must be small. But the meaning of 'small' is not precisely defined: no upper bound is given. On the simple example of the elastically supported piston in a waveguide, it appeared clearly that such an approximation is not valid in the neighbourhood of the *in vacuo* resonance frequency of the piston. The same restriction applies for the baffled plate in the neighbourhood of the critical frequency. It is reasonable to assume that the validity domain of a light fluid approximation depends mainly on the plate and the fluid physical characteristics, and not too much on the geometrical data. So, to determine the domain of application of such an approximation, we only need to examine the simple case of an infinite plate. As has been seen in Section 8.3, the Fourier transform of the solution is easily obtained. This Fourier transform can be expanded into a formal Taylor series of ε. The validity domain of the series truncated at order 1 is an excellent estimation of the validity domain of the approximation given by expressions (8.57, 8.58) established in the present subsection.

8.5. Finite-dimension Baffled Plate: Boundary Integrals Representation of the Solution and Boundary Integral Equations

In this section, the plate is excited by a harmonic force with density $f(M)$ and there is no acoustic source in the fluid domains. By using the Green's kernel of the *in vacuo* plate operator, the Green's representation of the plate displacement is obtained. It involves four fields: the *in vacuo* radiation of the external force f, and of the pressure difference $p^- - p^+$; and two *in vacuo* boundary layer potentials which enable us to account for the boundary conditions. As in the previous sections, the Green's representation of the pressure fields in terms of the plate displacement is introduced. Then the system of partial differential equations is replaced by a system of boundary integral equations: two over the plate domain Σ and two along its boundary $\partial\Sigma$. This system can be solved by numerical methods very similar to the method presented for the diffraction problems already studied (see chapter 6). For the case of a rectangular plate, we present a slightly different method: the unknown functions are approximated by truncated series of orthogonal polynomials; the system of boundary integral equations is approximated by a system of collocation equations which are equivalent to the Ritz–Galerkin equations. To illustrate the efficiency of this numerical technique, the response of the plate to the wall pressure of a turbulent flow is considered: it is shown that this problem reduces to solving the harmonic equations over a wide frequency bandwidth; a comparison between numerical results and experiment is shown.

For the sake of simplicity, the plate is supposed to be clamped along its boundary. The equations governing the system are thus

$$\begin{aligned}
&(\Delta + k^2)p^+Q = 0, \qquad Q \in \Omega^+ \\
&(\Delta + k^2)p^-Q = 0, \qquad Q \in \Omega^- \\
&(D\Delta^2 - \mu\omega^2)w(M) + P(M) = f(M), \qquad M \in \Sigma \\
&\partial_z p^-(M) = \text{Tr}\ \partial_z p^+(M) = \omega^2\mu_0 w(M), \qquad M \in \Sigma \\
&\qquad = 0, \qquad M \in \complement\Sigma \\
&w(M) = \partial_n w(M) = 0, \qquad M \in \partial\Sigma \\
&\text{Sommerfeld condition on } p^+ \text{ and } p^-
\end{aligned} \tag{8.59}$$

8.5.1. Green's kernel of the *in vacuo* infinite plate

This Green's function γ of the plate equation is the solution of

$$(D\Delta^2 - \mu\omega^2)\gamma = \delta \tag{8.60}$$

which satisfies the limit amplitude principle (which ensures that the principle of energy conservation is satisfied). We first replace ω by $\omega(1 + \iota\varepsilon)$, with $\varepsilon > 0$; and we introduce the parameter $\lambda_\varepsilon = \lambda(1 + \iota\varepsilon')$ defined by

$$\lambda_\varepsilon^4 = \frac{\mu\omega^2(1 + \iota\varepsilon)^2}{D}, \qquad \text{with } \lambda > 0 \text{ and } \varepsilon' > 0$$

The Fourier transform of the solution of equation (8.60) is written

$$\tilde{\gamma}_\varepsilon = \frac{1}{D(16\pi^4\xi^4 - \lambda_\varepsilon^4)} = \frac{1}{2D\lambda_\varepsilon^2}\left[\frac{1}{4\pi^2\xi^2 + \lambda_\varepsilon^2} - \frac{1}{4\pi^2\xi^2 - \lambda_\varepsilon^2}\right]$$

where ξ is the radial variable in the plane associated with the plane (x, y) through the Fourier transform. The function $1/(4\pi^2\xi^2 + \lambda_\varepsilon^2)$ is the Fourier transform of the Green's function of a Helmholtz equation with wavenumber λ_ε, that is $-\iota/4H_0(\lambda_\varepsilon r)$, r being the distance between M and the coordinates origin. Similarly, $1/(4\pi^2\xi^2+\lambda_\varepsilon^2)$ is the Fourier transform of $-\iota/4H_0(\iota\lambda_\varepsilon r)$. These two Hankel functions are finite at infinity: they both decrease exponentially as $r\longrightarrow\infty$. Thus, they define the unique bounded solution of equation (8.60) with λ replaced by λ_ε. By taking the limit for $\varepsilon\longrightarrow 0$, one gets

$$\gamma=-\frac{\iota}{8D\lambda^2}\left[H_0(\lambda r)-H_0(\iota\lambda r)\right] \tag{8.61}$$

The Green's kernel $\gamma(M, M')$ of the *in vacuo* plate operator is the plate response to a point unit force located at M'. It is given by (8.61) in which r is the distance between M and M'.

8.5.2. Green's representation of the fluid-loaded plate displacement

Let us introduce the bilinear form $a(w,\gamma^*)$ defined in (8.49) and perform two kinds of integration by parts. This gives

$$\begin{aligned}a(w,\gamma^*)=&\int_\Sigma D\Delta^2 w(M')\gamma(M,M')\,d\Sigma(M')+\int_{\partial\Sigma} D\ell_1(w(M'))\gamma(M,M')\\&-\operatorname{Tr}\partial_{n'}\Delta w(M')\gamma(M,M')+\ell_2(w(M'))\partial_{s'}\gamma(M,M')]\,ds(M')\end{aligned} \tag{8.62}$$

$$\begin{aligned}a(w,\gamma^*)=&\int_\Sigma D\Delta^2_{M'}\gamma(M,M')w(M')\,d\Sigma(M')+\int_{\partial\Sigma} D[\ell_{1M'}(\gamma(M,M'))\operatorname{Tr}w(M')\\&-\partial_{n'}\Delta_{M'}\gamma(M,M')\operatorname{Tr}w(M')+\ell_{2M'}(\gamma(M,M'))\partial_{s'}\operatorname{Tr}w(M')\,ds(M')\end{aligned} \tag{8.62'}$$

where the subscript M' in the second equality means that the derivatives are taken with respect to the coordinates of this point, and $\vec{n}'$ and $\vec{s}'$ are the unit normal and tangent vectors at M'; ℓ_1 and ℓ_2 are the boundary operators defined in section 8.2. The trivial equality

$$a(w,\gamma^*)-\mu\omega^2\langle w,\gamma\rangle=a(w,\gamma^*)-\mu\omega^2\langle w,\gamma\rangle$$

is written with (8.62) on one side and (8.62') on the other. Accounting for the equations satisfied by w and γ and for the boundary conditions verified by w, the Green's representation of the plate displacement is obtained as

$$\begin{aligned}w(M)=w_0(M)-&\int_\Sigma\gamma(M,M')P(M')d\Sigma(M')+\int_{\partial\Sigma} D[\ell_1(w(M'))\gamma(M,M')\\&-\operatorname{Tr}\partial_{n'}\Delta w(M')\gamma(M,M')+\ell_2(w(M'))\partial_{s'}\gamma(M,M')]\,ds(M')\end{aligned} \tag{8.63}$$

with $\quad w_0(M)=\langle\gamma(M,M'),f(M')\rangle=\int_\Sigma\gamma(M,M')f(M')\,d\Sigma(M')$

This representation depends on three boundary layers: $\ell_1(w(M'))$, $-\text{Tr}\ \partial_{n'}\Delta w(M')$ and $\ell_2(w(M'))$ which could be determined by writing the value on $\partial\Sigma$ of w, $\partial_n w$ and $\ell_2 w$ to get three boundary integral equations. But it is possible to reduce the system to two equations only. To this end, the last boundary integral is integrated by parts. Two cases must be considered:

- The contour $\partial\Sigma$ is continuously differentiable (no angular point), then

$$\int_{\partial\Sigma} \ell_2(w(M'))\partial_{s'}\gamma(M, M')\, ds(M') = -\int_{\partial\Sigma} \partial_s\ell_2(w(M'))\gamma(M, M')\, ds(M')$$

- The contour $\partial\Sigma$ has angular points M_i: let $\overline{\ell_2(w(M_i))}$ be the corresponding jump of $\ell_2(w(M))$, then

$$\int_{\partial\Sigma} \ell_2(w(M'))\partial_{s'}\gamma(M, M')\, ds(M') = -\int_{\partial\Sigma} \partial_s\ell_2(w(M'))\gamma(M, M')\, ds(M') + \sum_i \overline{\ell_2(w(M_i))}\gamma(M, M_i)$$

Introducing the explicit expressions of the boundary operators ℓ_1 and ℓ_2, the Green's representation of the fluid-loaded clamped plate, limited by a contour with angular points, has the form

$$\begin{aligned} w(M) = w_0(M) &- \int_{\Sigma} \gamma(M, M')P(M')\, d\Sigma(M') \\ &+ \int_{\partial\Sigma} D\{[\text{Tr}\ \Delta w(M') - (1-\nu)\ \text{Tr}\ \partial_{s^2} w(M')]\partial_{n'}\gamma(M, M') \\ &- [\text{Tr}\ \partial_{n'}\Delta w(M') + (1-\nu)\partial_s\ \text{Tr}\ \partial_n\partial_s w(M')]\gamma(M, M')\}\, ds(M') \\ &+ (1-\nu)\sum_i \overline{\text{Tr}\ \partial_n\partial_s w(M_i)}\gamma(M, M_i) \end{aligned} \quad (8.64)$$

This shows that the plate displacement depends on two boundary layers if the plate boundary has no angular point; when angular points are present, additional point forces must be introduced at each of them. This result can equally be written in a condensed form:

$$\begin{aligned} w(M) = w_0(M) &- \langle\gamma(M, M'), P(M')\rangle \\ &- \langle\gamma(M, M'), \chi_1 \otimes \delta'_{\partial\Sigma}(M')\rangle - \langle\gamma(M, M'), \chi_2 \otimes \delta_{\partial\Sigma}(M')\rangle \\ \chi_1 &= \text{Tr}\ \Delta w - (1-\nu)\text{Tr}\ \partial_{s^2} w \end{aligned}$$

$$\chi_2 = \text{Tr}\ \partial_{n'}\Delta w + (1-\nu)\partial_s\ \text{Tr}\ \partial_n\partial_s w - (1-\nu)\sum_i \overline{\text{Tr}\ \partial_n\partial_s w(M_i)}\delta_{M_i} \quad (8.64')$$

8.5.3. Boundary integral equations and polynomial approximation

Boundary integral equations

In the case of a regular contour, there are three unknown functions: the pressure jump P defined on the plate domain Σ, and the two layer densities χ_1 and χ_2 defined along the plate boundary $\partial\Sigma$. It is useful to introduce w as a fourth unknown function.

Expression (8.64) or (8.64′) provides a first integral equation on Σ. A second equation is given by the integral expression of P in terms of w:

$$P(M) = 2\omega^2\mu_0 \int_\Sigma w(M')\mathcal{G}_\omega(M, M')\, d\Sigma(M') \tag{8.65}$$

Two other equations along $\partial\Sigma$ are provided by the boundary conditions

$$\mathrm{Tr}\left\{ \int_\Sigma \gamma(M, M')P(M')\, d\Sigma(M') - \int_{\partial\Sigma} D\{\chi_1\partial_{n'}\gamma(M, M') - \chi_2\gamma(M, M')\}\, ds(M') \right\} = \mathrm{Tr}\ w_0(M), \qquad M \in \partial\Sigma \tag{8.66}$$

$$\mathrm{Tr}\ \partial_n\left\{ \int_\Sigma \gamma(M, M')P(M')\, d\Sigma(M') - \int_{\partial\Sigma} D\{\chi_1\partial_{n'}\gamma(M, M') - \chi_2\gamma(M, M')\}\, ds(M') \right\} = \mathrm{Tr}\ \partial_n w_0(M), \qquad M \in \partial\Sigma \tag{8.66'}$$

When $\partial\Sigma$ has angular points, additional equations are obtained by using expression (8.64) to evaluate the steps $\overline{\mathrm{Tr}\ \partial_n\partial_s w(M_i)}$. Because second order derivatives of the kernel γ are involved, some caution is required to evaluate these functions correctly. No more will be said on this; the study of the discontinuities of the layer potentials in plate theory is beyond the purpose of the present chapter.

Very often, equations (8.66, 8.66′) are used even for contours with angular points: the integral of the last term must be interpreted as a formal representation of the sum of the distribution dual products along each regular part $\partial\Sigma_i$ of the boundary $\partial\Sigma$:

$$\int_{\partial\Sigma} \partial_s\ \mathrm{Tr}\ \partial_n\partial_s w(M')\gamma(M, M')\, ds(M') = \sum_i \langle \gamma(M, M'), \partial_s\ \mathrm{Tr}\ \partial_n\partial_s w(M')\rangle$$

This last expression allows each layer density to be a distribution: in particular, it can include Dirac measures at each end of the contour elements $\partial\Sigma_i$. In a numerical procedure, the layer density is generally approximated by regular functions. This implies that the Dirac measures are approximated by regular functions too, which is quite correct: indeed, it is classically shown that a Dirac measure is the limit of, for example, a sequence of indefinitely derivable functions with compact support. Nevertheless, such a numerical approximation of the layer density will not converge to the exact solution in the neighbourhood of the contour angular points as fast as it does elsewhere.

Polynomial approximation

Among the various possible numerical methods to solve system (8.64, 8.65, 8.66, 8.66′), the polynomial approximation method is certainly one of the most powerful when rectangular domains are involved.

Let $\Psi_n(z)$ be any classical set of polynomials defined on the interval $]-1, +1[$ (Legendre, Chebyshev or Jacobi polynomials) and which satisfy an orthogonality relationship

$$\int_{-1}^{+1} \Psi_n(z)\Psi_m(z)\varpi(z)\, dz = \|\, \Psi_n \,\|^2\, \delta_n^m$$

where $\varpi(z)$ is a weight function.

Let the plate domain Σ be defined by $(-a < x < a;\ -b < y < b)$, with boundary $\partial\Sigma = (-a < x < a, y = \pm b) \cup (x = \pm a, -b < y < b)$. The plate displacement and the pressure jumps are approximated by truncated expansions:

$$w(x, y) \simeq \sum_{n=0,\, m=0}^{N,\, M} w_{nm}\Psi_n(x/a)\Psi_m(y/b)$$

$$P(x, y) \simeq \sum_{n=0,\, m=0}^{N,\, M} P_{nm}\Psi_n(x/a)\Psi_m(y/b)$$

The layer densities are also approximated by truncated expansions:

$$\chi_1(x, \pm b) \simeq \sum_{n=0}^{N} \chi_{1n}^{1\pm}\Psi_n(x/a), \qquad x \in]-a, +a[$$

$$\chi_1(\pm a, y) \simeq \sum_{m=0}^{M} \chi_{1m}^{2\pm}\Psi_m(y/b), \qquad y \in]-b, +b[$$

$$\chi_2(x, \pm b) \simeq \sum_{n=0}^{N} \chi_{2n}^{1\pm}\Psi_n(x/a), \qquad x \in]-a, +a[$$

$$\chi_2(\pm a, y) \simeq \sum_{m=0}^{M} \chi_{2m}^{2\pm}\Psi_m(y/b), \qquad y \in]-b, +b[$$

These expressions are introduced into the boundary integral equations. Then a collocation method is used. Let X_i $(i = 1, \ldots, N+1)$ be the zeros of $\Psi_{N+1}(x/a)$ and Y_j, $(j = 1, \ldots, M+1)$ those of $\Psi_{M+1}(y/b)$. The following collocation points are adopted: (X_i, Y_j) on Σ, X_i on $y = \pm b$ and Y_j on $x = \pm a$. It can be shown that, because of a fundamental property of orthogonal polynomials, the solution so obtained is identical to that given by the Ritz–Galerkin method based on the scalar product associated with the polynomials $\Psi_n(z)$. The main advantage is that it avoids the numerical evaluations of multiple integrals.

Let us show this result on a one-dimensional integral equation:

$$u(x) + \int_{-1}^{+1} u(y)K(x, y)\, dy = v(x), \qquad x \in]-1, +1[$$

The function $u(x)$ is approximated by a truncated series of orthogonal polynomials:

$$u(x) \simeq \sum_{n=0}^{N} u_n \Psi_n(x)$$

The Ritz–Galerkin method consists in minimizing the norm, as defined by the weighted scalar product associated with the Ψ_n, of the difference

$$\sum_{n=0}^{N} u_n \left[\Psi_n(x) + \int_{-1}^{+1} \Psi_n(y)K(x, y)\, dy \right] - v(x)$$

This is achieved if the following system of equations is satisfied:

$$\sum_{n=0}^{N} u_n \int_{-1}^{+1} \left[\Psi_n(x) + \int_{-1}^{+1} \Psi_n(y)K(x, y)\, dy \right] \Psi_m(x)\varpi(z)\, dx = \int_{-1}^{+1} v(x)\Psi_m(x)\varpi(z)\, dx$$

$$m = 0, 1, \ldots, N$$

Let x_i be the zeros of the polynomial $\Psi_{N+1}(x)$. The Gauss–Legendre quadrature formula associated with this polynomial leads to the replacement of the former system by:

$$\sum_{i=0}^{N} \sum_{n=0}^{N} u_n \int_{-1}^{+1} \left[\Psi_n(x_i) + \int_{-1}^{+1} \Psi_n(y)K(x_i, y)\, dy \right] \Psi_m(x_i)\Pi_i = \sum_{i=0}^{N} v(x_i)\Psi_m(x_i)\Pi_i$$

$$m = 0, 1, \ldots, N$$

where Π_i are known weights. This quadrature formula is exact as far as the functions to be integrated are polynomials of degree less than $N+1$. It is obvious that this last system is satisfied if the following collocation equations are satisfied:

$$\sum_{n=0}^{N} u_n \left[\Psi_n(x_i) + \int_{-1}^{+1} \Psi_n(y)K(x_i, y)\, dy \right] = \sum_{i=0}^{N} v(x_i), \qquad i = 0, 1, \ldots, N$$

Thus, the collocation equations are equivalent to the Ritz–Galerkin ones and they avoid an extra integration.

Remark. the integral of the product $\Psi_n(y)K(x_i, y)$ can be performed by any numerical method but it is much more consistent to use the quadrature formula associated with the polynomial $\Psi_{N+1}(y)$.

8.5.4. Response of a rectangular baffled plate to the wall pressure exerted by a turbulent boundary layer

The excitation of a vibrating structure by a flow occurs in many real life situations. This is in particular the case inside a fast car or a plane: in the neighbourhood of such a moving obstacle, vortices and turbulence appear which create a fluctuating pressure on its external boundary; this pressure induces vibrations of the elastic boundary (the vehicle walls) which then generates noise in the interior of the structure. The same phenomenon appears commonly with liquids: pipes carrying liquids generate noise; the performance of a pulled sonar is reduced due to the noise generated by the external flow inside the shell enclosing the sensors, etc.

Let us consider the simple problem of a thin baffled elastic plate; the notation is that of the former subsection. The fluid in the domain Ω^+ moves in the x direction, with a velocity U away from the plane $z=0$. It is assumed that close to this plane there exists a turbulent layer (for more details about turbulence, the reader must refer to classical fluid mechanics books). From the point of view of vibro-acoustics, it is sufficient to know that such a flow exerts on the plane $z=0$, and thus on the plate, a wall pressure which is correctly described as a random process which depends randomly on both the time variable and the space variables. Furthermore, the influence of the vibrations of the plate on the flow is negligible: indeed, the pressure induced by the plate motion is much lower than the turbulent wall pressure (by several tens of dB). Finally, as far as the plate vibration and the transmitted sound field are concerned, the motion of the fluid can be neglected in the vibro-acoustics equations (at least in a first approach). So, this subsection deals with the response of a baffled plate to a random excitation, the turbulent wall pressure being just an example among others.

Let $\tilde{f}(x, y; t)$ be a sample function of the random process which describes the turbulent wall pressure. It is characterized by a cross power spectrum density $S_f(M, M'; \omega)$ which is, roughly speaking, the mean value of $f(M; \omega)f^*(M'; \omega)$ where $f(M; \omega)$ is defined as a Fourier transform

$$f(M; \omega) = \int \tilde{f}(M; t)\mathrm{e}^{\iota\omega t}\, dt$$

Experiments show that $S_f(M, M'; \omega)$ depends on the space separation between the points M and M', that is it is a function of the two variables $X = x - x'$ and $Y = y - y'$. The space Fourier transform of the function $S_f(X, Y; \omega)$ is then defined by

$$\hat{S}_f(\xi, \eta; \omega) = \int_{\mathbb{R}^2} S_f(X, Y; \omega)\mathrm{e}^{-2\iota\pi(X\xi + Y\eta)}\, dX\, dY$$

Various models of the function $S_f(\xi, \eta; \omega)$ have been proposed, among which the best-known is due to Corcos. It is a product of three functions: the auto-spectrum $S_f(0, 0; \omega)$ (which can be either modelled analytically or measured); a function of $2\pi\xi/k$ and a function of $2\pi\eta/k$ (k is the acoustic wavenumber) which depends on the fluid and the flow characteristics. Figure 8.9 shows the aspect of $S_f(\xi, \eta; \omega)$ for fixed η and ω as a function of the ratio $2\pi\xi/k$: the fluid is air, the flow velocity is 130 m s^{-1} and the frequency is 1000 Hz.

The response of the system to a random excitation must be characterized by power cross spectra which have to be related to the power cross spectrum of the excitation. Such relationships are formally established as follows. For each sample function of the excitation process, the Fourier transform (w, p^-, p^+) of the system response satisfies equations (8.59). Let $\Gamma(M, Q; \omega)$ be the operator which expresses w in terms of the excitation, that is w is written

$$w(M; \omega) = \int_\Sigma \Gamma(M, Q; \omega) f(Q; \omega)\, d\Sigma(Q)$$

For the given sample function, the power cross spectrum of w is defined as

$$\begin{aligned} s_w(M, M'; w) &= w(M; w) w^*(M'; w) \\ &= \int_\Sigma \int_\Sigma \Gamma(M, Q; w) f(Q; w) f^*(Q'; w) \Gamma^*(M', Q'; w)\, d\Sigma(Q)\, d\Sigma(Q') \end{aligned}$$

The (mean) power cross spectrum density $S_w(M, M'; \omega)$ is the mean value of the former expression in which the only random quantity is $f(Q; \omega)$. Thus, by inverting the operations 'averaging' and 'integration', one gets

$$S_w(M, M'; \omega) = \int_\Sigma \int_\Sigma \Gamma(M, Q; \omega) S_f(Q, Q'; \omega) \Gamma^*(M', Q'; \omega)\, d\Sigma(Q)\, d\Sigma(Q')$$

The function $S_f(Q, Q'; \omega)$ is then replaced by its expression in terms of $\hat{S}_f(\xi, \eta; \omega)$:

$$S_f(Q, Q'; \omega) = \int_{\mathbb{R}^2} \hat{S}_f(\xi, \eta; \omega) e^{2\iota\pi(X\xi + Y\eta)}\, d\xi\, d\eta$$

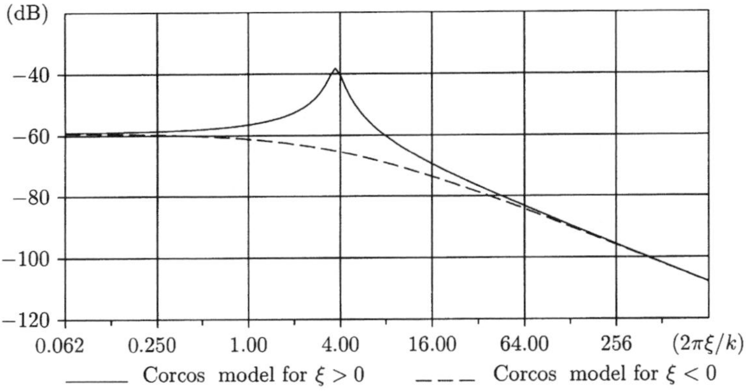

Fig. 8.9. Fourier transform of the power cross spectrum density of a turbulent layer wall pressure after the Corcos model, for fixed η (flow velocity = 130 m s^{-1}, frequency = 1000 Hz).

Introducing this last equality into the expression of $S_w(M, M'; \omega)$, one gets the following result:

$$S_w(M, M'; \omega) = \int_{\mathbb{R}^2} W(M, \xi, \eta; \omega)\hat{S}_f(\xi, \eta; \omega)W^*(M', \xi, \eta; \omega)\, d\xi\, d\eta$$

$$\text{with} \qquad W(M, \xi, \eta; \omega) = \int_{\Sigma} \Gamma(M, Q; \omega)\mathrm{e}^{2\iota\pi(X_Q\xi + Y_Q\eta)}\, d\Sigma(Q)$$

Similar expressions can be established for the power cross spectrum densities of the pressure fields.

This result shows that the statistic characteristics of the baffled plate response can be determined as soon as its response to the deterministic plate excitation $\mathrm{e}^{2\iota\pi(X\xi + Y\eta)}$ is known on a large domain of variations of the three parameters ξ, η and ω. The amount of computation is always important whatever the numerical method which is used. The method developed here, based on polynomial approximations of a system of boundary integral equations, is particularly efficient. As an illustration, numerical predictions have been compared to experimental results due to G. Robert. The plate dimensions are 0.30 m in the direction of the flow and 0.15 m in the other direction; its thickness is 0.001 m. It is made of steel characterized by $E = 1.96 \times 10^{11}$ Pa, $\nu = 0.3$, $\mu = 7.7\ \mathrm{kg\,m^{-2}}$. The fluid is air ($\mu_0 = 1.29\ \mathrm{kg\,m^{-3}}$, $c_0 = 340$ m/s) and the flow velocity is $130\ \mathrm{m\,s^{-1}}$. Figure 8.10 shows that there is a good agreement between the theoretical curve and the experimental one.

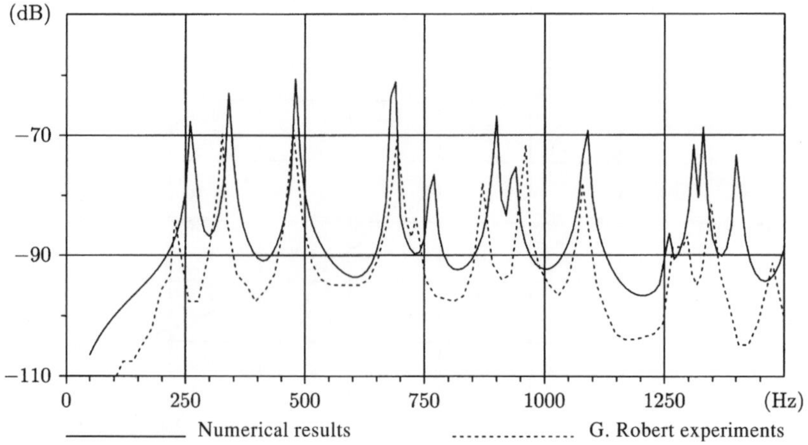

Fig. 8.10. Rectangular plate excited a turbulent flow of air: power density spectrum of a plate velocity (flow velocity at infinity = 130 m s^{-1}; Corcos model of wall pressure; point coordinates on the plate: $x/2a = 0.326$, $y/2b = 0.386$).

8.6. Conclusion

In this chapter, we have presented very simple examples of fluid/vibrating structure interactions, covering both aspects of the phenomenon: radiation and transmission of sound.

In real life, the structures involved are much more complicated. In buildings, plane walls are made of non-homogeneous materials, generally damped and very often with stiffeners; furthermore, double walls are very common. A plane fuselage is a very thin shell, with an array of stiffeners of various sizes, and covered with several layers of different visco-elastic materials. Analogous structures are used in the car, ship or train industries. A machine (machine tools as well as domestic equipment) is an assembly of plates, shells, bars, etc. A wide variety of materials are used.

Nevertheless, the phenomena of sound radiation or sound transmission follow the main rules which have been pointed out on the simple example of the elastically supported piston in a waveguide. Analytical approximations can be developed for such structures. The numerical method of prediction proposed here can be extended to those cases.

An alternative numerical approach is based on finite element approximations in which both the structure equations and the sound equation are approximated in the same way: the main difficulty is to create elements which account for the coupling between the solid and the fluid. Finally, mixed methods, in which the structure equations are approximated by finite elements and the acoustic equation by boundary elements have been developed successfully: here again, the coupling between finite elements and boundary elements is not quite a classical task.

It must be mentioned that the methods described here cannot be used for high frequencies. More precisely, they are limited to a frequency bandwidth within which the resonance frequency density remains rather small (this bandwidth depends, of course, of the elementary structures involved). Another limitation is the number of elementary substructures, which cannot be too large. For those cases of high frequencies or/and complex structures, a statistical method, when it can be used, is obviously much more suitable. A very well known approach of that class is called the statistical energy analysis, or, simply the S.E.A. method. When the underlying basic hypotheses are fulfilled by the structure, the predictions that it provides are quite reliable. This method, which could have been the topic of an additional chapter, is presented in many classical textbooks.

Bibliography

[1] CRIGHTON, D.G., 1989. The 1988 Rayleigh medal lecture: fluid loading – the interaction between sound and vibration. *J. Sound Vib.* 133, 1–27.

[2] FAHY, F., 1985. *Sound and structural vibration*. Academic Press, London.

[3] JUNGER, M.C. and FEIT, D., 1993. *Sound, structures, and their interaction*. Acoustical Society of America. (1972 and 1986 by the Massachusetts Institute of Technology.)

[4] FILIPPI, P.J.T. and MAZZONI, D., 1997. Response of a vibrating structure to a turbulent wall pressure: fluid-loaded structure modes series and boundary element method. In *Uncertainty*

Modeling in Finite Element, Fatigue and Stability of Structures. Series on Stability, Vibrations and Control of Systems. vol. 9, pp. 117–158. World Scientific Publishing.

[5] LANDAU, L. and LIFSCHITZ, E., 1967. *Théorie de l'élasticité.* Editions MIR, Moscow.

[6] LEISSA, A.W., 1973. *Vibrations of shells.* NASA Scientific and Technical Publications, Washington, D.C. 20546.

[7] LESUEUR, C., 1988. *Rayonnement acoustique des structures.* Collection de la Direction des Etudes et Recherches d'Electricité de France. Eyrolles, Paris.

[8] MORSE, Ph.M., 1948. *Vibration and sound.* McGraw-Hill, New York.

[9] NAYFEH, A., 1973. *Perturbation methods.* John Wiley and Sons, New York.

[10] MATTEI, P.-O., 1996. A two-dimensional Tchebycheff collocation method for the study of the vibration of a baffled fluid-loaded rectangular plate. *J. Sound Vib.* 196(4), 407–427.

[11] OHAYON, R. and SOIZE, C., 1998. *Structural acoustics and vibration.* Academic Press, London.

[12] ROBERT, G., 1984. Modélisation et simulation du champ excitateur induit sur une structure par une couche limite turbulente (réf. E.C.L. 84–02). Thèse, Ecole Centrale de Lyon, 36 avenue Guy de Collongue, B.P. 163, 69131 Ecully, France.

CHAPTER 9

Problems

Common Data for Problems 1 to 5

The first five following problems concern a parallelepipedic enclosure Ω, with boundary σ, defined by:

$$x \in \left] -\frac{a}{2}, +\frac{a}{2} \right[, \qquad y \in \left] -\frac{b}{2}, +\frac{b}{2} \right[, \qquad z \in \left] -\frac{c}{2}, +\frac{c}{2} \right[$$

The unit vector, normal to σ and pointing out to the exterior of Ω, is denoted by $\vec{n}$ (it is defined almost everywhere).

1. Eigenmodes for the Dirichlet Problem

Assume that the acoustic pressure satisfies a homogeneous Dirichlet condition along the boundary σ of the propagation domain. Calculate the set of eigenfrequencies and normalized eigenmodes (eigenmodes with unit $\mathbf{L}^2$-norm).

2. Forced Regime for the Dirichlet Problem

Let $f(x, y, z)$ be a harmonic ($e^{-\imath\omega t}$) source distribution located inside Ω and assume that a homogeneous Dirichlet boundary condition is imposed on σ. Establish the eigenmodes series expansion of the sound pressure $p(x, y, z)$ which satisfies the corresponding non-homogeneous Helmholtz equation.

3. Green's Function for the Helmholtz Equation Inside a Parallelepipedic Enclosure

Let $G_N(M, M')$, $G_D(M, M')$ and $G_R(M, M')$ be the Green's functions of the Neumann, Dirichlet and Robin problems respectively, which are defined by:

$$\begin{cases} (\Delta_{M'} + k^2)G_N(M, M') = \delta_{M'}(M), & M \in \Omega \\ \vec{n} \cdot \nabla G_N(M, M') = 0, & M \in \sigma \end{cases}$$

$$\begin{cases} (\Delta_{M'} + k^2)G_D(M, M') = \delta_{M'}(M), & M \in \Omega \\ G_D(M, M') = 0, & M \in \sigma \end{cases}$$

$$\begin{cases} (\Delta_{M'} + k^2)G_R(M, M') = \delta_{M'}(M), & M \in \Omega \\ \vec{n} \cdot \nabla G_R(M, M') - \dfrac{\iota k}{\zeta} G_R(M, M') = 0, & M \in \sigma \end{cases}$$

Find the eigenmodes series expansions of these Green's functions.

4. Green's Formula

Let $p_N(M)$, $p_D(M)$ and $p_R(M)$ be the solutions of the following boundary value problems:

$$\begin{cases} (\Delta + k^2)p_N(M) = f(M), & M \in \Omega \\ \vec{n} \cdot \nabla p_N(M) = g(M), & M \in \sigma \end{cases} \qquad (*)$$

$$\begin{cases} (\Delta + k^2)p_D(M) = f(M), & M \in \Omega \\ p_D(M) = g(M), & M \in \sigma \end{cases} \qquad (**)$$

$$\begin{cases} (\Delta + k^2)p_R(M) = f(M), & M \in \Omega \\ \vec{n} \cdot \nabla p_R(M) - \dfrac{\iota k}{\zeta} p_R(M) = g(M), & M \in \sigma \end{cases} \qquad (***)$$

Using the Green's function $G_N(M, M')$ (respectively $G_D(M, M')$, $G_R(M, M')$), show that $p_N(M)$ (respectively $p_D(M)$, $p_R(M)$) can be represented by a Green's formula similar to (2.114).

5. Green's Representations of the Solutions of the Neumann, Dirichlet and Robin Problems

Using the Green's representations established in Problem 4, give the solutions of the boundary value problems (*), (**) and (***).

6. Green's Kernel of the Helmholtz Equation in $\mathbb{R}^2$

Show that the function $G(M, S) = -(\iota/4)H_0^{(1)}(k \mid SM \mid)$ satisfies the following Helmholtz equation

$$(\Delta_M + k^2)G(S, M) = \delta_S(M)$$

in $\mathbb{R}^2$ and the Sommerfeld condition (3.2). This problem requires the knowledge of the fundamental properties of the Hankel function $H_0(x)$, in particular: its behaviour in the neighbourhood of $x = 0$ and the asymptotic series expansion for $x \rightarrow \infty$. The proof starts by a change of variables to get the coordinates origin at S and, then, cylindrical coordinates are used.

7. Singular Solutions of the Helmholtz Equation in $\mathbb{R}^2$

Show that the functions $H_n^{(1)}(k\rho) \exp(in\varphi)$ and $H_n^{(2)}(k\rho) \exp(in\varphi)$:

- satisfy a homogeneous Helmholtz equation in the complement of the coordinates origin;
- satisfy complex conjugate Sommerfeld conditions.

This problem requires the knowledge of the fundamental properties of the cylindrical Hankel functions: in particular, their behaviour in the neighbourhood of the origin and their asymptotic series expansion at infinity.

8. Singular Solutions of the Helmholtz Equation in $\mathbb{R}^3$

Show that the functions $h_n^{(1)}(k\rho)P_n^{|m|}(\cos\theta) \exp \iota m\phi$ and $h_n^{(2)}(k\rho)P_n^{|m|}(\cos\theta) \exp \iota m\phi$:

- satisfy a homogeneous Helmholtz equation in the complement of the coordinates origin;
- satisfy complex conjugate Sommerfeld conditions.

This problem requires the knowledge of the fundamental properties of the spherical Hankel functions: in particular, their behaviour in the neighbourhood of the origin and their asymptotic series expansion at infinity.

9. Expression of the Normal Derivative of a Double Layer Potential in $\mathbb{R}^3$

In subsection 3.1.5 it was shown that the normal derivative of a double layer potential in $\mathbb{R}^2$ has a value on the source support which can be expressed by convergent integrals (slightly restrictive conditions must be fulfilled).

Consider the double layer potential radiated by a source supported by a closed surface. The Green's kernel is written as the following sum:

$$-\frac{e^{\iota kr}}{4\pi r} = -\frac{1}{4\pi r} + \text{regular function}$$

The same steps as in 3.1.5 lead to (a) the singular term reduces to the solid angle from which the source support is seen; (b) under slightly restrictive conditions, the value, on the source support, of the normal derivative of the double layer potential is expressed with convergent integrals.

10. Green's Representation of the Exterior Dirichlet and Neumann Problems

In Theorem 3.4 it is stated that the B.I.E. deduced from the Green's representation of the pressure field can have real eigenwavenumbers, in particular for the Dirichlet and Neumann problems.

As stated in the theorem, show that, for these two boundary conditions, the second member, which corresponds to an incident field, is orthogonal to the eigenfunctions of the adjoint operator. (If $A(f) = \int_\sigma K(M, P)f(P)\, d\sigma(P)$ is an integral operator, its adjoint is defined by $\tilde{A}(f) = \int_\sigma K^*(M, P)f(M)\, d\sigma(M)$, where K^* is the complex conjugate of K.)

11. Interior Problem and Hybrid Potential Representation

Consider the Dirichlet and the Neumann boundary value problems for a bounded domain (interior problem) and look for the representation of the diffracted field by a hybrid layer potential. Show that the corresponding B.I.E. satisfy the same conditions of existence and uniqueness of the solution as the boundary value problem. (The proof is based on the remark that the hybrid layer potential satisfies a homogeneous Helmholtz equation and a Sommerfeld condition in the space region exterior to the propagation domain.)

12. Propagation in a Stratified Medium. Spatial Fourier Transform

We consider the following two-dimensional problem (O, y, z). Medium (1) corresponds to ($z < 0$ and $z > h$). Medium (2) corresponds to the constant depth layer ($0 < z < h$). Each medium (j) is characterized by a density ρ_j and a sound speed c_j. An omnidirectional point source $S = (0, s)$ is located in medium (2). The emitted signal is harmonic with an $(\exp(-\iota\omega t))$ behaviour. From questions (a) to (d), the parameters ρ_j and c_j are assumed to be constant.

(a) What is the system of equations for the sound pressures in (1) and (2)?
(b) A y-Fourier transform is applied to this system. What is the system obtained?
(c) What is the general form of the solutions of the equations of propagation in (1) and (2)? How many coefficients are to be determined? What are the boundary conditions to determine them?
(d) For which simple condition is the sound pressure the same at $M = (y, -a)$ and $M' = (y, h + a)$, for any a?
(e) If the sound speeds c_j are functions of z, how far can the previous method go? What is the difficulty? Answer the same questions if c_j are functions of y and z.

13. Asymptotic Expansions

Let us consider the integral:

$$J_1(z) = \frac{1}{\iota\pi}\int_0^{\pi} \exp\,(\iota z \cos\,\theta)\cos\,\theta\; d\theta$$

J_1 is the Bessel function of order 1. Find the asymptotic behaviour of J_1 when z is real and tends to infinity. Can a method of integration by parts be applied? Why? What is the first term of the asymptotic expansion obtained through the method of stationary phase?

14. Parabolic Approximation

Let us consider the two-dimensional Helmholtz equation with a constant wavenumber:

$$(\Delta + k^2)p(x, z) = \delta_S(x, z) \tag{9.1}$$

$S = (0, s)$ is an omnidirectional point; the signal is harmonic ($\exp\,(-\iota\omega t)$). The aim of this exercise is to compare the exact solution of this equation (with Sommerfeld conditions) and the expansion obtained from the parabolic approximation.

(a) What is the parabolic equation obtained from (9.1) if p is written as $p(x, z) = \psi(x, z)\exp\,(\iota kx)$?
(b) The parabolic equation for ψ can be solved by using a z-Fourier transform. What is the expression for ψ?
(c) Compare the exact solution and its approximation. For which conditions is the approximation valid?

15. Method of Images

Let us consider the part of the plane $(0, x, y)$ corresponding to $(x > 0, y > 0)$. S is an omnidirectional point source, located at $(x_s > 0, y_s > 0)$. The boundaries $(x > 0, y = 0)$ and $(x = 0, y > 0)$ are described by a homogeneous Dirichlet condition and a homogeneous Neumann condition respectively.

(a) What is the expression of the sound pressure p? Is it an exact representation (check that it satisfies the right Helmholtz equation and the boundary conditions).
(b) If A is the amplitude of the source, what is the level measured on the wall at $M = (x > 0, 0)$?
(c) Is the expression of p still valid if the boundaries are of finite length L? Comment.

16. Integral Equation and Fourier Transform

Let us consider the sound propagation in the half-plane $(z > 0)$. The boundary $(z = 0)$ is characterized by a reduced specific normal impedance. An omnidirectional point source is located at $S = (y = 0, z_s > 0)$.

(a) What is the y-Fourier transform of the sound pressure $p(y, z)$?
(b) What is the integral equation obtained for p by using the Green's formula and the classical Green's kernel?
(c) By applying the y-Fourier transform to the integral equation, give the expression of $\hat{p}(\xi, z)$. Check that both methods lead to the same expression.

17. Born Method

Let us consider the equation:

$$p''(z) + k^2 n^2(z)p(z) = O \qquad \text{for any real } z$$

with

$$n^2(z) = \begin{cases} 1 + a \tanh{(kz/2s)} & \text{if } z > 0 \\ 1 & \text{if not} \end{cases}$$

a and s are parameters. a is 'small'. $p(z)$ represents the sound pressure emitted by an incident plane wave.

(a) $p(z)$ can be written:

$$p(z) = \begin{cases} \exp{(\iota kz)} + R_E \exp{(-\iota kz)} & \text{if } z < 0 \\ T_E \exp{(\iota kz)}L(z) & \text{if } z > 0 \end{cases}$$

$L(z)$ is known explicitly but its expression is of no interest here. Why can $p(z)$ be expressed like this? What are the expressions of the coefficients R_E and T_E?
(b) An approximate expression of p can be written

$$p(z) \simeq \exp{(\iota kz)} \left(1 + \sum_{n=1}^{\infty} a^n \rho_n(z) \right) \qquad \text{for } z \geqslant 0$$

Functions $\rho_n(z)$ can be expressed as

$$\rho_n(z) = \exp{(-\iota kz)}F_n(z)/2\iota k$$

What is the equation satisfied by each F_n? Give a representation of F_n as a function of ρ_{n-1}.

18. Diffraction by a Thin Screen

Let p be the sound pressure diffracted by an infinitely thin screen described by a homogeneous Neumann condition. Show that the Green's formula applied to p and the classical Green's kernel leads to a double layer potential on one side of the screen.

19. Integral Equation

Let p be the sound pressure emitted by a point source above the plane $(z = 0)$. The plane is described by an impedance condition. Write the integral representations of p obtained by applying Green's formula to p and the following Green's kernels:

(a) G_1 solution of the Helmholtz equation in the whole space and satisfying Sommerfeld conditions.
(b) G_2 satisfying the homogeneous Neumann condition on $(z = 0)$.
(c) G_3 satisfying the same impedance condition on $(z = 0)$.
(d) Write the corresponding integral equations and comment. What are the advantages of each one?

20. Method of Wiener–Hopf

Let us consider the following two-dimensional problem. In the half-plane $(y > 0)$, an incident plane wave impinges on the boundary $(y = 0)$ which is described by a reduced normal impedance ζ_1 on $(x < 0)$ and ζ_2 on $(x > 0)$. ζ_1 and ζ_2 are assumed to be constants.

(a) By using Green's formula and partial Fourier transforms, write a Wiener–Hopf equation equivalent to the initial differential problem. The solution can be found in ref. [24] of chapter 5.
(b) For the particular case of $\zeta_1 = 0$ and ζ_2 tends to infinity, find the expression of the Fourier transform of the sound pressure.

21. Neumann Condition

Give the expression of the sound pressure emitted by an omnidirectional point source above a plane described by the homogeneous Neumann condition. Show by a detailed calculation that this expression is the solution of the Helmholtz equation and satisfies the boundary condition. Can this solution be used to find the solution in the case of a non-homogeneous Neumann condition?

22. Integral Equations in an Enclosure

A point source $S = (x_s, y_s)$ is located in a rectangular enclosure $(0 < x < a, 0 < y < b)$. Each boundary Σ_j is characterized by an impedance α_j.

(a) Write an integral equation to evaluate the sound pressure by applying Green's formula to the sound pressure and the classical Green's kernel.

(b) Let us assume that α_2, α_3 and α_4 have the same value β. Which kernel, denoted L, must be chosen to obtain the following integral representation for p:

$$p(M) = p_{inc}(M) + \int_{\Sigma_1} K(M, P)p(P)\, d\Sigma(P)$$

What are the expressions of p_{inc} and K versus L. How can this kernel L be calculated? Is it easier to obtain than p?

23. Propagation in a Waveguide

Let us consider an infinite tube of rectangular cross section ($0 \leqslant x \leqslant a$; $0 \leqslant z \leqslant d$) with the axis parallel to the y-axis. The boundaries ($0 \leqslant x \leqslant a$, $z = 0$) and ($x = 0$, $0 \leqslant z \leqslant d$) are described by a homogeneous Dirichlet condition. The other two are described by a homogeneous Neumann condition.

(a) Give the expressions of the eigenfrequencies and eigenmodes in x and z. Give the general expressions of the sound pressure if the source is an incident plane wave.
(b) What is the condition for which all the modes are attenuated except the first one?
(c) If a point source is located in the tube, give the expression of the sound pressure.

24. Propagation in a Layer

Let us consider the sound propagation in a layer of fluid of constant thickness. A point source is located in the layer. Describe the methods which can be used to obtain an exact or approximate expression of the sound pressure. Comment on their respective advantages and conditions of validity (frequency band, source, sound speed, boundary conditions, ...).

25. Fourier Transform and Separation Method

Let us consider the case of propagation between two parallel planes. One plane is described by a homogeneous Dirichlet condition, the other one by a homogeneous Neumann condition. Show that the method of spatial Fourier transform and the method of separation of variables lead to the same solution.

26. Integral Equations

Let us consider an integral equation on a straight line written with the classical Green's kernel. Check that the elements A_{ij} of the matrix of the equivalent linear system are functions of $|i-j|$. Which property of the kernel is responsible for this?

27. Geometrical Theory of Diffraction

Show that when the sound speed profile of a medium is a linear function of depth z, the trajectories of the rays are circular. This case is studied in ref. [18] of chapter 5.

28. Elastically Supported Piston in a Waveguide

Let us consider a system similar to that of Section 8.1: an elastically supported piston is located at $x=0$; the half-guide $x<0$ contains a fluid characterized by a density ρ and a sound velocity c; the half-guide $x>0$ contains a different fluid characterized by ρ' and c'. Do the same analysis as in Section 8.1 and comment on the results for the two cases $\rho c > \rho' c'$ and $\rho c < \rho' c'$.

29. Roots of the Dispersion Equation

Consider the dispersion equation (8.33″).

(1) Prove that it has two complex roots which are symmetrical with respect to the imaginary axis.
(2) Using the light fluid approximation, evaluate the roots for $k > \lambda$.

30. Infinite Fluid-loaded Plate with Two Different Fluids (a)

The system is almost the same as in Section 8.3: an infinite plate separates the space into Ω^-, which contains a fluid characterized by (ρ, c), and Ω^+, which contains a fluid characterized by (ρ', c'). Write the dispersion equation and analyse the positions of the roots. Assuming that both fluids are gases, calculate the roots by a light fluid approximation.

31. Infinite Fluid-loaded Plate with Two Different Fluids (b)

The system is almost the same as in Section 8.3: an infinite plate separates the space into Ω^- which contains a fluid characterized by (ρ, c), and Ω^+ which contains a fluid characterized by (ρ', c'). As in subsection 8.3.2, analyse the reflection and the transmission of a plane wave $p^+(x, y, z) = e^{\iota k(x \sin\theta - z\cos\theta)}$. Comment on the results for the two cases $\rho c > \rho' c'$ and $\rho c < \rho' c'$.

32. Fluid-loaded Baffled Plate: Eigenmodes

(1) Prove that the eigenmodes defined by equation (8.51) satisfy the orthogonality relationship

$$\langle U_n, U_m^* \rangle - 2\varepsilon\beta_\omega(U_n, U_m^*) = 0 \qquad \text{for} \quad m \neq n$$

$$\text{or} \qquad a(U_n, U_m^*) = 0 \qquad \text{for} \quad m \neq n$$

(2) Establish the expansions (8.52).

33. Fluid-loaded Baffled Plate: Light Fluid Approximation

Consider a rectangular baffled plate with a simply supported boundary.

(1) Using the method of separation of variables, determine the set of the *in vacuo* resonance frequencies and resonance modes.
(2) Using the light fluid approximation, calculate the fluid-loaded baffled plate resonance frequencies and resonance modes (first order approximation).
(3) Assuming that the system is excited by an incident plane wave $p^+(x, y, z) = e^{ik(x \sin\theta - z \cos\theta)}$, give the expressions of the plate displacement and the reflected and transmitted acoustic pressure fields.
(4) Write the corresponding computer program.

Mathematical Appendix: Notations and Definitions

Dominique Habault and Paul J.T. Filippi

Introduction

The main aim of this appendix is to present some mathematical definitions and theorems in order to help the reader who is not familiar enough with some of the expressions used in this book. It also contains the main notations used in most of the chapters.

The mathematical terms used in this book are often related to mathematical analysis or functional analysis. This is why we first present some ideas on how this theory applies in acoustics and vibrations.

To describe a physical phenomenon, the usual procedure is to model it by differential or integral equations and boundary conditions. The mathematical equations are then approximated and solved numerically. The numerical solution is obtained from calculations on computers. Then the following questions arise:

- How do we write properly a system of equations for functions which can be discontinuous or not defined everywhere, especially when these functions must be differentiated?
- How do we approximate the 'exact' equations by equations which can be solved numerically? How 'close' are these approximations from the exact system?
- Do both exact and approximate systems of equations have solutions? What are the properties of the solutions (uniqueness, smoothness, finite energy, continuity, differentiation, ...)?

These questions are essential and computers are not able to provide answers. Functional analysis can answer these questions. From the properties of the operators involved in the equations and the boundary conditions and from the properties of the source terms (right-hand sides of the equations), we can deduce the properties of the solutions and determine the accuracy of approximations.

In order to do this, function spaces have been defined. Hilbert spaces and Sobolev spaces are two of them. To be useful, these function spaces must include discontinuous functions or generalized functions like the Dirac distribution. In these spaces are defined norms and inner products with which it is possible to 'measure' the distance between two functions (a solution and its approximation, for instance). The definition of the distance depends on the space the functions belong to.

The definitions and theorems presented here have been chosen among the basic results of the mathematical theory. To provide a better understanding of the text, they are illustrated whenever possible by examples chosen in acoustics.

Section A.1 is a list of notations used in this book. Section A.2 recalls briefly the definitions of some classical mathematical terms. Section A.3 is a presentation of some of the most relevant function spaces in mechanics. Section A.4 is devoted to the theory of distributions, also called generalized functions.

Let us finally underline that this text is by no means a rigorous presentation of mathematical results. However, it has largely taken advantage of the mathematical books ([1] to [8]) listed at the end of the appendix. For rigorous and complete presentations, the reader is highly recommended to refer to these books.

A.1. Notations Used in this Book

The following notations are used in most chapters.

$k = \omega/c$: acoustic wave number; ω: angular frequency; c: sound speed.
$\lambda = 2\pi/k$: acoustic wavelength.
∇: gradient or divergence operator in 1, 2 or 3 dimensions.
Δ: Laplace operator in 1, 2 or 3 dimensions.
ι: complex number such that $\iota^2 = -1$ and $\Im(\iota) > 0$.
δ: Dirac measure or Dirac distribution.

Harmonic signals: Harmonic signals are assumed to behave as exp $(-\iota\omega t)$ with respect to time.

Helmholtz equation: the Helmholtz equation is often written with a Dirac measure as source term:

$$(\Delta + k^2)p(M) = \delta_S(M)$$

or, if $M = (x, y, z)$ and $S = (x_0, y_0, z_0)$:

$$(\Delta + k^2)p(x, y, z) = \delta(x - x_0)\delta(y - y_0)\delta(z - z_0)$$

or

$$(\Delta + k^2)p(x, y, z) = \delta_{x_0}(x)\delta_{y_0}(y)\delta_{z_0}(z)$$

These notations, although very common in mechanics, are not quite correct and should be replaced by

$$(\Delta + k^2)p(x, y, z) = \delta_{x_0}(x) \otimes \delta_{y_0}(y) \otimes \delta_{z_0}(z)$$

The notation $\otimes$ represents the tensor product of distributions. It is defined in Section A.4.

Normal derivative: If $\vec{n}$ is a unit vector normal to a surface (or a curve) Γ, at point P, the notations are

$$\frac{\partial p}{\partial \vec{n}} \equiv \partial_n p \equiv \vec{n}.\nabla p \quad \text{or} \quad \frac{\partial G(M, P)}{\partial \vec{n}(P)} \equiv \partial_{n(P)} G(M, P) \equiv \vec{n}(P).\nabla_P G(M, P)$$

Sommerfeld conditions: These are defined in Chapter 3, section 3.1.
Order relations: These can be simply formulated by:

$f(x) = \mathcal{O}(g(x))$ when x tends to x_0 if $f(x)$ behaves as $g(x)$ when x tends to x_0
$f(x) = o(g(x))$ when x tends to x_0 if $f(x)$ decreases faster than $g(x)$ when x tends to x_0

or more rigorously:

$f(x) = \mathcal{O}(g(x))$, when x tends to x_0, means that there exists a constant A and a neighbourhood D of x_0 such that $|f(x)| \leqslant A\,|g(x)|$ for every x in D.
$f(x) = o(g(x))$, when x tends to x_0, means that for any $\varepsilon > 0$ there exists a neighbourhood D_ε of x_0 such that $|f(x)| \leqslant \varepsilon\,|g(x)|$ for every x in D_ε.

See also [9].

A.2. Classical Definitions

The functions considered in this book are defined on a domain Ω included in or equal to the n-dimensional infinite space $\mathbb{R}^n$, with $n = 1$, 2 or 3. x is a point of this space and is written $x = (x_1, \ldots, x_n)$ in what follows. $f(x)$ is a real or complex number.

Definitions

- A *bounded domain* is a domain which can be included in a sphere of finite radius R.
- In this book, the boundary Γ of Ω is said to be *smooth* if it is possible to define a normal vector at any point of the boundary.
- f has a *compact support* if $f = 0$ outside a closed and bounded subset of $\mathbb{R}^n$.
- f is *continuous* at point x_0 if the limit of $f(x)$ when $x \to x_0$ is equal to $f(x_0)$.
- f is *continuous on the set* $\mathcal{A}$ if it is continuous at any point of $\mathcal{A}$.
- f is *piecewise continuous* if it is continuous on N subsets $\mathcal{A}_i$ of $\mathcal{A}$, with $\mathcal{A}_i$ not equal to a single point.
- f is *bounded* on $\mathcal{A}$ if there exists a real positive number a such that $|f(x)| \leqslant a$ for any x in $\mathcal{A}$.
- f is k-times *continuously differentiable* if all its partial derivatives of order $j \leqslant k$ exist and are continuous. The space of all such functions f is denoted $C^k(\Omega)$. The same kind of definition holds for an infinitely differentiable function (space $C^\infty(\Omega)$).
- f is *singular* at x_0 if $f(x_0)$ is not defined or infinite.

Example. The Green's function of the Helmholtz equation

$$G(S, M) = -\frac{\exp(\iota k r(S, M))}{r(S, M)}$$

is singular at $M = S$. It becomes infinite when the observation point M approaches the source S. This leads to singular integral equations (see Section A.5). ■

- f is *square integrable* on $\mathscr{A}$ if $\int_{\mathscr{A}} |f(x)|^2 \, dx$ exists and is finite (see Section A.3.3).
- Let f be an integrable function on the interval $[a, b]$ of $\mathbb{R}$, except in the neighbourhood of a point c of the interval. Then $\int_a^b f(x)\, dx$ is defined as a *Cauchy principal value*, vp, if

$$\lim_{\varepsilon \to 0} \left(\int_a^{c-\varepsilon} f(x)\, dx + \int_{c+\varepsilon}^{b} f(x)\, dx \right)$$

exists.

Example

$$\text{vp} \int_a^b \frac{dx}{x} = \log \left| \frac{b}{a} \right| \qquad \text{if } a < 0 \text{ and } b > 0$$

A.3. Function Spaces

The most useful spaces in mechanics are the Hilbert and Sobolev spaces, because their properties are well adapted to the study of the operators and functions involved in the equations.

With such spaces, it is possible to define operations on functions, distances between functions (through inner products and norms) and also convergence properties which are quite essential when approximations are obtained.

A.3.1. Space $\mathscr{D}$ and Space $\mathscr{D}'$

Definition. $\mathscr{D}$ (or $\mathscr{D}(\mathbb{R}^n)$) is the space of all functions φ which are infinitely differentiable on $\mathbb{R}^n$ and have a compact support. This means that a function φ belongs to the space $\mathscr{D}$ if and only if

- all the derivatives of φ exist and are continuous;
- there exists a bounded domain K of $\mathbb{R}^n$ such that φ is equal to zero outside K.

$\mathscr{D}$ is a linear space (also called a vector space). It is not empty. For example, let us define φ by

$$\varphi(x) = \begin{cases} \exp\left(-\dfrac{1}{1 - r^2} \right) & \text{if } r < 1 \\ 0 & \text{if not} \end{cases}$$

with x a point of $\mathbb{R}^n$ and $r = \left(\sum_{i=1}^{n} x_i^2\right)^{1/2}$.

φ is equal to zero outside the ball of radius 1. It is obviously differentiable for $r<1$ and $r>1$. For $r=1$, all the derivatives are equal to zero. Then all its derivatives are continuous everywhere on $\mathbb{R}^n$.

Theorem. Any continuous function f, with a bounded support K, can be uniformly approximated by a function φ of $\mathcal{D}$. This means that, for any $\varepsilon>0$, there exists a function φ such that

$$\sup_{x\in\mathbb{R}^n} |f(x)-\varphi(x)| \leqslant \varepsilon$$

Definition. $\mathcal{D}'$ is the dual of $\mathcal{D}$. This means that $\mathcal{D}'$ is the space of all continuous linear functions defined on $\mathcal{D}$. $\mathcal{D}'$ is the space of distributions which are defined in Section A.4.

A.3.2. Space $\mathcal{S}$

Definition. $\mathcal{S}$ is the space of all functions $f(x)$ indefinitely differentiable and which decrease faster than x^{-k} for any k, when x tends to infinity. It is easy to see that $\mathcal{D}\subset\mathcal{S}$.

Theorem. If f belongs to $\mathcal{S}$, then its Fourier transform exists and also belongs to $\mathcal{S}$.
This is then a convenient space to define Fourier transforms.

Two kinds of Fourier transforms are used in acoustics: time Fourier transforms and space Fourier transforms. In this book, they are defined by

$$\hat{h}(\omega)=\int_{-\infty}^{+\infty} h(t)\exp(+\iota\omega t)\,dt \quad\text{and}\quad \hat{f}(\xi)=\int_{\mathbb{R}^n} f(x)\exp(-2\iota\pi\xi x)\,dx$$

Because of the first definition, the time dependence for harmonic signals is chosen as $\exp(-\iota\omega t)$.

A.3.3. Hilbert spaces

Let us consider again functions defined on Ω. Ω represents the space $\mathbb{R}^n$ or a subspace of $\mathbb{R}^n$.

A Hilbert space is a particular type of linear space in which an inner product is defined. The exact definition of a Hilbert space is not given here. Too many notions, although simple, are required.

One of the most commonly used spaces is $\mathbf{L}^2(\Omega)$. The inner product between two functions f and g of $\mathbf{L}^2(\Omega)$ is defined by

$$\langle f, g\rangle_2 \equiv \int_\Omega f(x)g^*(x)\,dx$$

where $g^*(x)$ is the complex conjugate of $g(x)$.

Definition. $\mathbf{L}^2(\Omega)$ is the space of functions f such that:

$$\|f\|_2 \equiv (\langle f, f \rangle_2)^{1/2} \equiv \left(\int_\Omega |f(x)|^2 \, dx \right)^{1/2} < +\infty$$

$\|f\|_2$ is the norm defined in $\mathbf{L}^2(\Omega)$. $\mathbf{L}^2$ is the space of functions with finite energy. This is why it is very well adapted to problems in mechanics.

Definition. Let V be a normed linear space of functions. A system of functions $(b_j(x), j=0, 1, \ldots)$ is *a basis* for V if any function f of V has a unique representation:

$$f = \sum_{j=0}^{+\infty} \alpha_j b_j$$

where α_j are scalars.

The b_j are linearly independent.

Remark. $\sum_{j=0}^{+\infty} \alpha_j b_j(x) = 0$ for any x in Ω implies that $\alpha_j = 0$ for any j. This is one of the properties used in the geometrical theory of diffraction and the W.K.B. method (see Chapters 4 and 5).

Example 1. $b_j(x) \equiv x^j$ is a basis in $\mathbf{L}^2(]-1, +1[)$. This basis is not orthogonal, because $\langle x^j, x^k \rangle$ is not zero for any $j \neq k$. Orthogonal polynomials such as Legendre polynomials form an orthogonal basis.

Example 2. The modes often form an orthogonal basis. For example, the modes, which are solutions of the homogeneous Helmholtz equation in a rectangular enclosure and satisfy homogeneous Dirichlet boundary conditions, form a basis. This implies that any solution of a Helmholtz equation and Dirichlet conditions in a rectangular enclosure can be written as a series of these modes. This series is called the modal representation of the solution. This is the property used in the separation of variables method.

Definition. A sequence of functions f_k of V *converges* to f in V when k tends to infinity if $\|f - f_k\|_V$ tends to zero. This is called a 'strong' convergence.

Example 3. The $\mathbf{L}^2$-convergence is obtained if f_k and f belong to $\mathbf{L}^2$ and if $\|f - f_k\|_2$ tends to zero. The modal series is $\mathbf{L}^2$-convergent to the solution.

Definition. A sequence of functions f_n of V *weakly converges* to f of V when k tends to infinity if $\lim_{k \to +\infty} \langle f_k, b_j \rangle = \langle f, b_j \rangle$ for any j.

Similarly, if A is an operator on functions of V, $\langle Au, v \rangle = \langle f, v \rangle$ for any v in V is a weak formulation of the equation $Au = f$.

Example 4. The variational formulation is a weak formulation. The example for the Dirichlet problem is given in the next section. A weak formulation often corresponds to the principle of conservation of energy.

Remark. The strong convergence implies the weak convergence.

A.3.4. Sobolev spaces

With Sobolev spaces, it is possible to take into account in the definition of the norm the properties of a function and its derivatives. These spaces are then quite useful in the theory of partial differential equations.

Definition. If s is a positive integer, the $H^s(\Omega)$ space (also denoted $W^{s,2}(\Omega)$ in some mathematics books) is defined by

$$H^s(\Omega) = \{f \text{ of } \mathbf{L}^2(\Omega) \text{ such that } D^\alpha f \text{ belongs to } \mathbf{L}^2(\Omega) \text{ for any integer } \alpha = 0, 1, \ldots, s\}$$

Example 1. $H^0(\Omega)$ is obviously equal to $\mathbf{L}^2(\Omega)$. $H^1(\Omega)$ contains all the functions of $\mathbf{L}^2(\Omega)$ such that their first-order partial derivatives are also in $\mathbf{L}^2(\Omega)$. If f belongs to $H^1(\Omega)$, f and its first-order derivatives are of finite energy. This space is a convenient space to study solutions of the Helmholtz equation as shown in the last example of this section. ■

The previous definition has been extended quite naturally to negative and also noninteger s by introducing the Fourier transform $\hat{f}(\xi)$ of $f(x)$.

Definition. If s is a real number, $H^s(\mathbb{R}^n)$ is the space of functions f such that $(1 + |\xi|^2)^{s/2}\hat{f}(\xi)$ is in $\mathbf{L}^2(\mathbb{R}^n)$. The inner product is defined by

$$\langle f, g \rangle = \int_{\mathbb{R}^n} (1 + |\xi|^2)^s \hat{f}(\xi)\hat{g}^*(\xi)\, d\xi$$

The spaces $H^s(\mathbb{R}^n)$ are Hilbert spaces.

With this generalization, it is possible to find a function space which the Dirac distribution belongs to, namely $H^{-1/2}(\mathbb{R}^n)$.

Example 2. Dirac distribution and Helmholtz equation. In acoustics, the Dirac distribution is extensively used for two reasons:

- It can be used to model an omnidirectional point source.
- If the solution is known when the right-hand side member of the Helmholtz equation is a Dirac distribution, then the solution is known for any source term, because of the convolution property of the Helmholtz equation (see Section A.5).

It is then essential to be able to determine the conditions for which the system of equations

$$(\Delta + k^2)G(S, M) = \delta_S(M) \text{ in } \mathbb{R}^n \text{ or } \Omega$$

with appropriate boundary conditions and/or Sommerfeld conditions has a solution and to which space this solution belongs. Knowing the properties of the operator defined by the system and that the Dirac distribution is in the $H^{-1/2}(\mathbb{R}^n)$ space, it is possible to deduce the properties of the solution (existence, uniqueness, behaviour, ...). ■

It is also possible to define the limit (or the trace) of a function $f(M)$ defined on Ω when M tends to a point P of the boundary Γ of Ω.

Definition. The term 'trace' is used here for the limit of $f(M)$ when the point M of Ω tends to a point P of the boundary Γ. If f is a continuous function, the trace is equal to the value of f on Γ. But the notion of trace extends to less smooth functions and to distributions. With the help of Sobolev spaces, it is possible to define traces which are not functions defined at any point of Γ.

Example 3. Layer potentials. Let $G(S, M)$ be the Green's function for the Helmholtz equation.
Then the simple layer operator K defined by

$$K\mu(M) \equiv \int_\Gamma \mu(P')G(M, P')\, d\sigma(P')$$

relates μ of $H^{-1/2}(\Gamma)$ to $K\mu$ of $H^1(\Omega)$.

Its value on Γ is given by

$$L\mu(P) \equiv \mathrm{Tr}(K\mu(M)) = \lim_{M \to P} (K\mu(M))$$

and relates μ of $H^{-1/2}(\Gamma)$ to $L\mu$ of $H^{1/2}(\Gamma)$.

Similarly, the double layer operator K' defined by

$$K'\mu(M) \equiv \int_\Gamma \mu(P') \frac{\partial G(M, P')}{\partial \vec{n}(P')}\, d\sigma(P')$$

relates μ of $H^{-1/2}(\Gamma)$ to $K'\mu$ of $H^0(\Omega)$.

Its value on Γ is given by

$$L'\mu(P) \equiv \mathrm{Tr}(K'\mu(M)) = \lim_{M \to P} (K'\mu(M))$$

and relates μ of $H^{-1/2}(\Gamma)$ to $L'\mu$ of $H^{-1/2}(\Gamma)$.

From the properties of the operators K, K$'$, L and L$'$, (compact, adjoint, ...), it is then possible to determine the conditions of existence and uniqueness of solutions of integral equations. ■

Example 4. Dirichlet problem. Let Ω be a bounded domain of $\mathbb{R}^n$ and V the Sobolev space $H_0^1(\Omega)$. $H_0^1(\Omega)$ is the space of all functions of $H^1(\Omega)$ which are equal to 0 on the boundary Γ of Ω. The Dirichlet problem consists in finding the solution u in V of the system

$$c^2\Delta u + \omega^2 u = -f \text{ in } \Omega \qquad \text{and} \qquad u = 0 \text{ on } \Gamma$$

It can be replaced by the weak formulation

$$a(u, v) = \mathcal{L}(v) \qquad \text{for any } v \text{ in } V$$

where a and $\mathcal{L}$ are defined by

$$a(u, v) \equiv \left(\int_\Omega c^2 \nabla u \cdot \nabla v^* - \omega^2 u v^* \right) \qquad \text{and} \qquad \mathcal{L}(v) \equiv \int_\Omega f v^*$$

In $a(u, v)$, the first term corresponds to a potential energy and the second term to a kinetic energy.

The equation is solved by a numerical procedure. First, the solution u in V is approximated by a function u_h in a subset V_h of V, such that

$$a(u_h, v_h) = \mathcal{L}(v_h) \qquad \text{for any } v_h \text{ in } V_h$$

Because of the properties of the operators and spaces, it can be shown that there exists a solution u_h in V_h. Its convergence to the exact solution u in V depends on the properties of functions v_h.

Furthermore, if the $(v_{h_i}, i = 1, ..., N)$ is an orthogonal basis of V_h, u_h can be written

$$u_h = \sum_{i=1}^{N} U_{h_i} v_{h_i}$$

and the previous equation leads to a simple linear system of order N:

$$BU_h = F_h$$

with matrix B and vectors U_h and F_h defined by

$$B_{ij} \equiv a(v_{h_i}, v_{h_j}); \qquad U_{h_i} \equiv \int_\Omega u v_{h_i}^* \qquad \text{and} \qquad F_{h_i} = \mathcal{L}(v_{h_i})$$

Theorem. If $s > (n/2) + k$, then $H^s(\mathbb{R}^n) \subset C^k(\mathbb{R}^n)$.

This relates the properties of functions of Sobolev spaces to their differentiation properties.

A.4. Distributions or Generalized Functions

The theory of distributions has been developed by L. Schwartz [7]. Distributions are also called generalized functions (see for example, [8] for mathematical theorems and [10] for applications in acoustics).

A.4.1. Distributions

Definition. T is a *distribution* if T is a continuous linear function defined on $\mathfrak{D}$.

This means that T associates to a function φ of $\mathfrak{D}$ a complex number $T(\varphi)$. $T(\varphi)$ is also written $\langle T, \varphi\rangle$ and the following three properties hold:

$$\langle T, \varphi_1 + \varphi_2\rangle = \langle T, \varphi_1\rangle + \langle T, \varphi_2\rangle$$

$$\langle T, \lambda\varphi\rangle = \lambda\,\langle T, \varphi\rangle \qquad \text{if } \lambda \text{ is a complex number}$$

If φ_j converges to φ when j tends to $+\infty$, then $\langle T, \varphi_j\rangle$ converges to $\langle T, \varphi\rangle$. ■

How can this general definition be related to applications in acoustics? Here are some illustrative examples.

Example 1. Let f be a locally integrable function, that is a function integrable on any bounded domain. T_f defined by

$$\langle T_f, \varphi\rangle \equiv \int_{\mathbb{R}^n} f(x)\varphi(x)\, dx$$

is a distribution.

Here f is a function and T_f is the associated distribution. For simplicity (although it is not rigorous), f is often written instead of T_f and f is called a distribution.

The notation is then $\langle T_f, \varphi\rangle$ or $\langle f, \varphi\rangle$. ■

Example 2. The Dirac measure or Dirac distribution. This is defined by

$$\langle T_\delta, \varphi\rangle \qquad \text{or} \qquad \langle \delta, \varphi\rangle \equiv \varphi(0)$$

It is easy to check that it satisfies the properties of a distribution. It corresponds to the definition of a monopole source at point 0 in mechanics.

If a is a point of $\mathbb{R}^n$, $\langle \delta_a, \varphi\rangle \equiv \varphi(a)$ is also a distribution, called the Dirac measure at point a. ■

Example 3. Derivatives. If T is defined by $\langle T, \varphi\rangle \equiv D\varphi(a)$ where D is any partial differential operator, then T is a distribution.

$\langle T, \varphi\rangle \equiv -\varphi'(a)$ corresponds to a dipole source at point a. ■

Example 4. Distribution of monopoles on a surface Γ, *with density* ρ. This is defined by

$$\langle T, \varphi\rangle \equiv \int_\Gamma \rho(x)\varphi(x)\, d\sigma(x)$$ ■

Example 5. Distribution of normal dipoles on a surface Γ, *with a density* ρ. This is defined by

$$\langle T, \varphi\rangle \equiv -\int_\Gamma \rho(x)\,\frac{\partial\varphi(x)}{\partial\vec{n}}\, d\sigma(x)$$

where $\vec{n}$ is a unit vector normal to Γ. ■

A.4.2. Derivation of a distribution

Definition. The partial derivative of a distribution T with respect to variable x_i is defined by:

$$\left\langle \frac{\partial T}{\partial x_i}, \varphi \right\rangle \equiv - \left\langle T, \frac{\partial \varphi}{\partial x_i} \right\rangle$$

and more generally

$$\langle D^p T, \varphi \rangle \equiv (-1)^{|p|} \langle T, D^p \varphi \rangle$$

where

$$D^p \equiv \left(\frac{\partial}{\partial x_1} \right)^{p_1} \left(\frac{\partial}{\partial x_2} \right)^{p_2} \cdots \left(\frac{\partial}{\partial x_n} \right)^{p_n} \qquad \text{with } |p| = p_1 + p_2 + \cdots + p_n$$

Particular case: In the case of a distribution T_f associated to a continuously differentiable function f, one has:

$$\langle T_f, \varphi \rangle \equiv \int_{\mathbb{R}^n} f(x)\varphi(x)\, dx \qquad \text{and} \qquad \left\langle \frac{\partial T_f}{\partial x_i}, \varphi \right\rangle \equiv - \int_{\mathbb{R}^n} f(x) \frac{\partial \varphi(x)}{\partial x_i}\, dx_1 \ldots dx_n$$

By integrating by parts, it can be shown that in the last right-hand side term the integral with respect to x_i can be written

$$- \int_{-\infty}^{+\infty} f(x) \frac{\partial \varphi(x)}{\partial x_i}\, dx_i = f(x)\varphi(x)]_{x_i=-\infty}^{x_i=+\infty} + \int_{-\infty}^{+\infty} \frac{\partial f(x)}{\partial x_i} \varphi(x)\, dx_i$$

The first term on the right-hand side is equal to zero since φ is zero outside a bounded domain (φ is in $\mathcal{D}$). Then the right-hand side is equal to $\langle T_{\partial f/\partial x_i}, \varphi \rangle$.

For distributions associated to a function, the rules of derivation correspond to the classical rules. ■

Example 1. One-dimensional Heaviside function Y

$$Y(x) = \begin{cases} 1 & \text{if } x > 0 \\ 0 & \text{if not} \end{cases}$$

T_Y (or Y) is the distribution defined by

$$\langle Y, \varphi \rangle \equiv \int_0^{+\infty} \varphi(x)\, dx$$

and then

$$\langle Y', \varphi \rangle \equiv - \int_0^{+\infty} \varphi'(x)\, dx$$

This last integral is obviously equal to

$$-\varphi(x)]_0^{+\infty} = \varphi(0) = \langle \delta, \varphi \rangle$$

Then the derivative of Y is the Dirac distribution: $Y' = \delta$. ■

Example 2. Functions discontinuous at $x = 0$ in $\mathbb{R}$. Let us consider a function f infinitely differentiable for $x > 0$ and for $x < 0$. It is assumed that all the derivatives have a limit on the $(x < 0)$ side, called the lower limit, and have a limit on the $(x > 0)$ side, called the upper limit, when x tends to zero.

Let us note $\sigma_m = f^{(m)}(0_+) - f^{(m)}(0_-)$, the difference between the upper and lower limits of the mth-derivative of f at $x = 0$. Let us calculate the derivative of the distribution T_f defined by

$$\langle T_f, \varphi \rangle \equiv \int_{\mathbb{R}} f(x)\varphi(x)\, dx$$

Then

$$\langle T'_f, \varphi \rangle = -\langle T_f, \varphi' \rangle = -\int_{\mathbb{R}} f(x)\varphi'(x)\, dx = -\int_{-\infty}^{0} f(x)\varphi'(x)\, dx - \int_{0}^{+\infty} f(x)\varphi'(x)\, dx$$

By integrating by parts each of the last two integrals, it is shown that

$$\langle T'_f, \varphi \rangle = (f(0_+) - f(0_-))\varphi(0) - \int_{-\infty}^{+\infty} f'(x)\varphi(x)\, dx = \sigma_0\varphi(0) + \langle f', \varphi \rangle$$

This means that if f' is used instead of T'_f the derivative of f in the distribution sense is $f' = \sigma_0\delta + \{f'\}$ where $\{f'\}$ is the classical derivative of f in the domains $(x > 0)$ and $(x < 0)$.

Theorem. More generally,

$$f^{(m)} = \{f^{(m)}\} + \sigma_0\delta^{(m-1)} + \sigma_1\delta^{(m-2)} + \cdots + \sigma_{m-1}\delta \qquad \text{for } \geqslant 1$$

These results can be generalized to the case of functions f infinitely differentiable on $\mathbb{R}^n$ except on a smooth hypersurface Γ. For the Laplace operator, it leads to

$$\Delta f = \{\Delta f\} + \left(\left\{ \frac{\partial f^+}{\partial \vec{n}} \right\} - \left\{ \frac{\partial f^-}{\partial \vec{n}} \right\} \right) \delta_\Gamma + (f^+ - f^-)\delta'_\Gamma$$
$$= \{\Delta f\} + \sigma_{\vec{n}}\delta_\Gamma + \sigma_0\delta'_\Gamma$$

where $\vec{n}$ is the unit vector normal to Γ. δ'_Γ is defined by

$$\langle \delta'_\Gamma, \varphi \rangle = -\int_\Gamma \frac{\partial \varphi}{\partial \vec{n}}$$

The (+) (resp. (−)) sign corresponds to the side of the normal $\vec{n}$ (resp. to the opposite side to the normal $\vec{n}$). σ_0 and $\sigma_{\vec{n}}$ are respectively the jump of f and the jump of $\partial f/\partial\vec{n}$ across the surface Γ.

Application to the Green's formula. If the surface Γ is the boundary of a volume Ω, if f is equal to zero outside Ω, then

$$\begin{aligned}\langle\Delta f, \varphi\rangle = \langle f, \Delta\varphi\rangle &\equiv \int_\Omega f\Delta\varphi \, d\Omega \\ &= \langle\{\Delta f\}, \varphi\rangle + \left\langle \frac{\partial f}{\partial\vec{n}_i}\delta_\Gamma, \varphi\right\rangle + \langle f\delta'_\Gamma, \varphi\rangle \\ &\equiv \int_\Omega \varphi\Delta f \, d\Omega + \int_\Gamma \frac{\partial f}{\partial\vec{n}_i}\varphi \, d\Gamma - \int_\Gamma f\frac{\partial\varphi}{\partial\vec{n}_i} \, d\Gamma\end{aligned}$$

or

$$\int_\Omega (f\Delta\varphi - \varphi\Delta f) \, d\Omega + \int_\Gamma \left(f\frac{\partial\varphi}{\partial\vec{n}_i} - \varphi\frac{\partial f}{\partial\vec{n}_i}\right) d\Gamma = 0$$

In these integrals, $d\Omega$ and $d\Gamma$ respectively represent the element of integration on Ω and on Γ. $\vec{n}_i$ is the unit vector, normal to Γ and interior to Ω.

The theory of distributions is then an easy and rigorous way to obtain Green's formula.

A.4.3. Tensor product of distributions

In this book, the following notations are used:

$$\delta_{x_s, y_s, z_s}(x, y, z) \quad \text{or} \quad \delta(x - x_s, y - y_s, z - z_s) \quad \text{or} \quad \delta_S(M)$$

All these notations are improper and should be replaced by

$$\delta_{x_s}(x) \otimes \delta_{y_s}(y) \otimes \delta_{z_s}(z)$$

This is called a tensor product.

Definition. Let S_x of $\mathcal{D}'_x$ and T_y of $\mathcal{D}'_y$ be two distributions. (To make things clearer, variables x and y are introduced as subscripts.) Then, the *tensor product* of these two distributions is the distribution $W_{x,y}$ of $\mathcal{D}'_{x,y}$ such that

$$\langle W, \varphi\rangle \equiv \langle S_x, \langle T_y, \varphi(x, y)\rangle\rangle$$

This means in particular that if S_x and T_y are defined by

$$\langle S_x, u\rangle \equiv \int f(x)u(x) \, dx \qquad \text{and} \qquad \langle T_y, v\rangle \equiv \int g(y)v(y) \, dy$$

then

$$\langle W_{xy}, \varphi \rangle \equiv \int f(x) \left(\int g(y)\varphi(x, y)\, dy \right) dx$$

Example. The 2-dimensional delta distribution is defined by

$$\langle \delta_a(x) \otimes \delta_b(y), \varphi(x, y) \rangle \equiv \langle \delta_a(x), \varphi(x, b) \rangle \equiv \varphi(a, b)$$

Similarly, the 3-dimensional delta distribution is defined by

$$\langle \delta_a(x) \otimes \delta_b(y) \otimes \delta_c(z), \varphi(x, y, z) \rangle \equiv \varphi(a, b, c)$$

A.5. Green's Kernels and Integral Equations

In this book, the terms 'Green's function' or 'Green's kernel' of a Helmholtz equation are used equivalently to refer to the solutions of the following systems

$$(\Delta + k^2)G(S, M) = \delta_S(M) \text{ in } \mathbb{R}^n \text{ and Sommerfeld conditions}$$

or

$$(\Delta + k^2)G(S, M) = \delta_S(M) \text{ in } \Omega$$

with boundary conditions on Γ and Sommerfeld conditions if Ω is not bounded.

The terms 'elementary kernel' and 'elementary solution' refer to any solution of a Helmholtz equation with $\delta_S(M)$ as second member.

The Helmholtz equation with constant coefficients is a convolution equation. This means in particular that if f is the solution of one of the two systems written above with a source term $Q(M)$ instead of the Dirac distribution, then f is given by

$$f(M) = (G * Q)(M) \equiv \int_{\mathbb{R}^n} Q(S)G(S, M)\, d\sigma(S)$$

Then if G is known, the solution f is known for any source term Q.

Definition. Let p be defined on the domain Ω, with boundary Γ. Let $K(M, M')$ be any kernel.

Terms of the form

$$p(M) = \varepsilon p_0(M) + \int_\Gamma \mu(P')K(M, P')\, d\sigma(P')$$

where M is a point of Ω and P' a point of Γ are called integral representations of p. ε is generally a constant; it can be equal to zero. p_0 is a known function.

Terms of the form

$$\varepsilon\mu(P) + \int_\Gamma \mu(P')K(P, P')\,d\sigma(P') = g(P)$$

where P and P′ are points of Γ are called integral equations for μ defined on Γ.

Because Green's functions $G(S, M)$ for the Helmholtz equation are singular (i.e. tend to infinity when M tends to S), the integral equations are singular. It has been necessary to develop a theory of singular integral operators to study the properties of the solutions of integral equations. A very large number of articles and books have been published on this subject (see [6] for example).

In this book, we do not go through this theory, which is far beyond our scope. The term 'singular' refers to the integrability of the kernels. Three types of singularities are encountered:

- *Weakly singular integral.* This is the case of a single layer potential, which is perfectly defined since the kernel is integrable in the Riemann sense.
- *Cauchy principal value.* This is the case of a double layer potential or of the derivative of a single layer potential.
- *Hyper-singular integral.* This is the case of the derivative of a double layer potential. The integral must be defined as a finite part in the sense of Hadamard.

Bibliography

[1] COURANT, R. and HILBERT, D., 1953. *Methods of mathematical physics*, vol. 1, and 1962, *Methods of mathematical physics*, vol. 2. Wiley, New York.

[2] DAUTRAY, R. and LIONS, J.L., 1992. *Mathematical analysis and numerical methods for science and technology*. Springer-Verlag, New York.

[3] JOHN, F., 1986. *Partial differential equations*, 4th edn, Applied Mathematical Sciences Series, vol. 1. Springer-Verlag, New York.

[4] KRESS, R., 1989. *Linear integral equations*, Applied Mathematical Sciences Series, vol. 82. Springer-Verlag, New York.

[5] LEBEDEV, L.P., VOROVICH, I.I. and GLADWELL, G.M.L., 1996. *Functional analysis: Applications in mechanics and inverse problems*, Solid Mechanics and Its Applications Series, vol. 41. Kluwer Academic, Dordrecht.

[6] MIKHLIN, S.G., 1965. *Multidimensional singular integrals and integral equations*. Pergamon Press, Oxford.

[7] SCHWARTZ, L., 1965. *Méthodes mathématiques pour les sciences physiques*. Hermann, Paris.

[8] YOSIDA, K., 1980. *Functional analysis*, Classics in Mathematics Series. Springer-Verlag, New York.

[9] ERDELYI, A., 1956. *Asymptotic expansions*. Dover, New York.

[10] CRIGHTON, D.G., DOWLING, A.P., FFOWCS-WILLIAMS, J.E., HECKL, M. and LEPPINGTON, F., 1992. *Modern methods in analytical acoustics*, Springer Lecture Notes. Springer-Verlag, New York.

Index

asymptotic expansion
  elementary kernel 161
  Kirchhoff approximation 169
  layer potentials 163
  of integrals
    stationary phase method 164
    steepest descent method 166
    Watson's lemma 164

Bessel equation 73
Bessel function 73
  Padé approximants 125
Boundary Integral Equations (B.I.E.) 86
  and Green's representation 110
  Boundary Element Method (B.E.M.), collocation equations 191
  Boundary Element Method (B.E.M.), Galerkin equations 194
  fluid-loaded baffled plate 283
  ground effect problems 132
  polynomial approximation 284

cylindrical coordinates 71

Dirichlet condition 43, 47

eigenfrequency 47, 49
  numerical approximation by B.E.M. 195
eigenmodes 47, 49
  of a fluid-loaded plate 274
  orthogonality 73, 274
  series 54, 55, 60, 74

fluid-loaded plate
  coincidence frequency 267
  critical frequency 262
  dispersion equation 262
  eigenvalues and eigenmodes 273
  Green's representation of the solution 281
  non-dimensional equation 272
  resonance frequencies and resonance modes 274
forced regime 54, 55, 80

free oscillation 48
  of a piston in a waveguide 256

Geometrical Theory of Diffraction 137
  diffracted ray 140
  incident ray 138
  inhomogeneous medium 177
  ray method 176
  reflected ray 139
  theoretical basis 174
Green's function
  free space 91
  of a duct 221
Green's representation 81, 84, 86, 104
  Boundary Integral Equation 112, 114, 132
  exterior problem 113
  fluid-loaded baffled plate 283
  interior problem 112
ground effect
  homogeneous ground 122
  inhomogeneous ground 132
  Laplace type integral 125
  surface wave 125
ground impedance
  measurement 128
  models 131
guided waves
  infinite plane waveguide 214
  W.K.B. approximation 209

Hankel function 75
harmonics
  cylindrical 95
  spherical 95
Helmholtz equation 43

inhomogeneous medium 149
  Born and Rytov approximations 173
  continuously varying index 151
  Geometrical Theory of Diffraction 154, 177
  layered medium 150
  parabolic approximation 155, 179
  W.K.B. method 153, 171

Laplace transform 59
layer potential
  discontinuities 99
  double 85, 97
  far-field asymptotic expansion 163
  ground effect representation 124
  hybrid 107, 114
  simple 85, 97
light fluid approximation
  baffled plate 276
  infinite plate 264
limit absorption principle 103
limit amplitude principle 104

method of images 80, 174

Neumann condition 42, 47
Neumann problem 49, 54, 71

parabolic approximation 179
piston in a waveguide 224, 226
  radiation of sound 250
  transmission of sound 253
point source
  isotropic 93
  multipolar 94

reflection coefficient 78, 126
resonance frequency 52
resonance mode 52, 65
reverberation time 67, 70
Robin condition 44, 47
Robin problem 52, 55, 78

Sabine absorption coefficient 70
Sabine formula 71
separation of variables 49, 52, 71, 75, 136
space Fourier Transform 123
specific normal impedance 44, 123, 131
steepest descent method 166

thin plate
  energy equation 258
  partial differential equation 259
thin screen 108
  diffraction by a 143
  Kirchhoff approximation 169
transient 58, 65

wave equation 42
waveguide
  circular duct 220
  cut-off frequency 219
  Green's function 221
  impulse response 228
  rectangular duct 218
  with absorbing walls 236
Wiener–Hopf method 182
  example of application 135